Handbook of New Concepts in Water Purification

Handbook of New Concepts in Water Purification

Contributors

Javier Miguel Ochando-Pulido, Antonio Martinez-Ferez et al.

AURIS
Reference

www.aurisreference.com

Handbook of New Concepts in Water Purification

Contributors: Javier Miguel Ochando-Pulido, Antonio Martinez-Ferez et al.

Published by Auris Reference Limited

www.aurisreference.com

United Kingdom

Handbook of New Concepts in Water Purification

ISBN: 978-1-78154-829-5
British Library Cataloguing in Publication Data
A CIP record for this book is available from the British Library

Printed in the United Kingdom

Exclusively distributed by CBS Publishers & Distributors Pvt. Ltd.

Sales & Distribution Rights only for India, Pakistan, Bangladesh, Sri Lanka, Nepal and Bhutan. This book is not to be sold outside these territories.

Contents

List of Abbreviations

AOPs	Advanced oxidation processes
AGMD	Air Gap Membrane Distillation
APDC	Ammonium pyrrolidine dithiocarbamate
ASV	Anodic Stripping Voltammetry
AAS	Atomic Absorption Spectrometry
AFM	Atomic force microscopy
BOD	Biological Oxygen Demand
BDD	Boron-doped diamond
CNTs	Carbon nanotubes
CRM	Certified Reference Materials
COD	Chemical oxygen demand
CVD	Chemical Vapor Deposition
CBM	Coal bed methane
CSG	Coal Seam Gas
CBNG	Coal-bed natural gas
DDT	Dichlorodiphenyltrichloroethane
DCMD	Direct Contact Membrane Distillation
ESI	Electrospray ionization
EPA	Environmental Protection Agency
FAAS	Flame atomic absorption spectrometry
FBC	Fluidized bed crystallizer
GOR	Gain output ratio
HCL	Hollow cathode lamp
IDS	Iminodisuccinic acid
ICC	Indian Childhood Cirrhosis
ICP	Inductively coupled plasma
IARC	International Agency for Research on Cancer
IDA	International Desalination Association
IUPAC	International Union of Pure and Applied Chemistry
KBHD	Kay Bailey Hutchison Desalination Plant
LIBS	Laser induced breakdown spectroscopy
LNG	Liquefied natural gas
LDT	Liquid decay test
LEP	Liquid entry pressure
LOD	Lowers limit of detection
MBR	Membrane bioreactor
MCR	Membrane chemical reactor
MDC	Membrane distillation crystallization
MES	Mercaptoethane sulfonate

MGDA	Methylglycinediacetic acid
MENA	Middle East and North Africa
MLSS	Mixed liquor suspended solids
MED	Multiple-effect distillation
MSF	Multi-stage flash distillation
MWCNT	Multi-walled carbon nanotubes
OMW	Olive mill wastewaters
OCP	Open circuit potential
OVD	Outside vapor deposition
ppm	Parts per million
PMRs	Photocatalytic membrane reactors
PACl	Poly-aluminium chloride
PAHs	Polyaromatic hydrocarbons
PAHs	Polycyclic aromatic hydrocarbons
PLE	Polymeric ligand exchanger
PID	Proportional–integral–derivative
REEs	Rare earth elements
RMP	Reactive Metal Penetration
SEM	Scanning electron microscopy
SEIs	Secondary election images
STP	Sewage treatment plants
SAR	Sodium adsorption ratio
SPE	Solid phase extraction
SAGD	Steam assisted gravity drainage
SGMD	Sweep Gas Membrane Distillation
TGS	Tight gas sands
UNIDO	United Nations Industrial Development Organization
VMD	Vacuum Membrane Distillation
V-MEMD	Vacuum-Multi-Effect-Membrane-Distillation
WHPCO	Wet hydrogen peroxide catalytic oxidation
XRD	X-ray diffraction

List of Contributors

Javier Miguel Ochando-Pulido
Department of Chemical Engineering, University of Granada, Granada 18071, Spain

Antonio Martinez-Ferez
Department of Chemical Engineering, University of Granada, Granada 18071, Spain

Tsuyoshi Ochiai
Kanagawa Academy of Science and Technology, KSP building East 407, 3-2-1 Sakado, Takatsu-ku, Kawasaki, Kanagawa 213-0012, Japan
Photocatalysis International Research Center, Tokyo University of Science, 2641 Yamazaki, Noda, Chiba 278-8510, Japan

Shoko Tago
Kanagawa Academy of Science and Technology, KSP building East 407, 3-2-1 Sakado, Takatsu-ku, Kawasaki, Kanagawa 213-0012, Japan

Mio Hayashi
Kanagawa Academy of Science and Technology, KSP building East 407, 3-2-1 Sakado, Takatsu-ku, Kawasaki, Kanagawa 213-0012, Japan

Akira Fujishima
Kanagawa Academy of Science and Technology, KSP building East 407, 3-2-1 Sakado, Takatsu-ku, Kawasaki, Kanagawa 213-0012, Japan
Photocatalysis International Research Center, Tokyo University of Science, 2641 Yamazaki, Noda, Chiba 278-8510, Japan

Hiromasa Tawarayama
Optical Communications R&D Laboratories, Sumitomo Electric Industries, Ltd., 1 Taya-cho, Sakae-ku, Yokohama 244-8588, Japan

Toshifumi Hosoya
R&D General Planning Division, Sumitomo Electric Industries, Ltd., 1-1-3, Shimaya, Konohana-ku, Osaka 554-0024, Japan

Marjan S. Ranđelović
University of Niš, Faculty of Science and Mathematics, Department of Chemistry, Niš,, Serbia

Milovan M. Purenović
University of Niš, Faculty of Science and Mathematics, Department of Chemistry, Niš,, Serbia

Aleksandra R. Zarubica
University of Niš, Faculty of Science and Mathematics, Department of Chemistry, Niš,, Serbia

Ramiro Escudero
Instituto de Investigaciones Metalúrgicas, Universidad Michoacana de San Nicolás de Hidalgo, Morelia, Michoacán, México

Francisco J. Tavera
Instituto de Investigaciones Metalúrgicas, Universidad Michoacana de San Nicolás de Hidalgo, Morelia, Michoacán, México

Eunice Espinoza
Instituto de Investigaciones Metalúrgicas, Universidad Michoacana de San Nicolás de Hidalgo, Morelia, Michoacán, México

Adina Elena Segneanu
National Institute for Research and development in Electrochemistry and Condensed Matter –INCEMC Timisoara, Romania

Carmen Lazau
National Institute for Research and development in Electrochemistry and Condensed Matter –INCEMC Timisoara, Romania

Paula Sfirloaga
National Institute for Research and development in Electrochemistry and Condensed Matter –INCEMC Timisoara, Romania

Paulina Vlazan
National Institute for Research and development in Electrochemistry and Condensed Matter –INCEMC Timisoara, Romania

Cornelia Bandas
National Institute for Research and development in Electrochemistry and Condensed Matter –INCEMC Timisoara, Romania

Ioan Grozescu
National Institute for Research and development in Electrochemistry and Condensed Matter –INCEMC Timisoara, Romania

Cristina Orbeci
Politehnica University Bucuresti, Romania

Chunli Zheng
School of Energy and Power Engineering, Xi'an Jiaotong University, China

Ling Zhao
College of Environment & Resources of Inner Mongolia University, China

Zhimin Fu
College of Environment & Resources of Inner Mongolia University, China

Xiaobai Zhou
The Environmental Monitoring Center of Jiangsu Province, Nanjing, China

An Li
School of Petrochemical Engineering, Lanzhou University of Technology, China

Bulent Sen
University of Firat, Faculty of Fisheries Department of Aquatic Basic Sciences Elazığ, Turkey

Feray Sonmez
University of Firat, Faculty of Fisheries Department of Aquatic Basic Sciences Elazığ, Turkey

Ozgur Canpolat
University of Firat, Faculty of Fisheries Department of Aquatic Basic Sciences Elazığ, Turkey

Mehmet Tahir Alp
University of Mersin, Faculty of Fisheries Department of Aquatic Basic Sciences Mersin, Turkey

Mehmet Ali Turan Kocer
Mediterranean Fisheries Research Production and Training Institue, Antalya,
Turkey

Preface

Water purification is the process of removing undesirable chemicals, biological contaminants, suspended solids and gases from contaminated water. The text *Handbook of New Concepts in Water Purification* presents a review on optimization of conventional drinking water treatment plant that eventually proposing a method to maximize process efficiency with less risks. In first chapter, a review on the actual state of the art on the treatment of olive mill wastewaters (OMW) by membrane technologies has been addressed. Second chapter focuses on TiO_2-impregnated porous silica tube and its application for compact air- and water-purification units. In third chapter, new composite materials in the technology for drinking water purification from ionic and colloidal pollutants have been described. The advances in possible water treating improvements are issued in fourth chapter. Waste water treatment methods have been introduced in fifth chapter. In sixth chapter, the treatment technologies for organic wastewater have been reviewed. The relationship of algae to water pollution and waste water treatment has been discussed in seventh chapter. In eighth chapter, the novel and common methods for the determination of trace heavy metals in waste water and their removal processes are explained. Selective removal of heavy metal ions from waters and waste waters using ion exchange methods have been described in last chapter.

Chapter 1

ON THE RECENT USE OF MEMBRANE TECHNOLOGY FOR OLIVE MILL WASTEWATER PURIFICATION

Javier Miguel Ochando-Pulido and Antonio Martinez-Ferez
Department of Chemical Engineering, University of Granada, Granada 18071, Spain

ABSTRACT

Many reclamation treatments as well as integrated processes for the purification of olive mill wastewaters (OMW) have already been proposed and developed but not led to completely satisfactory results, principally due to complexity or cost-ineffectiveness. The olive oil industry in its current status, composed of little and dispersed factories, cannot stand such high costs. Moreover, these treatments are not able to abate the high concentration of dissolved inorganic matter present in these highly polluted effluents. In the present work, a review on the actual state of the art concerning the treatment and disposal of OMW by membranes is addressed, comprising microfiltration (MF), ultrafiltration (UF), nanofiltration (NF), and reverse osmosis (RO), as well as membrane bioreactors (MBR) and non-conventional membrane processes such as vacuum distillation (VD), osmotic distillation (OD) and forward osmosis (FO). Membrane processes are becoming extensively used to replace many conventional processes in the purification of water and groundwater as well as in the reclamation of wastewater streams of very diverse sources, such as those generated by agro-industrial activities. Moreover, a brief insight into inhibition and control of fouling by properly-tailored pretreatment processes upstream the membrane operation and the use of the critical and threshold flux theories is provided.

INTRODUCTION

The olive oil sector has represented since several decades one of the most important industries in the Countries of the Mediterranean River Basin. Spain, Italy, Portugal, Greece and the Northern African countries—Syria, Algeria,

Turkey, Morocco, Tunisia, Libya, Lebanon—cope with the major production worldwide (Figure 1). Other countries such as France, Serbia and Montenegro, Macedonia, Cyprus, Egypt, Israel, and Jordan also present a considerable annual olive oil yield, according to data from the International Olive Oil Council (IOOC, 2013−2014) [1]. Moreover, olive oil production is also becoming an emergent agro-food industry in China and other countries such as the USA, Australia, and the Middle East. It is very worth highlighting the case of China, which exhibits favorable edaphoclimatic conditions for the growth of olive trees, and is expected to develop a significant olive oil production potential in the near future. Hence, the treatment of the olive mill effluents is already a task of global concern and it is no longer a problem limited to a specific region.

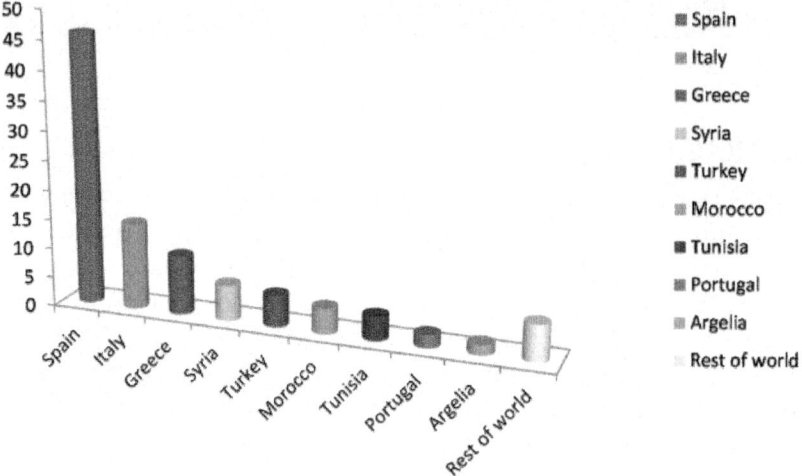

Figure 1: Olive oil production worldwide (data from International Olive Oil Council, IOOC 2014) [1].

Indeed, in recent decades a deep growth of the olive oil industrial sector has been experienced as a result of the modernization of olive oil mills, in response to the increasing demand of olive oil worldwide. Taking the case of Spain, there are more than 1,700 olive mills currently authorized and operating [1]. The production of olive oil employs a very significant number of people and is one of the main industrial activities, as in other countries of the Mediterranean Basin and Southern Europe. More than 1,400,000 tons of olive oil were produced in Spain during the 2013–2014 campaign, 70% of which were obtained in Andalucía where there are 850 olive mills, which yielded a production of 1,022,000 tons of olive oil as well as 4,778,451 tons of table olives, highlighting the provinces of Jaén, with more than 515,000 tons of olive oil and 2,229,000 tons of olives; Granada with 118,000 tons of olive oil and 509,383 tons of olives; and Córdoba with 219,000 tons of olive oil and 1,123,800 tons of olives [1].

The significant boost of this industrial sector in the last years has brought an undesired side-effect: the amounts of olive mill wastewaters (OMW) by-produced have also significantly increased, especially as a result of the change of the antique batch press method for the continuous centrifugation-based olive oil production processes currently used, which ensure much higher productivity (Table 1 and Table 2). On one hand, these continuous systems guarantee a higher yield in recovering olive oil from the olives, up to 21%, but on the other they lead to an increased production of wastewater streams due to water injection in the centrifugation processes [2]. An average-sized modern olive oil factory currently generates between 10 and 15 m^3 daily of OMW derived from the proper extraction process, and also 1 m^3 of wastewater derived from the washing of the olives per processed ton (olive washing wastewater, OWW) [2]. This raises several million cubic meters of these effluents annually, which represents a huge amount of these highly contaminant wastewaters as well as of potable water consumption.

Table 1: Characteristics of the effluents of batch and continuous olive oil extraction processes [3]

Process	Effluent	COD, g/L	BOD₅, g/L	TSS, g/L	pH	EC, mS/cm	TPh, g/L
Olive cleaning	OWW	0.8–2.2	0.3–1.5	8–18	5.5–6.6	2.5–3.0	0–0.1
Batch press	OMW-P	130–130	90–100	10–12	4.5–5.0	2.0–5.0	1.0–2.4
Three phase	OMW-3	30–220	5–45	5–35	3.5–5.5	2.0–7.9	0.3–7.5
Two phase	OMW-2	4–18	0.8–6.0	2–7	3.5–6.0	1.5–2.5	0.1–1.0

COD: chemical oxygen demand; BOD₅: biochemical oxygen demand; TSS: total suspended solids; EC: electric conductivity; TPh: total phenolic compounds.

Table 2. Effluents flowrates of continuous olive oil extraction processes [2]

Effluent, L/kg	3-phase extraction	2-phase extraction
Washing of olives (OWW)	0.06	0.05
Horizontal centrifuge	0.90	0
Vertical centrifuge	0.20	0.15
Cleaning	0.05	0.05
Total	1.21	0.25

The current necessity of maximizing the production processes too often excludes the planning of the environmental protection. Wastewater treatment for ulterior uses in multiple applications contributes to sustainable water consumption and the regeneration of the public hydraulic domain and the ecosystems. In this scenario, the European Directive 2000/60/CE took the lead in establishing the legal framework to confer the utmost protection to water, highlighting the use of regenerated wastewater. Moreover, European Environmental Regulations will become more stringent in virtue of the 'H2020 Horizon' (http://ec.europa.eu/programmes/horizon2020/).

Direct discharge of OMW has been reported to cause strong odor nuisance, soil contamination, inhibition of plant growth, hindrance of self-purification processes, underground leaks, and water body pollution, as well as severe impacts to the aquatic fauna and to the ecological status [3,4,5]. Discharge of untreated OMW to the ground fields and superficial water bodies is actually prohibited in Spain, whereas in Italy and other European countries only partial discharge on suitable terrains is allowed. Because of the presence of high levels of organic pollutants and refractory compounds, direct disposal of these effluents to the municipal sewage treatment systems is also prohibited. Legal limits are established in order to prevent difficulties to the municipal sewer treatment plants, which operate on the basis of biomasses that must be maintained alive.

OMW exhibit a series of characteristics that make their reclamation by conventional physicochemical treatments utterly difficult. The presence of phytotoxic recalcitrant pollutants—such as phenolic compounds, organic and long chain fatty acids, tannins and organohalogenated contaminants— makes these effluents resistant to biological degradation and thus inhibits the efficiency of biological processes. Furthermore, the physicochemical composition of OMW is extremely variable as it depends on several factors such as the extraction process, edaphoclimatic and cultivation parameters, as well as the type, quality and maturity of the processed olives, among others. OMW typically exhibit intense violet-dark colour, acid pH, strong odor, high saline toxicity reflected in the electroconductivity values, and very heavy organic pollutants load.

A focus on Table 1 shows that the pollutants concentration levels in the effluent from the washing of the olives (OWW) stands normally below the standard limits for discharge on superficial suitable terrains (e.g., data from the Guadalquivir Hydrographical Confederation, 2006–2014: TSS < 500 mg/L and COD < 1000 mg/L) (Table 3) [6]. However, concentration values might exceed the established standards, and this fact depends on the water flowrate used in the olive washing machines during the fruit cleaning procedure.

Table 3: Parametric standard limits for discharge of OMW in suitable terrains (data from the Guadalquivir Hydrographical Confederation, 2006–2014) [6]

Parameter	2002	2003	2004	2006	2014
pH	6–9	6–9	6–9	6–9	6–9
TSS, mg/L	600	500	500	500	500
COD, mg/L	2,500	2,000	1,500	1,000	1,000
BOD_5, mg/L	2,000	-	-	-	-

On the other hand, major organic pollutants load is commonly measured in the effluent coming out of the centrifuges (OMW-2 and OMW-3), and most of them are phytotoxic compounds recalcitrant to biological degradation. Therefore, the presence of these substances would be hardly reflected by biochemical oxygen demand (BOD_5) measurements, thus the chemical oxygen demand (COD) seems a more appropriate parameter together with total phenolic compounds (TPh) concentration, as reflected in Table 3.

Taking into consideration that in the two-phase extraction process water injection is only practiced in the final vertical centrifugation step, the volume of liquid effluent derived from the production process (OMW-2) is reduced by one-third on average with respect to the amount required for the three-phase system. Moreover, much of the organic matter remains in the solid waste, which contains higher moisture than the pomace from the three-phase system (60–70% in two-phase systems vs. 30–45% in three-phase ones, OMW-3) and hence OMW-2 exhibits a lower degree of pollutants, too: the measured chemical oxygen demand (COD) in OMW-2 is commonly in the range 4–18 g/L in contrast with up to 30–220 g/L in case of OMW-3. Inorganic compounds including chloride, sulfate, and phosphoric salts of potassium, calcium, iron, magnesium, sodium, copper, and traces of other elements are other common traits of OMW [3,4,5].

The disposal of the solid waste is not within the scope of the present work, which is focused only towards the management of the problem related to the reclamation of the liquid effluents. Some examples of solutions already proposed for the management of the pomace waste are adsorption of heavy metals [7,8], dyes [9], and phenols [10], as well as composting [11] or biogas production [12], among others.

The small size and geographical dispersion of olive oil mills as well as the seasonality of olive oil production represent the main obstacles for the implementation of cost-effective management processes for these effluents. A centralized treatment of OMW does not seem feasible currently, thus an effective and simple solution is needed for these small plants.

Many reclamation treatments and integrated processes for OMW have already been proposed and developed but not led to completely satisfactory results, such as lagooning or natural evaporation and thermal concentration [13], composting [14], treatments with clay or with lime [15,16], physico-chemical procedures such as coagulation-flocculation [17,18], and biosorption [19], advanced oxidation processes including ozonation [20], Fenton´s reaction [21,22] and photocatalysis [23,24], electrochemical treatments [25,26], and hybrid processes [27,28,29,30].

Furthermore, because of the significant salinity of OMW, reflected in their high electroconductivity (EC) values, conventional physicochemical treatments are not effective. These treatments are not able to abate the high concentration of dissolved ionic concentration present in these effluents, which present hazardous salinity levels according to the guidelines established by the Food and Agricultural Organization (F.A.O.) for irrigation purposes (Table 4).

In order to attempt the complete depuration of OMW, membrane technology offers compact modular nature, high efficiency, and moderate investment and maintenance expenses. Membrane processes are becoming extensively used to replace many conventional processes in the purification of water and groundwater as well as in the reclamation of wastewater streams of very diverse sources, such as those generated by agro-industrial activities [31,32,33,34,35,36,37]. The availability of new membrane materials, designs, module concepts and know-how has promoted credibility among investors.

In the present work, a review on the actual state of the art of the treatment and disposal of the OMW from both the two-phase and three-phase systems by membranes will be addressed, comprising microfiltration (MF), ultrafiltration (UF), nanofiltration (NF), and reverse osmosis (RO), as well as membrane bioreactors (MBR) and other non-conventional membrane processes.

Table 4: Irrigation water quality as a function of its salinity (Food and Agriculture Organization, FAO)

EC, dS/cm	Water quality	Risk due to salinity
0–1	Excellent to good	Low to mid
1–3	Good to marginal	High
> 3	Marginal to unacceptable	Very high

MEMBRANE TREATMENT OF THREE-PHASE OLIVE MILL WASTEWATER

The biggest technical drawback for the implementation of membrane technologies in wastewater treatment plants is the high fouling potential.

Membrane fouling—which can be caused by colloids, soluble organic compounds, and microorganisms that are typically not well removed with conventional pretreatment methods—increases the feed pressure and requires frequent membrane cleaning procedures. Concerning this, properly-tailored pretreatment processes are required in order to avoid high fouling rates, which would rapidly lead to zero flux conditions if no pretreatment is conducted on the raw effluent upstream the membrane operation.

Turano *et al.* [38] proposed an integrated centrifugation-UF process for the treatment of OMW from three-phase centrifugation-based olive mills in Italy. The preliminary centrifugation step achieved the removal of suspended solids, and the centrifuge supernatant was conducted to UF (polysulfone, MWCO 17 kDa). It was reported up to 55% COD reduction was achieved after the proposed centrifugation pretreatment, whereas 80% removal of suspended solids concentration and 90% COD reduction was achieved at the outlet of the integrated process. Moreover, centrifugation is a simple and mechanical operation, which does not need addition of chemicals, and it is available in the production line.

Paraskeva *et al.* [39] investigated the treatment and complete fractionation of OMW coming from a three-phase-decanter olive oil mill in Greece, using a combination of different membrane processes. The authors tested UF followed by NF and/or RO in a batch mode, keeping constant the composition of the initial raw wastewater.

Prior to UF, screening with an 80 µm polypropylene filter was performed to remove suspended solids. The UF membrane was made of ceramic material (zirconia, mean pore size 100 nm) in multichannel configuration. The recovery ratio was fixed between 80–90% of the initial OMW volume, at 15–35 °C operating temperature and TMP between 1.0 and 2.25 bar. UF succeeded in the separation of high molecular weight constituents including suspended solids, as well as the condensation of solid lipid components (up to 90%) and a large amount of the phenolic compounds (~50%).

Polymeric membranes in spiral wound configuration were used for either NF (200 Da MWCO) or RO (100 Da MWCO) tests, in order to further treat the UF permeate. In NF tests, the temperature was kept constant at 20 °C, and the best TMP (20 bar) led to satisfactory permeate flowrate between 100–120 L/h. Following the NF step, phenols were removed to an extent exceeding 95% of the initial value, but even better efficiency was achieved by applying RO after NF, which enabled a significant conductivity, salinity, and turbidity decrease and nearly 30 L/h permeate flowrate. Best performances were found to occur at 35 °C and high-pressure values (TMP = 40 bar), which allowed reaching a turbidity value of 14 NTU and a decrease of up to 98.9% of the raw water conductivity, with a recovery between 75–80% of the initial OMW volume. Finally, they stated not only that the chemical composition of the post-treated effluent was suitable for disposal in aquatic receptors or for irrigation purposes, but also that both the inorganic part of OMW (N, P, Mg, K, metal traces) and the organic fraction (hydrocarbons, nitrogenous compounds, organic acids, polyalcohols) may be used as plant nutrients, perhaps in combination with other inorganic or organic fertilizers such as manure or sludge from biological treatments of other types of wastes [40,41]. On the other hand, polyphenols and fats ought to be separated, as described before, due to their phytotoxic properties [42,43]. Nonetheless, no data on the permeate flux decline regarding the operating conditions was reported and fouling was not discussed by the authors.

Stoller and Chianese [44,45,46] studied the purification of the wastewater stream generated in the olive washing procedure (OWW) to comply with discharge standards in municipal sewers (Italy). In contrast with OMW, this effluent presents a moderate organic pollutants load, but high concentration of suspended solids. Therefore, the authors proposed a treatment process comprising an initial solid/liquid (S/L) separation stage, followed by batch UF and NF in series (composite thin-film spiral-wound membranes). As S/L separation operation they addressed the use of coagulation-flocculation with polyelectrolytes: aluminum sulfate (AS) or aluminum hydroxide (AH). The two pretreatment processes performed similarly with regard to COD and BOD_5 rejection efficiencies, but higher productivity was observed in the subsequent membranes-in-series process after flocculation with AS.

In a following research work, Stoller and Bravi [34] applied the same coagulant-flocculants to pretreat three-phase OMW before batch MF, UF, and NF membranes in sequence, and a final RO step. In addition, they examined two other pretreatment processes: photocatalysis (PC) with nanometric titanium dioxide anatase powders irradiated by UV light, and aerobic digestion (AD). To sum up, all pretreatment processes successfully tested down to RO provided final permeate streams with COD equal to 456 mg/L, 242 mg/L, and 385 mg/L (AS, BIO+AS and PC, respectively), complying with irrigation quality standards. However, UV/TiO_2 photocatalysis performed more efficiently, showing the highest membrane productivity within the shortest residence time (24 h). Coagulation-flocculation residence time was 72 h for both coagulant-flocculants tested. A significant reduction of polyphenols and final dry matter was attained with both coagulants, but with AH the process had to be stopped at the UF step since cost-effective permeate flow rates were not observed due to quick fouling build-up on the membrane. Otherwise, much longer residence time (seven days) was needed for biodigestion.

Following this, Stoller [47] conducted a deeper study on how flocculation as pretreatment affected the performance of MF, UF, NF, and RO membranes in the treatment of three-phase OMW, by examining the particle size distribution in the effluent of each stage. Stoller underlined the effect created by a secondary flocculation induced by the AS flocculant-derived salts accumulating near the membrane surface, which gathered particles together into bigger aggregates that could be more easily carried away by the tangential flow, thus sensibly reducing fouling.

In sum, in these studies Stoller underlined the key importance of the pretreatment and highlighted that higher pollutants reductions, e.g., in the COD value, do not guarantee the utmost suitability of the adopted pretreatment process. The fact that the pretreatment brings off pollutant particle size (d_p) far away from that of the membrane pores (D_p) is of paramount importance to further enhance the steady-state flux values. In conclusion, a non-linear relationship between the critical flux value and the pore-blocking particles density was remarked.

In another study, Akdemir and Ozer [48] proposed a membrane treatment process based on UF (MWCO 30–100 kDa) preceded by pH adjustment (acidic or alkaline) and cartridge filtration (20 µm). In detail, the pretreatment reached 63% COD removal. These authors reported that the highest permeate flux productivity obtained with the 100k Da UF membrane could be attained upon operating conditions of 200 L/h flow rate and an operating pressure equal to 4 bar. However, an operating pressure of 1 bar was observed to be more adequate in order to preserve the UF membrane from fouling issues.

Finally, the COD, TOC, SS, and oil and grease concentrations in the stream exiting the proposed treatment procedure were equal to 6.4 g/L, 2.5 g/L, 320 mg/L and 270 mg/L, respectively, corresponding to removal ratios of 92.3%, 92.7%, 97.1%, and 98.9%. However, the pursued compliance with Turkish standards for wastewater discharge into sewers was not achieved.

Coskun *et al.* [49] investigated the treatment of three-phase OMW (Turkey) previously centrifuged, then filtered via UF membranes followed by NF, and finally RO membranes. The fluxes reached values of up to 21.2–28.3 L/m²h for the NF membranes whereas 12.6–15.5 L/m²h for the RO membranes, respectively. The maximum COD removal efficiencies obtained at 10 bar ranged from 59.4–79.2% for the NF membranes, whereas between 96.2% and 96.3% for the RO membranes, respectively. For these latter RO membranes, the conductivity removal efficiencies obtained at 25 bar ranged between 93.2% and 94.8%, respectively. One aspect not considered by these authors was the effect of the fouling issues on the steady-state and long-run performance of the used membranes, which was not reported but should be further investigated.

Zirehpour *et al.* [50] examined an integrated MF-UF–NF membrane system for the purification of three-phase OMW in Iran. The effluent was pre-filtered by three-step MF in series, with nominal pore sizes of 50, 5 and 0.2 μm, in concentration mode, subsequently followed by two and three UF and NF membranes, respectively.

One positive aspect of this study is that the authors did perform an analysis of the fouling behavior of the used UF and NF membranes, by assessment of the flux recovery ratio and degree of the total flux loss during volume reduction factor experiments. The commercial UF membrane provided higher permeate flux than the lab-made one, but it was pointed that the antifouling properties and rejection efficiency of the latter were significantly better. A specific arrangement of the integrated membrane system was concluded to be the UF membrane followed by two-step NF membranes in series, the first NF step providing high flux while the second one providing high rejection.

Di Lecce *et al.* [51] recently investigated the fractionation of OMW (three-phase, Italy) using a two-step MF (tubular polypropylene) and NF (polyamide thin-film composite spiral wound) membrane process at pilot scale. Previously, the OMW samples were filtered through a cotton fabric filter to reduce the suspended solid concentration from 5.4% to 3.8% (w/w). Results revealed a rejection of the NF membrane towards COD, dry matter, phenolic compounds and antioxidant activity higher than 98%, independently of the volume concentration factor. Finally, the purified NF permeate stream obtained presented COD and phenolic contents values very close to those requested for discharge into surface waters. Nevertheless, no dynamic behavior of the membrane system was reported by the authors, nor the permeate flux profile vs. the operation time, nor the fouling build-up rates on the used membranes, which are crucial prints in order to balance the economic viability of the proposed treatment process.

MEMBRANE TREATMENT OF TWO-PHASE OLIVE MILL WASTEWATER

It is worth pointing out that the existing studies report mainly membrane treatment processes for OMW exiting olive mills operating with the three-phase olive oil production technology, but few membrane studies on OMW generated by olive factories working with the two-phase extraction technology can be found. Also, the existing research works are focused principally on the use of UF and NF membranes, yet there is a gap of knowledge on RO purification of OMW.

Ochando-Pulido *et al.* [35,36,37] studied the reclamation of two-phase OMW (Spain) by RO on a bench-scale, to achieve the quality to recirculate the final effluent to the olives washing machines of the manufacture process to finally close the loop. The raw OMW was previously subjected to an advanced oxidation process (AOP) based on homogeneous Fenton-like reaction [21,22]. This AOP achieved high abatement of pesticides as well as polyphenols and tannins, besides other organic non-humic and humic contaminants associated to OMW. These compounds are resistant to biological degradation and phytotoxic, and have also demonstrated development of fouling on the membranes. The authors highlighted high and stable permeate flux was provided upon recirculation of a fraction of the permeate stream. Under those conditions, 100% suspended solids, phenols and iron removal was achieved, in addition to 99.4% and 98.2% overall COD and conductivity rejection efficiencies, respectively.

On the other hand, Ochando-Pulido *et al.* [3] studied a batch membranes-in-series processes, in detail UF followed by NF and RO, for the reclamation of two-phase olive mill wastewater (OMW-2). In this work, a pH-T flocculation process followed by photocatalysis with ferromagnetic-core titanium dioxide under ultraviolet irradiation (UV/TiO$_2$) was examined as pretreatment. The adoption of this treatment sequence helped in reducing the required membrane area, equal to 104.6 m^2 and 81.4 m^2 for the UF and NF membranes, respectively, leading to a limited need of overdesign of the membrane plant. In addition, the use of the applied UV/TiO$_2$ photocatalysis process enhanced the productivity and longevity of both membranes, and permitted achieving a final treated effluent stream compatible for irrigation purposes.

Furthermore, Ochando-Pulido *et al.* [52] proposed another integrated batch UF and NF membranes-in-series process for the simultaneous treatment of the two main effluents generated in olive mills operating with the two-phase technology, in particular wastewater derived from the washing of the olives (OWW) and from the olive oil washing during the vertical centrifugation (OMW). Beforehand, the raw effluent, that is, 1:1 v/v mixture of OWW and OMW, designed as OWMW-2, was pretreated by pH-temperature flocculation followed by photocatalysis with TiO$_2$ nanoparticles under UV irradiation. Significant and stable fluxes were observed on both UF and NF membranes, 15.5 and 22.2 L/m^{2}h, respectively. Finally, the treatment line just comprising UF preceded by pH-T flocculation and UV/TiO$_2$ photocatalysis provided an effluent compatible for irrigation and permitted the minimum membrane plant dimension, with only six total modules.

THE USE OF MEMBRANE BIOREACTORS FOR OMW RECLAMATION

The main handicaps in the treatment of OMW by membrane bioreactors (MBRs) are basically, on one hand, the high residence times required, and on the other hand, the high load of recalcitrant organic compounds present in this type of effluent, that typically hinder the efficiency of biological processes; moreover, the high organic matter concentration in OMW also multiplies the fouling propensity of MBR systems, principally because it represents a good culture medium for deleterious bio-fouling, which appears in the form of gel layers (mainly caused by extracellular polymeric substances) and pore blocking caused by microorganisms as well as the organic particles themselves.

The feasibility of the use of MBRs for the treatment of OMW (three-phase, Tunisia) was examined in a research study by Dhaouadi and Marrot [53]. These authors used an external ceramic UF MBR with 0.1 μm MWCO. The experimental study was carried out with various diluted OMW solutions continuously fed to the reactor, in order to permit the employed biomass to become gradually acclimated. It is worth highlighting that, by using backpulse combinations (1 s/1 min), a significant and stable permeate flux (100 L/hm^2) could be successfully attained, with zero suspended solids and almost no phenolic compounds. Moreover, backpulsing enabled maintaining a constant permeate flux over a period of several days.

In another study, Conidi et al. [54] proposed an integrated membrane treatment for three-phase OMW in Italy, comprising MF (0.2 μm) followed by UF (10 kDa MWCO) membranes and a final MBR, to achieve the reclamation and valorization of OMW from a three-phase olive oil production process in Italy. The MF membrane permitted the removal of suspended solids, after which low molecular weight polyphenolic compounds could be recovered in the UF permeate stream.

The novelty of this study resides in the fact that the UF permeate was finally driven to a multiphase biocatalytic MBR, where the isomer of oleuropein-aglycon was isolated from the phenolic fraction produced during the former oleuropein hydrolysis reaction step, achieving a maximum conversion of 45.7%. This may give a sensible boost to the treatment of OMW, counterbalancing the energy costs.

NON-CONVENTIONAL MEMBRANE PROCESSES

Russo [55] studied the reclamation of three-phase OMW (Italy) by preliminary MF followed by two-stepped UF (6 kDa followed by 1 kDa membranes) and final RO operation. Best productivities upon lowest fouling were 50 L/hm^2 for

VRF equal to 3 for the ceramic MF membrane, whereas between 10–15 L/hm^2 for UF with polymeric membranes of the MF permeate, and up to 35 L/hm^2 when ulfrafiltering the UF permeate with 1 kDa ceramic membranes. Finally, 20–25 L/hm^{2}were yielded by the RO membrane.

The author reported high concentration of low molecular weight (LMW) polyphenols (349 ppm from initial 55 ppm in the raw OMW samples) in the MF permeate, 76% being hydroxytyrosol. Finally, the RO retentate stream contained enriched and purified LMW polyphenols, proposed by Russo for food, pharmaceutical or cosmetic industries, whereas MF and UF retentates were suggested as fertilizers or for production of biogas in anaerobic reactors.

Garcia-Castello *et al.* [56] proposed another treatment method for three-phase OMW in Italy, comprising MF, NF, and finally vacuum membrane distillation (VMD) or osmotic distillation (OD). MF ensured 91% and 26% reduction of suspended solids and total organic carbon (TOC), respectively, as well as 78% recovery of the initial content of polyphenols in the permeate stream. Further, NF achieved the recovery of most polyphenols in the permeate, which was enriched by ulterior treatment by osmotic distillation (OD) or vacuum membrane distillation (VMD).

A final solution containing about 0.5 g/L free LMW polyphenols, with hydroxytyrosol representing 56% of the total amount, was produced by using calcium chloride dihydrate solution as brine. The authors highlighted the interest of these formulations for food, cosmetics, and pharmaceuticals.

Recently, Gebreyohannes *et al.* [57] investigated the treatment of three-phase OMW (Italy) by forward osmosis (FO) with a cellulose acetate membrane. They proposed a single-step FO operated with 3.7 m MgCl$_2$ draw solution and 6 cm/s crossflow velocity. This treatment provided a volume reduction of 71%, complete decolorization of the permeate, and more than 98% rejection to OMW components, including biophenols and ions, thus making FO more attractive from the point of view of its cost-efficiency. Moreover, the authors studied a MBR-based pre-treatment prior to FO, which reduced pectins concentration by 92.3%, thus resulting in 30% flux enhancement.

The authors addressed the fouling issues occurring on the used membrane, and reported that 95% pure water permeability could be recovered by applied a cleaning cycle based on osmotic back-flushing, after continuous OMW dehydration tests carried out for 200 h.

As pointed by these authors, various studies have shown that fouling occurring in FO is in most part reversible, as a result of the low foulant compaction given the negligible operating pressure gradient. Hence, FO holds a great potential to treat wastewater streams such as OMW, which has high fouling propensity. However, still low permeate productivities were reported in this research paper, and a certain permeate flux loss due to membrane fouling upon increasing the volume recovery was always noticed. Therefore, further investigation should be performed, but FO may be a promising technique for OMW reclamation.

FOULING MITIGATION AND CONTROL IN MEMBRANE TREATMENT OF OMW

In order to successfully implement a membrane process for a specific application at industrial scale, the prediction of the performance of the selected membrane is mandatory. Here an additional difficulty is given by concentration polarization and membrane fouling phenomena occurring over the membrane during the operation time, which alter the membrane output continuously.

Fouling mitigation and control is mandatory to ensure the competitiveness and economic efficiency of membrane technology. This is especially relevant in the case of membrane processes applied for the treatment and reclamation of wastewater streams like OMW, where the added value of the product, that is purified water, is low, and thus key are costs.

Fouling reduces the membrane performances in time and this leads to energy costs increments and premature substitution of the membrane modules. Since wastewater treatment must imply low operating costs, the substitution of the membrane modules cannot be performed frequently. Consistent fouling inhibition methods may assure this result, thus making membrane processes for OMW streams treatment both technically and economically feasible.

One of the possible solutions to increase the reliability of a process is the use of stable control systems. In the case of most membrane processes, this is performed by simple control strategies that do not include knowledge and control of fouling. The complete lack of advanced control systems in membrane technologies, capable of taking fouling issues into account, limits the reliability of the technology and represents one key problem to be solved to permit its further maturation. In order to achieve this result, the fouling behaviour of the system must be estimated *a priori*.

In the last years, Stoller and Ochando have successfully applied the concepts of the critical and threshold flux for fouling and process control during membrane treatment of OMW from both two-phase and three-phase systems [58,59,60,61,62,63].

The concept of the critical flux was theoretically proven and physically explained by Bacchin *et al.* [64], who proposed the first theoretical model giving explanation to membrane transport phenomena of colloidal particles. They gave a first definition of the critical flux, stating that it is the performance point above which the repulsive barrier is overcome, and below which no fouling occurs. Afterwards, Field *et al.* [65] gave an empirical approach of the concept of the critical flux for MF membranes, defining it as the permeate flux which can be successfully attained without incurring in fouling formation during the operation time. Later on, this concept was also extended to UF and NF membranes [66].

Nevertheless, some authors noted that critical flux behavior is not always observed strictly in all membrane separation processes, pointing to the idea that it might not be possible to completely inhibit fouling during the operation of some liquid-liquid membrane systems, and the case of wastewater treatments was indicated as a clear example [67,68,69]. These researchers noticed that fouling was unavoidable to a certain extent at every operating condition, and thus the concept of the threshold flux was introduced [66]. In contrast with the critical flux, the threshold flux instead makes reference to the maximum permeate flux at which fouling builds up at a very low and constant rate, and above which the rate of fouling becomes exponentially increased. In other words, the threshold flux divides a low fouling region from a high fouling region of pressure-driven membrane processes. Recently, Stoller and Ochando [69] verified the validity of this theory in the treatment of OMW by UF and NF membranes.

Long-term studies on the treatment of OMW are also relevant to examine the efficacy and cost-efficiency of membrane processes applied for the reclamation of these effluents in the long run. In this regard, Stoller published a very valuable work on a three-year long experience of effective fouling inhibition by threshold flux based optimization methods on a NF membrane module for OMW treatment [70]. The used NF membrane module could be successfully operated for three years during pilot-scale work. This could be accomplished by the adoption of appropriate fouling inhibition control, relaying on both critical flux measurements and the development of an optimized operation method. The critical flux theory was successfully applied to this system, but was not capable to explain the observed fouling behavior of the examined membrane, which could be explained instead by the threshold flux model.

Moreover, Stoller *et al.* reported in a recent paper [59] a reliable method for the conversion of critical flux measurement data into threshold flux measurements of a NF membrane module in the treatment of three-phase OMW. Stoller highlighted the important need to develop methods capable of measuring threshold fluxes quickly, and to convert critical flux measurement data into threshold flux data. In a recent work, both critical and threshold flux concepts, which share many common aspects, were merged for simplification purpose by Stoller and Ochando into a new concept, the boundary flux, introduced in year 2014 [71]. The authors underlined that the knowledge of real-time boundary flux values is a key factor to design stable control systems for membrane processes, since operation within sub-boundary flux conditions avoids irreversible fouling and thus premature technical (and economical) failures. The boundary flux values are sensibly influenced by those parameters affecting the critical and threshold fluxes, that is: hydrodynamics, temperature, membrane properties, time, and feedstock characteristics. This concept was successfully applied in the treatment of OMW [72,73].

In particular, in the case of batch membrane treatments used to reduce the required membrane area, the fact that the bulk becomes increasingly concentrated during operation increases the difficulty in setting the adequate operating conditions. In these cases, Stoller and Ochando [71,72,73] point that the boundary flux value estimated at the end of the batch should be adopted as the target flux value initially. A deeper analysis on the concept of the boundary flux for both batch and continuous membrane operations can be found in the Boundary Flux Handbook [71].

In view of the available research papers on fouling issues in the treatment of OMW by membranes, future investigation on the fouling mechanisms occurring during membrane treatments of OMW is needed in order to fully understand the interactions among the effluent particles and the membrane surface. In this regard, the critical and threshold models, gathered in the boundary flux theory, seems a good approach which should be further studied.

CONCLUSIONS

In the present work, a review on the actual state of the art on the treatment of olive mill wastewaters (OMW) by membrane technologies is addressed. The study focuses on microfiltration (MF), ultrafiltration (UF), nanofiltration (NF), and reverse osmosis (RO), as well as membrane bioreactors (MBR) and other non-conventional membrane processes such as vacuum distillation (VD), osmotic distillation (OD), and forward osmosis (FO).

A wide range of reclamation treatments as well as integrated processes for the treatment of OMW have already been proposed and developed by scientists and engineers, but not led to completely satisfactory results due to cost-ineffectiveness. The olive oil industry in its current status, composed of little and dispersed factories, cannot bear such high costs. Furthermore, these treatments are not able to abate the high concentration of dissolved ionic concentration present in these effluents.

Membrane processes are becoming extensively used to replace many conventional processes in the purification of water and groundwater as well as in the reclamation of wastewater streams of very diverse sources, such as those generated by agro-industrial activities. However, only a limited number of studies have been published so far regarding OMW treatment by membrane technologies.

As it can be asserted from the present review, there are still some unresolved problems that slow down large-scale membrane applications with respect to OMW management, despite the promising perspectives. A brief insight into inhibition and control of fouling by properly-tailored pretreatment processes upstream the membrane operation and the use of the critical and threshold flux theories is also highlighted as important to ensure the cost-effectiveness of the treatment of OMW by membrane processes when transferred to the industrial scale. With respect to this, several membrane materials, configurations and pore sizes have been tested up to now, and also different pretreatments prior to membrane operations. These pretreatments provide different organic concentration abatement and also shift the particle size distribution of the colloidal and suspended matter differently. The latter is also a relevant factor: according to the pore blocking model, particles of certain sizes tend to cause fouling issues in a short time, which are those having a size similar to that of pores.

As future tasks, research on cost-effective OMW pretreatments, either new ones or more efficient combinations of those already existing, should keep on to ensure steady state efficiencies, and further research on optimized operating conditions for each integrated pretreatment-membrane operation, as well as improvement of the existing membrane materials and new ones, which may help in enhancing the process' economic efficiency. In this regard, the boundary flux theory seems a good approach to be further studied.

ACKNOWLEDGMENTS

Spanish Ministry of Science and Innovation is gratefully acknowledged for funding the project CTQ2010-21411: Depuration of wastewater from olive oil industry for its reutilization in the process. European project PHOTOMEM (contract no.FP7-SME-2011, grant 262470) and European project ETOILE (contract no. FP7-SME-2007-1, grant 222331) are also very much acknowledged.

AUTHOR CONTRIBUTIONS

Javier Miguel Ochando Pulido performed the data collection and elaboration of the manuscript, while Antonio Martínez Férez is acknowledged for revising the manuscript.

REFERENCES

1. International Olive Oil Council (IOOC). Available online: http://www.internationaloliveoil.org/ (accessed on 1 June 2015).

2. Mendoza, A.; Hidalgo-Casado, F.; Ruiz-Gómez, M.A.; Martínez-Román, F.; Moyano-Pérez, M.J.; Cert-Ventulá, A.; Pérez-Camino, M.C.; Ruiz-Méndez, M.V. Characteristics of olive oils from First and second centrifugation. *Oil Grease* 1996, *47*, 163–181.

3. Ochando-Pulido, J.M.; Hodaifa, G.; Victor-Ortega, M.D.; Rodriguez-Vives, S.; Martinez-Ferez, A. Effective treatment of olive mill effluents from two-phase and three-phase extraction processes by batch membranes in series operation upon threshold conditions. *J. Hazard. Mater.* 2013, *263*, 168–176.

4. Voreadou, K. Olive Mill Wastewater: A Trial for the Water Ecosystems. In Proceedings of Symposium on Treatment of Wastes from Olive Mills; Greek Agrotechnical Society: Heraklion, Greece, 1989.

5. Ntougias, S.; Gaitis, F.; Katsaris, P.; Skoulika, S.; Iliopoulos, N.; Zervakis, G.I. The effects of olives harvest period and production year on olive mill wastewater properties—Evaluation of Pleurotus strains as bioindicators of the effluent's toxicity. *Chemosphere* 2013, *92*, 399–405.

6. Guadalquivir Hydrographical Confederation. Available online: http://www.chguadalquivir.es/opencms/portalchg/index.html (accessed on 1 June 2015).

7. Baccar, R.; Bouzid, J.; Feki, M.; Montiel, A. Preparation of activated carbon from Tunisian olive-waste cakes and its application for adsorption of heavy metal ions. *J. Hazard. Mater.* 2009, *162*, 1522–1529.

8. Bouzid, J.; Elouear, Z.; Ksibi, M.; Feki, M.; Montiel, A. A study on removal characteristics of copper from aqueous solution by sewage sludge and pomace ashes. *J. Hazard. Mater.* 2008, *152*, 838–845.

9. Akar, T.; Tosun, I.; Kaynak, Z.; Ozkara, E.; Yeni, O.; Sahin, E.N.; Akar, S.T. An attractive agro-industrial by-product in environmental cleanup: Dye biosorption potential of untreated olive pomace. *J. Hazard. Mater.* 2009, *166*, 1217–1225.

10. Stasinakis, A.S.; Elia, I.; Petalas, A.V.; Halvadakis, C.P. Removal of total phenols from olive-mill wastewater using an agricultural by-product, olive pomace. *J. Hazard. Mater.* 2008, *160*, 408–413.

11. Haddadin, M.S.Y.; Haddadin, J.; Arabiyat, O.I.; Hattar, B. Biological conversion of olive pomace into compost by using Trichoderma harzianum and Phanerochaete chrysosporium. *Bioresour. Tech.* 2009, *100*, 4773–4782.

12. Tekin, A.R.; Coşkun Dalgıç, A. Biogas production from olive pomace. *Resour. Conserv. Recycl.* 2000, *30*, 301–313.

13. Annesini, M.; Gironi, F. Olive oil mill effluent: ageing effects on evaporation behavior. *Water Res.* 1991, *25*, 1157–1960.

14. Papadimitriou, E.K.; Chatjipavlidis, I.; Balis, C. Application of composting to olive mill wastewater treatment.*Environ. Technol.* 1997, *18*, 10–107.

15. Aktas, E.S.; Imre, S.; Esroy, L. Characterization and lime treatment of olive mill wastewater. *Water Res.* 2001, *35*, 2336–2340.

16. Al-Malah, K.; Azzam, M.O.J.; Abu-Lail, N.I. Olive mills effluent wastewater post-treatment using activated clay. *Sep. Purif. Technol.* 2000, *20*, 225–234.

17. Nieto, L.M.; Hodaifa, G.; Rodríguez, S.; Giménez, J.A.; Ochando, J. Flocculation-sedimentation combined with chemical oxidation process. *Clean Soil Air Water* 2011, *39*, 949–955.

18. Sarika, R.; Kalogerakis, N.; Mantzavinos, D. Treatment of olive mill effluents. Part II. Complete removal of solids by direct flocculation with poly-electrolytes. *Environ. Int.* 2005, *31*, 297–304.

19. Hodaifa, G.; Eugenia-Sánchez, M.; Sánchez, S. Use of industrial wastewater from olive-oil extraction for biomass production of Scenedesmus obliquus. *Bioresour. Technol.* 2008, *99*, 1111–1117.

20. Cañizares, P.; Paz, R.; Sáez, C.; Rodrigo, M.A. Costs of the electrochemical oxidation of wastewaters: a comparison with ozonation and Fenton oxidation processes. *J. Environ. Manag.* 2009, *90*, 410–420.

21. Nieto, L.M.; Hodaifa, G.; Rodríguez, S.; Giménez, J.A.; Ochando, J. Degradation of organic matter in olive oil mill wastewater through homogeneous Fenton-like reaction. *Chem. Eng. J.* 2011, *173*, 503–510.

22. Hodaifa, G.; Ochando-Pulido, J.M.; Rodriguez-Vives, S.; Martinez-Ferez, A. Optimization of continuous reactor at pilot scale for olive-oil mill wastewater treatment by Fenton-like process. *Chem. Eng. J.* 2013, *220*, 117–124.

23. De Caprariis, B.; Di Rita, M.; Stoller, M.; Verdone, N.; Chianese, A. Reaction-precipitation by a spinning disc reactor: Influence of hydrodynamics on nanoparticles production. *Chem. Eng. Sci.* 2012, *76*,

73–80.

24. Sacco, O.; Stoller, M.; Vaiano, V.; Ciambelli, P.; Chianese, A.; Sannino, D. Photocatalytic degradation of organic dyes under visible light on *n*-doped photocatalysts. *Int. J. Photoenergy* 2012, *2012*, 626759:1–626759:8.

25. Papastefanakis, N.; Mantzavinos, D.; Katsaounis, A. DSA electrochemical treatment of olive mill wastewater on Ti/RuO$_2$ anode. *J. Appl. Electrochem.* 2010, *40*, 729–737.

26. Tezcan Ün, Ü.; Altay, U.; Koparal, A.S.; Ogutveren, U.B. Complete treatment of olive mill wastewaters by electrooxidation. *Chem. Eng. J.* 2008, *139*, 445–452.

27. Cañizares, P.; Martinez, L.; Paz, R.; Saéz, C.; Lobato, J.; Rodrigo, M.A. Treatment of Fenton-refractory olive oil mill wastes by electrochemical oxidation with boron-doped diamond anodes. *J. Chem. Technol. Biotechnol.* 2006, *81*, 1331–1337.

28. De Heredia, J.B.; Garcia, J. Process integration: Continuous anaerobic digestion–ozonation treatment of olive mill wastewater. *Ind. Eng. Chem. Res.* 2005, *44*, 8750–8755.

29. Grafias, P.; Xekoukoulotakis, N.P.; Mantzavinos, D.; Diamadopoulos, E. Pilot treatment of olive pomace leachate by vertical-flow constructed wetland and electrochemical oxidation: An efficient hybrid process. *Water Res.* 2010, *44*, 2773–2780.

30. Lafi, W.K.; Shannak, B.; Al-Shannag, M.; Al-Anber, Z.; Al-Hasan, M. Treatment of olive mill wastewater by combined advanced oxidation and biodegradation. *Sep. Purif. Technol.* 2009, *70*, 141–146.

31. Iaquinta, M.; Stoller, M.; Merli, C. Optimization of a nanofiltration membrane for tomato industry wastewater treatment. *Desalination* 2009, *245*, 314–320.

32. Luo, J.; Ding, L.; Wan, Y.; Jaffrin, M.Y. Threshold flux for shear-enhanced nanofiltration: experimental observation in dairy wastewater treatment. *J. Membr. Sci.* 2012, *409*, 276–284.

33. Bódalo, A.; Gómez, J.L.; Gómez, E.; Máximo, F.; Hidalgo, A.M. Application of reverse osmosis to reduce pollutants present in industrial wastewater. *Desalination* 2003, *155*, 101–108.

34. Stoller, M.; Bravi, M. Critical flux analyses on differently pretreated olive vegetation wastewater streams: Some case studies. *Desalination* 2010, *250*, 578–582.

35. Ochando-Pulido, J.M.; Hodaifa, G.; Martinez-Ferez, A. Permeate recirculation impact on concentration polarization and fouling on RO

purification of olive mill wastewater. *Desalination* 2014, *343*, 169–179.

36. Ochando-Pulido, J.M.; Rodriguez-Vives, S.; Martinez-Ferez, A. The effect of permeate recirculation on the depuration of pretreated olive mill wastewater through reverse osmosis membranes. *Desalination* 2012, *286*, 145–154.

37. Ochando-Pulido, J.M.; Hodaifa, G.; Rodriguez-Vives, S.; Martinez-Ferez, A. Impacts of operating conditions on reverse osmosis performance of pretreated olive mill wastewater. *Water Res.* 2012, *46*, 4621–4632.

38. Turano, E.; Curcio, S.; De Paola, M.G.; Calabrò, V.; Iorio, G. An integrated centrifugation–ultrafiltration system in the treatment of olive mill wastewater. *J. Membr. Sci.* 2002, *206*, 519–531.

39. Paraskeva, C.A.; Papadakis, V.G.; Tsarouchi, E.; Kanellopoulou, D.G.; Koutsoukos, P.G. Membrane processing for olive mill wastewater fractionation. *Desalination* 2007, *213*, 218–229.

40. Harvey, P.J.; Campanella, B.F.; Castro, P.M.L.; Harms, H.; Lichtfouse, E.; Schaffner, A.R.; Smreck, S.; Werck-Reichhart, D. Phytoremediation of polyaromatic hydrocarbons, anilines and phenols. *Environ. Sci. Pollut. Res.* 2002, *9*, 29–47.

41. Bais, H.P.; Walker, T.S.; Stermitz, F.R.; Hufbauer, R.A.; Vivanco, J.M. Enantiomeric-dependent phytotoxic and antimicrobial activity of (+/−)-catechin. A rhizosecreted racemic mixture from spotted knapweed. *Plant Physiol.* 2002,*128*, 1173–1179.

42. Capasso, R.; Evidente, A.; Schivo, L.; Orru, G.; Marcialis, M.A.; Cristinzio, G. Antibacterial polyphenols from olive oil mill waste waters. *J. Appl. Bacteriol.* 1995, *79*, 393–398.

43. Garcia, I.; Jimenez Pena, P.R.; Bonilla Venceslada, J.L.; Martin, A.; Martin Santos, M.A.; Ramos Gomez, E. Removal of phenol compounds from olive oil mill wastewater using Phanerochaete chrysosporium, Aspergillus niger, Aspergillus tereus and Geotrichum candidum. *Proc. Biochem.* 2000, *35*, 751–758.

44. Stoller, M.; Chianese, A. Optimization of membrane batch processes by means of the critical flux theory. *Desalination*2006, *191*, 62–70.

45. Stoller, M.; Chianese, A. Influence of the adopted pretreatment process on the critical flux value of batch membrane processes. *Ind. Eng. Chem. Res.* 2007, *46*, 2249–2253.

46. Stoller, M.; Chianese, A. Technical optimization of a batch olive wash wastewater treatment membrane plant.*Desalination* 2006, *200*, 734–736.

47. Stoller, M. On the effect of flocculation as pretreatment process and

particle size distribution for membrane fouling reduction. *Desalination* 2009, *240*, 209–217.

48. Akdemir, E.O.; Ozer, A. Investigation of two ultrafiltration membranes for treatment of olive oil mill wastewater.*Desalination* 2009, *249*, 660–666.

49. Coskun, T.; Debik, E.; Demir, N.M. Treatment of olive mill wastewaters by nanofiltration and reverse osmosis membranes. *Desalination* 2010, *259*, 65–70.

50. Zirehpour, A.; Jahanshahi, M.; Rahimpour, A. Unique membrane process integration for olive oil mill wastewater purification. *Sep. Purif. Technol.* 2012, *96*, 124–131.

51. Di Lecce, G.; Cassano, A.; Bendini, A.; Conidi, C.; Giorno, L.; Gallina, T. Characterization of olive mill wastewater fractions treatment by integrated membrane process. *J. Sci. Food Agric.* 2014, *94*, 2935–2942.

52. Ochando-Pulido, J.M.; Verardo, V.; Segura-Carretero, A.; Martinez-Ferez, A. Technical optimization of an integrated UF/NF pilot plant for conjoint batch treatment of two-phase olives and olive oil washing wastewaters. *Desalination*2015, *364*, 82–89.

53. Dhaouadi, H.; Marrot, B. Olive mill wastewater treatment in a membrane bioreactor: process stability and fouling aspects. *Environ. Technol.* 2010, *31*, 761–770.

54. Conidi, C.; Mazzei, R.; Cassano, A.; Giorno, L. Integrated membrane system for the production of phytotherapics from olive mill wastewaters. *J. Membr. Sci.* 2014, *454*, 322–329.

55. Russo, C. A new membrane process for the selective fractionation and total recovery of polyphenols, water and organic substances from vegetation waters (VW). *J. Membr. Sci.* 2007, *288*, 239–246.

56. Garcia-Castello, E.; Cassano, A.; Criscuoli, A.; Conidi, C.; Drioli, E. Recovery and concentration of polyphenols from olive mill wastewaters by integrated membrane system. *Water Res.* 2010, *44*, 3883–3892.

57. Gebreyohannes, A.Y.; Curcio, E.; Poerio, T.; Mazzei, R.; Di Profio, G.; Drioli, E.; Giorno, L. Treatment of olive mill wastewater by forward osmosis. *Sep. Purif. Technol.* 2015, *147*, 292–302.

58. Stoller, M. Effective fouling inhibition by critical flux based optimization methods on a NF membrane module for olive mill wastewater treatment. *Chem. Eng. J.* 2011, *168*, 1140–1148.

59. Stoller, M.; Bravi, M.; Chianese, A. Threshold flux measurements of a nanofiltration membrane module by critical flux data conversion.

Desalination 2013, *315*, 142–148.

60. Stoller, M.; de Caprariis, B.; Cicci, A.; Verdone, N.; Bravi, M.; Chianese, A. About proper membrane process design affected by fouling by means of the analysis of measured threshold flux data. *Sep. Purif. Technol.* 2013, *114*, 83–89.

61. Ochando-Pulido, J.M.; Stoller, M.; Bravi, M.; Martinez-Ferez, A.; Chianese, A. Batch membrane treatment of olive vegetation wastewater from two-phase olive oil production process by threshold flux based methods. *Sep. Purif. Technol.* 2012, *101*, 34–41.

62. Ochando-Pulido, J.M.; Hodaifa, G.; Victor-Ortega, M.D.; Martinez-Ferez, A. Fouling control by threshold flux measurements in the treatment of different olive mill wastewater streams by membranes-in-series process. *Desalination* 2014, *343*, 162–168.

63. Ochando-Pulido, J.M.; Hodaifa, G.; Martinez-Ferez, A. Threshold flux measurement of an ultrafiltration membrane module in the treatment of two-phase olive mill wastewater. *Chem. Eng. Res. Des.* 2014, *92*, 769–777.

64. Bacchin, P.; Aimar, P.; Sanchez, V. Influence of surface interaction on transfer during colloid ultrafiltration. *J. Membr. Sci.* 1996, *115*, 49–63.

65. Field, R.W.; Wu, D.; Howell, J.A.; Gupta, B.B. Critical flux concept for microfiltration fouling. *J. Membr. Sci.* 1995, *100*, 259–272.

66. Field, R.W.; Pearce, G.K. Critical, sustainable and threshold fluxes for membrane filtration with water industry applications. *Adv. Colloid Interface Sci.* 2011, *164*, 38–44.

67. Le-Clech, P.; Chen, V.; Fane, T.A.G. Fouling in membrane bioreactors used in wastewater treatment. *J. Membr. Sci.* 2006, *284*, 17–53.

68. Luo, J.; Ding, L.; Wan, Y.; Jaffrin, M.Y. Threshold flux for shear-enhanced nanofiltration: experimental observation in dairy wastewater treatment. *J. Membr. Sci.* 2012, *409*, 276–284.

69. Stoller, M.; Ochando-Pulido, J.M. Going from a critical flux concept to a threshold flux concept on membrane processes treating olive mill wastewater streams. *Procedia Eng.* 2012, *44*, 607–608.

70. Stoller, M. A three-year long experience of effective fouling inhibition by threshold flux based optimization methods on a NF membrane module for olive mill wastewater treatment. *Membranes* 2013, *32*, 37–42.

71. Stoller, M.; Ochando Pulido, J.M. *The boundary flux handbook: A comprehensive database of critical and threshold flux values for membrane practitioners*; Elsevier: Amsterdam, Netherlands, 2015.

72. Ochando-Pulido, J.M.; Stoller, M. Boundary flux optimization of a nanofiltration membrane module used for the treatment of olive mill wastewater from a two-phase extraction process. *Sep. Purif. Technol.* 2014, *130*, 124–131.

73. Stoller, M.; Ochando-Pulido, J.M.; Di Palma, L.; Martínez-Férez, A. Membrane process enhancement of 2-phase and 3-phase olive mill wastewater treatment plants by photocatalysis with magnetic-core titanium dioxide nanoparticles. *J. Ind. Eng. Chem.* 2015. in press.

Chapter 2

TIO₂-IMPREGNATED POROUS SILICA TUBE AND ITS APPLICATION FOR COMPACT AIR- AND WATER-PURIFICATION UNITS

Tsuyoshi Ochiai[1,2], Shoko Tago[1], Mio Hayashi[1], Hiromasa Tawarayama[3], Toshifumi Hosoya[4] and Akira Fujishima[1,2]

[1]Kanagawa Academy of Science and Technology, KSP building East 407, 3-2-1 Sakado, Takatsu-ku, Kawasaki, Kanagawa 213-0012, Japan

[2]Photocatalysis International Research Center, Tokyo University of Science, 2641 Yamazaki, Noda, Chiba 278-8510, Japan

[3]Optical Communications R&D Laboratories, Sumitomo Electric Industries, Ltd., 1 Taya-cho, Sakae-ku, Yokohama 244-8588, Japan

[4]R&D General Planning Division, Sumitomo Electric Industries, Ltd., 1-1-3, Shimaya, Konohana-ku, Osaka 554-0024, Japan

ABSTRACT

A simple, convenient, reusable, and inexpensive air- and water-purification unit including a one-end sealed porous amorphous-silica (a-silica) tube coated with TiO_2 photocatalyst layers has been developed. The porous a-silica layers were formed through outside vapor deposition (OVD). TiO_2 photocatalyst layers were formed through impregnation and calcinations onto a-silica layers. The resulting porous TiO_2-impregnated a-silica tubes were evaluated for air-purification capacity using an acetaldehyde gas decomposition test. The tube (8.5 mm e.d. × 150 mm) demonstrated a 93% removal rate for high concentrations (*ca.* 300 ppm) of acetaldehyde gas at a single-pass condition with a 250 mL/min flow rate under UV irradiation. The tube also demonstrated a water purification capacity at a rate 2.0 times higher than a-silica tube without TiO_2 impregnation. Therefore, the tubes have a great potential for developing compact and in-line VOC removal and water-purification units.

INTRODUCTION

Photocatalytic environmental purification, particularly VOC removal, has received increased attention owing to its low cost and enduring stability. However, popularly used photocatalysts and photocatalytic filters are significantly limited in their application due to relatively low purification efficiency and difficulty in handling the powder. Thus, although extensive research has been conducted on photocatalytic air purification, the difficulty in creating a practical air purifier has rendered it ineffective for implementation in real-world industrial technology. We have reported various methods for the design and application of a TiO_2 photocatalyst to maximize its photocatalytic abilities [1,2,3,4]. Recently, we have succeeded in the simple fabrication of novel one-end sealed porous TiO_2-coated amorphous-silica (a-silica) tubes with large porosity using the outside vapor deposition (OVD) method [5]. The tube was evaluated through *Escherichia coli* removal and Qβ phage inactivation testing. The impregnation method was used to fill TiO_2 precursor deep into the pores of one-end sealed porous a-silica tubes. The porous a-silica tubes were assayed for their VOC removal ability through an acetaldehyde decomposition test. In addition, the water purification ability of these tubes was preliminarily evaluated through the methylene blue decolorization test. These tests revealed more efficient materials, with emphasis on their ability to remove VOC.

RESULTS AND DISCUSSION

Characterization

The average bulk density and average porosity of the porous tubes were 0.84 g/cm^3 and 0.62, respectively. SEM images of the surface, secondary election images (SEIs) of the cross-section, and high-magnification SEIs of the cross-section of the porous TiO_2-impregnated a-silica tube are shown in Figure 1. Figure 1e–g shows high-magnification SEIs of cross-section of modified TiO_2 particles on the a-silica particles. White, gray, and black areas in Figure 1e–g represent TiO_2 particles, a-silica particles, and the resin intruding the pore, respectively. The estimated TiO_2 particle size is several tens of nanometers, which is smaller than the TiO_2 grain size in the TiO_2-coated a-silica tubes fabricated using the OVD method (several hundreds of nanometers) [5]. TiO_2 exists on the surface of a-silica skeleton even if it is located deep within a silica pore. However, in the deeper parts of the silica pores, the amount of observed TiO_2 declined. TiO_2 impregnation onto the porous silica tubes increase their pressure drops slightly but maintains a breathability sufficient to let the air or water pass through the tubes during purification or decomposition (Figure 2). The pore diameter of the tubes in this research can be estimated to 0.4 μm as the same as the previous research [5].

The Raman spectra of the TiO_2-impregnated a-silica tube and the TiO_2-coated a-silica tube by the OVD method are shown in Figure 3. The Raman spectrum of the TiO_2-coated a-silica tube by the OVD method is similar to the spectrum of the TiO_2 nanopowders with 60 wt. % of anatase content [5,6]. Repeating the heat process with a burner in the OVD method seemed to lead TiO_2 phase to rutile crystals. In contrast, the Raman bands of the TiO_2-impregnated a-silica tube at 142, 194, 396, 514, and 639 cm^{-1} are nearly identical to the spectrum of the anatase phase [7]. Thus, Raman spectroscopy indicates that the TiO_2 particles in the TiO_2-impregnated a-silica tube consisted of anatase crystals. Anatase crystals with exposed high-energy facets, including (001) and (010) facets, have attracted significant attention because of their high photocatalytic property [8,9]. The combination of TiO_2 particle size and crystal phase of the TiO_2-impregnated a-silica tube are more effective than the TiO_2-coated a-silica tube by the OVD method alone for photocatalytic capacity.

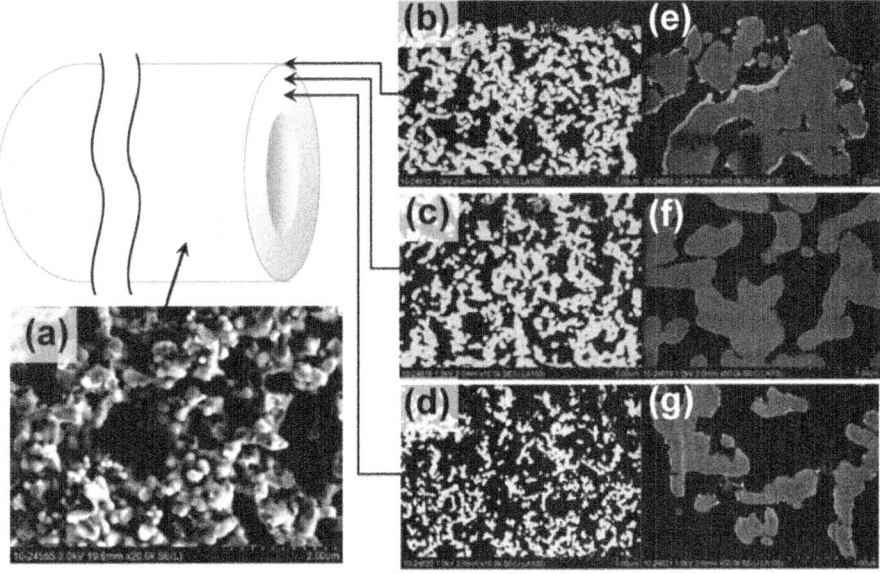

Figure 1: SEM images of the surface (a), secondary election images (SEIs) of the cross-section (b–d), and high-magnification SEIs of cross-section (e–g) of the TiO_2-impregnated a-silica tube. Cross-section images were obtained at 0 (b,e), 0.4 (c,f), 0.8 (d,g) mm from the surface.

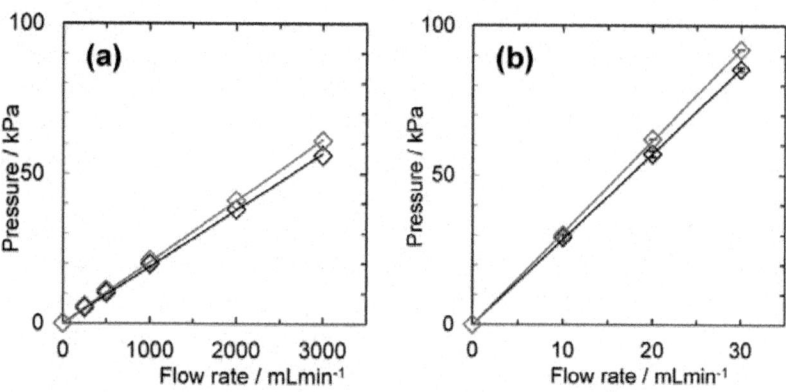

Figure 2: Pressure drops of the porous TiO_2-impregnated a-silica tube under the air (a) or water (b) flow. Black: a-silica tube without TiO_2 impregnation; red: TiO_2-impregnated a-silica tube.

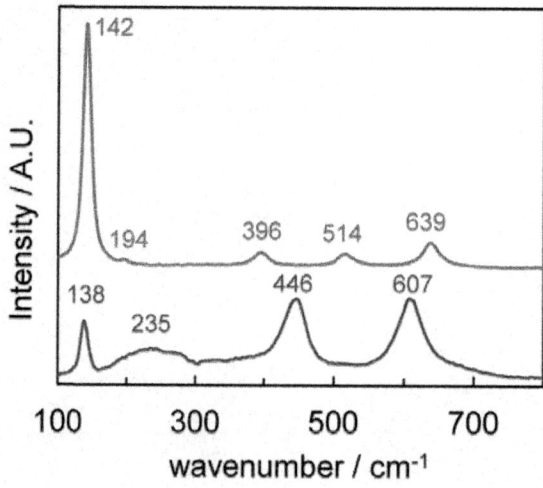

Figure 3: Raman spectra of the TiO_2-impregnated a-silica tube (red) and the TiO_2-coated a-silica tube by the outside vapor deposition (OVD) method (blue).

Results of Air- and Water-Purification Test

Figure 4a shows a typical data set of acetaldehyde removal (red) and CO_2 generation (green) by the TiO_2-impregnated a-silica tube with UV-C irradiation. The tube decomposed 100 ppm of acetaldehyde almost completely with a single-pass condition at a 250 mL/min flow rate. Under the same condition, the tube showed 93%, 78%, and 68% removal of 300, 700, and 1000 ppm of acetaldehyde, respectively (Figure 4b red). On the other hand, the a-silica tube covered with TiO_2/Ni-foam showed 89%, 46%, and 30% removal of 300, 700, and 1000 ppm of acetaldehyde, respectively (Figure 4b black). The TiO_2-coated a-silica tube by the OVD method could not remove high concentrations of acetaldehyde (Figure 4b blue). The significant difference between the TiO_2-impregnated a-silica tube and the TiO_2-coated a-silica tube by the OVD method may be caused by the particle size and crystal phase of the TiO_2. The high photocatalytic property of the anatase phase and smaller particle size of TiO_2 of the TiO_2-impregnated a-silica tube led to an effective decomposition of gaseous compounds [8,9]. However, the removal ratio of the tube was slightly decreased during long-term treatment (Figure 5). The data indicated that any type of deactivation process may occur. Now we are attempting to establish a re-activation method of the tube using a simple method such as washing or heating.

The methylene blue decomposition property of the TiO_2-impregnated a-silica tube exceeded that of the a-silica tube without TiO_2 impregnation during the experiments in which water passed through the tubes repeatedly (Figure 6). Both the decolorization behaviors occurred within the UV light and showed a similarity to the first order reaction equation. The reaction rate constant (k_1) of the TiO_2-impregnated a-silica tube (0.28, Figure 6 red) is 2.0 times higher than the k_1 of the a-silica tube without TiO_2 impregnation (0.14, Figure 6 white). These preliminarily evaluation indicate the potential for photocatalytic water purification ability of the tube [10].

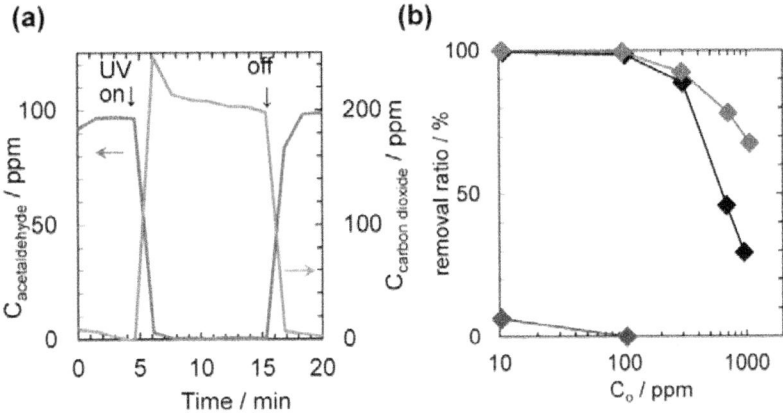

Figure 4: (a) Typical data set of acetaldehyde removal (red) and CO_2 generation

(green) by the TiO_2-impregnated a-silica tube with UV-C irradiation. (b) Removal ratio *vs.* initial concentration of acetaldehyde by the TiO_2-impregnated a-silica tube (red), a-silica tube covered with TiO_2/Ni-foam (black), TiO_2-coated a-silica tube by the OVD method (blue).

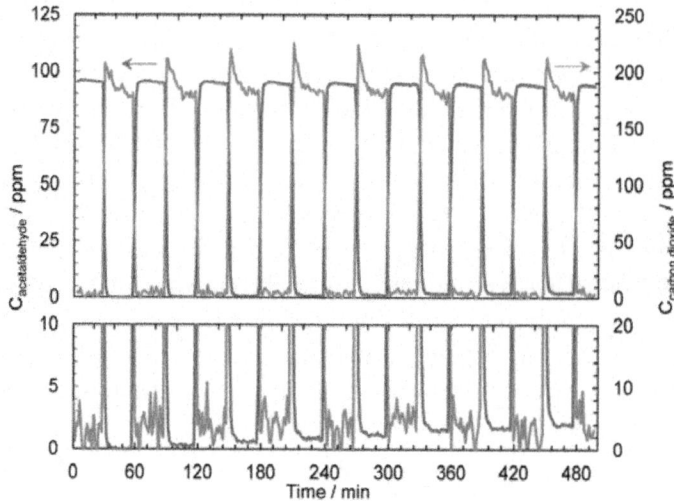

Figure 5: Data set of acetaldehyde removal (red) and CO_2 generation (green) by the TiO_2-impregnated a-silica tube with a 30/30-min on/off cycle of UV-C irradiation.

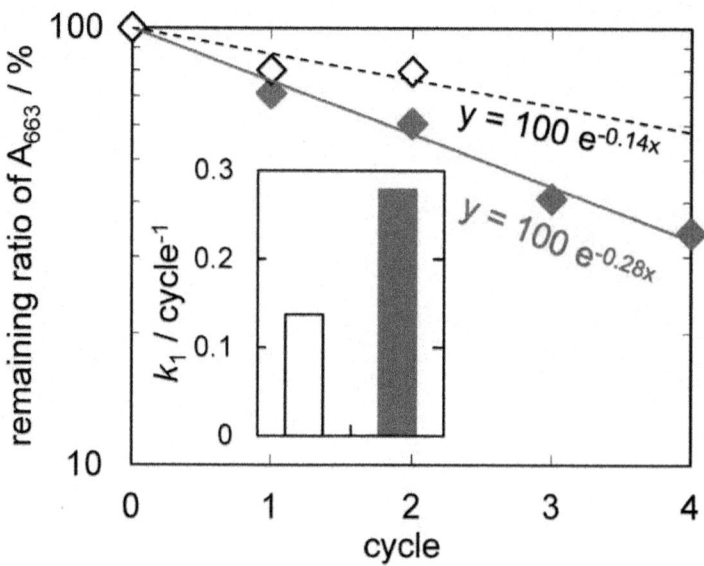

Figure 6: Plot showing the results of methylene blue decomposition test of porous

TiO_2-impregnated a-silica tube (red) and a-silica tube without TiO_2 impregnation (white). Inset: Reaction rate constants (k_1) of the tubes.

EXPERIMENTAL SECTION

Figure 7 shows the method of fabricating the porous TiO_2-impregnated a-silica tube. A-silica tubes with an external diameter of 8.6 mm, a thickness of 1.3 mm, and a length of 300 mm were fabricated using the OVD method [5,11] (Figure 7a). One-end sealed porous tubes were obtained by pulling out the rod target from the soot body (Figure 7b). Then the tube was soaked in a 1 M titanium(IV) isopropoxide/ethanol solution, pulled out of the solution, vacuumed to dry, heated at 550 °C for 1 h, and dried again after soaking in milli-Q under the effect of ultrasonic treatment (Figure 7c). The porous structure of the tubes was observed using an FE-SEM (S-4800, Hitachi, Tokyo, Japan). Samples for cross-section observation were prepared by embedding the tubes in resin and then polishing them with a cross-section polisher (SM-09010, JEOL, Tokyo, Japan). For the structural characterization of the TiO_2 layer, Raman spectroscopy excited by 532 nm Nd:YAG laser (LabRAM HR-800, HORIBA JOVIN YVON, Longjumeau, France) was used. The pressure drops caused by the TiO_2 modification over the tubes were also measured.

Photographs of the TiO_2-impregnated porous a-silica tubes in air and water purification using decomposition tests of acetaldehyde and methylene blue are shown in Figure 8. For a continuous single-pass condition, a prescribed concentration of acetaldehyde gas was introduced into the TiO_2-impregnated porous a-silica tube at a flow rate of 250 mL/min and was exhausted after the reaction (Figure 8b). The TiO_2-impregnated porous a-silica tube was inserted into a quartz glass tube (27 mm i.d. × 30 cm length) and irradiated by a UV-C lamp. Acetaldehyde and CO_2 concentrations in the quartz glass tube were analyzed simultaneously and continuously by photo-acoustic infrared spectroscopy using an Innova AirTech Instruments Multi-gas Monitor Type 1412 with suitable optical filters (Ballerup, Denmark). For comparison, the a-silica tube without TiO_2 impregnation, the TiO_2-coated a-silica tube by the OVD method [5], and the a-silica tube covered with conventional TiO_2-impregnated Ni-foam (TiO_2/Ni-foam) [12] were also evaluated. A made-to-measure helical UV-C lamp (Kyokko Denki Co., Ltd., Tokyo, Japan) was used as the UV light source. The UV intensity at 254 nm at the surface of the porous tube was measured using a UV-radiometer UVR-300 with a sensor head UD-250 (Topcon Corp., Tokyo, Japan).

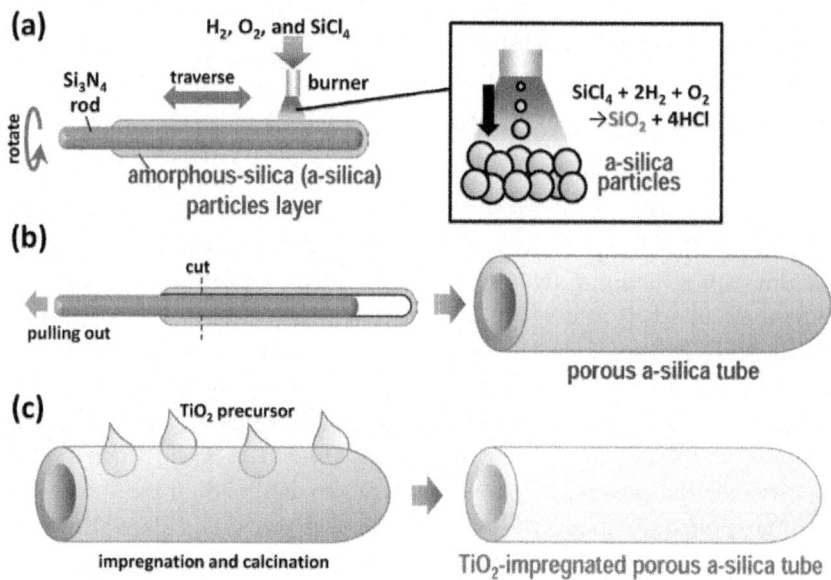

Figure 7: Schematic of the method of fabricating porous TiO$_2$-impregnated a-silica tubes.

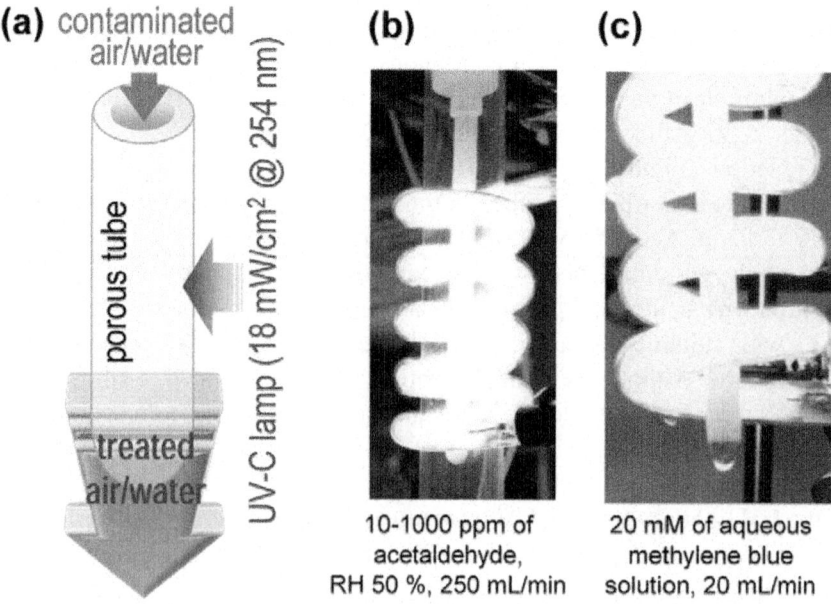

Figure 8: Schematic (a) and photographs of the air- (b) and water- (c) purification test for the TiO$_2$-impregnated a-silica tube.

The methylene blue decomposition test of the TiO_2-impregnated porous a-silica tube was carried out by passing 50 mL of 20 mM aqueous methylene blue solution through the tube at a flow rate of 20 mL/min with UV-C irradiation. The solution was then stored in a beaker (Figure 8c). The remaining ratio of methylene blue was calculated by a decreased absorbance at 663 nm using UV-visible spectrophotometer 2450 (Shimadzu Co., Kyoto, Japan). Then the stored and treated solution was passed through the tube again. Pseudo first order reaction rate constants (k_1) were calculated from the remaining ratio as a function of cycle number.

CONCLUSIONS

A convenient air and water purification unit that uses a TiO_2-impregnated porous a-silica tube was investigated. The tubes showed a possibility for air and water purification. In particular, VOC decomposition property was outstanding with a condition of high concentration acetaldehyde (78% at 700 ppm) and single-pass process. Moreover, a-silica glass can be welded to fused silica glass (Figure 9). Therefore, the tubes have a great potential for compact and in-line VOC removal and water-purification units.

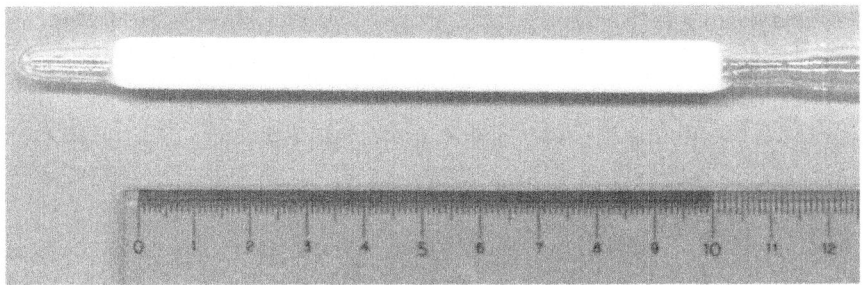

Figure 9: Porous a-silica tube welded to a fused silica glass tube.

REFERENCES

1. Ochiai, T. Environmental and medical applications of TiO_2 photocatalysts and boron-doped diamond electrodes.*Electrochemistry* 2014, *82*, 720–725.

2. Nakata, K.; Kagawa, T.; Sakai, M.; Liu, S.; Ochiai, T.; Sakai, H.; Murakami, T.; Abe, M.; Fujishima, A. Preparation and photocatalytic activity of robust Titania monoliths for water remediation. *ACS Appl. Mater. Interfaces* 2013, *5*, 500–504.]

3. Liu, B.; Nakata, K.; Sakai, M.; Saito, H.; Ochiai, T.; Murakami, T.; Takagi,

K.; Fujishima, A. Hierarchical TiO_2 spherical nanostructures with tunable pore size, pore volume, and specific surface area: Facile preparation and high-photocatalytic performance. *Catal. Sci. Technol.* 2012, *2*, 1933–1939.

4.   Reddy, K.R.; Nakata, K.; Ochiai, T.; Murakami, T.; Tryk, D.A.; Fujishima, A. Facile fabrication and photocatalytic application of Ag nanoparticles-TiO_2 nanofiber composites. *J. Nanosci. Nanotechnol.* 2011, *11*, 3692–3695.

5.   Ochiai, T.; Tago, S.; Tawarayama, H.; Hosoya, T.; Ishiguro, H.; Fujishima, A. Fabrication of a porous TiO_2-coated silica glass tube and its application for a handy water purification unit. *Int. J. Photoenergy* 2014.

6.   Oh, S.-M.; Ishigaki, T. Preparation of pure rutile and anatase TiO_2 nanopowders using RF thermal plasma. *Thin Solid Films* 2004, *457*, 186–191.

7.   Balachandran, U.; Eror, N.G. Raman spectra of Titanium dioxide. *J. Solid State Chem.* 1982, *42*, 276–282.

8.   Liu, M.; Piao, L.; Zhao, L.; Ju, S.; Yan, Z.; He, T.; Zhou, C.; Wang, W. Anatase TiO_2 single crystals with exposed {001} and {110} facets: Facile synthesis and enhanced photocatalysis. *Chem. Commun.* 2010, *46*, 1664–1666.

9.   Zhu, J.; Wang, S.; Bian, Z.; Xie, S.; Cai, C.; Wang, J.; Yang, H.; Li, H. Solvothermally controllable synthesis of anatase TiO_2 nanocrystals with dominant {001} facets and enhanced photocatalytic activity. *CrystEngComm* 2010, *12*, 2219–2224.

10. Ochiai, T.; Hoshi, T.; Slimen, H.; Nakata, K.; Murakami, T.; Tatejima, H.; Koide, Y.; Houas, A.; Horie, T.; Morito, Y.; *et al.* Fabrication of TiO_2 nanoparticles impregnated Titanium mesh filter and its application for environmental purification unit. *Catal. Sci. Technol.* 2011, *1*, 1324–1327.

11. Petit, V.; le Rouge, A.; Béclin, F.; Hamzaoui, H.E.; Bigot, L. Experimental study of SiO_2 soot deposition using the outside vapor deposition method. *Aerosol Sci. Technol.* 2010, *44*, 388–394.

12. Ochiai, T.; Fukuda, T.; Nakata, K.; Murakami, T.; Tryk, D.; Koide, Y.; Fujishima, A. Photocatalytic inactivation and removal of algae with TiO_2-coated materials. *J. Appl. Electrochem.* 2010, *40*, 1737–1742.

Chapter 3

NEW COMPOSITE MATERIALS IN THE TECHNOLOGY FOR DRINKING WATER PURIFICATION FROM IONIC AND COLLOIDAL POLLUTANTS

Marjan S. Ranđelović[1], Aleksandra R. Zarubica[1] and Milovan M. Purenović[1]

[1]University of Niš, Faculty of Science and Mathematics, Department of Chemistry, Niš,, Serbia

INTRODUCTION

Composite materials (composites) are inherently heterogeneous and represent a defined combination of chemically and structurally different constituent materials, ensuring the required properties such as mechanical strength, stiffness, low density, or other specific characteristics depending on their purpose. Therefore, composite material is a system composed of two or more physically distinct phases whose combination produces a synergistic effect and aggregate properties that are different from those of its constituents. Favorable characteristics of composite materials were known to the people even in the period BC (before Christ-Century) and were used in order to improve the quality of human daily life. For example, it is known that in the ancient period, people made bricks that were reinforced with straw, and thus secured greater longevity and durability of their buildings. The incorporation of the straw improves the strength, toughness and thermal insulation properties of these composites. In principle, the degree of reinforcement (volume fraction of straw) and the level of alignment of the straw stalks (and their lengths) may be adjusted so that not only the properties but their anisotropy may be optimised differently in various parts of the structure [1]. Significant development and application of composites began in the second half of the 20th century, wherein their diversity and areas of application are constantly increasing. Development of composite materials is resulted mainly from the increasing need for materials with better mechanical characteristics that would be used as components in various constructions. For this purpose, such composites should have an adequate strength, stiffness, good oxidation resistance and low weight. Intensive study of composite materials and their processing methods has caused that these materials replace metals and alloys and become indispensable in the manufacture of parts for automobiles, spacecrafts, sports equipment etc. In terms of exploiting modern engineering composites this remains a central principle. Modern composites can be said to have "designed micro- and nanostructures" which means that the constituents of composites have much more finely divided structures and tend to have sizes in the micrometre or nanometre range. Basic factors affecting properties of composites are as follows:

- Properties of phases;
- Amount of phases;
- Bonding and the interface between the phases;
- Size, distribution and shape (particles, flakes, fibers, laminates) of the dispersed phase - reinforcement;
- Orientation of the dispersed phase - reinforcement (random or preferred).

Good bonding (adhesion) between matrix and dispersed phase provides a high level of mechanical properties of the composite via the interface. In addition, interfaces are responsible for numerous processes of electron transfers and play crucial role in redox processes, heterogeneous catalysis, adsorption etc. Usually, there are three forms of interface between the two phases within the composite:

1. Direct bonding with no intermediate layer. In this case adhesion ("wetting") is provided by either covalent bonding or van der Waals force;
2. Intermediate layer in form of solid solution of the matrix and dispersed phases constituents;
3. Intermediate layer (interphase) in form of a third bonding phase (adhesive).

Current challenges in the field of composite materials are associated with the extension of their application area from structural composites to functional and multifunctional composites. In this respect, a great improvement of composite materials through processing has been made enabling the development of composite materials for electrical, thermal and other functional applications that are relevant to current technological needs. Examples of functions are joining, repair, sensing, actuation, deicing (as needed for aircraft and bridges), energy conversion (as needed to generate clean energy), electrochemical electrodes, electrical connection, thermal contact improvement and heat dissipation (*i.e.*, cooling, as needed for microelectronics and aircrafts) [2]. Modern processing includes the use of additives (which may be introduced as liquids or solids), the combined use of fillers at the micrometer and nanometer scales, the formation of hybrids, the modification of the interfaces in a composite and control over the microstructure. Therefore, it can be said that the development of composite materials for current technological needs must be application driven and process oriented. The conventional composites engineering approach, which is focused on mechanics and purely structural applications, is in contrast to mentioned modern practice.

On the contemporary level of science development it is known that materials of certain characteristics can be obtained only by strictly defined procedures of processing and depend on their chemical composition and structure. Since composites are heterogeneous systems, as already has been noted, the matrix is of great importance whose structure and chemical composition determine the most dominant features of the composite as a unit. However, it should be noted here that the composite does not possess properties of a single component but exhibits qualitatively new features, because of which it is considered as a new material. In addition to the dominant use of composites as structural elements, important application of composite materials is in the water purification technologies. In this field of application, composites usually have the role of adsorbent, electrochemically active materials, catalysts, photocatalysts etc. Bearing in mind that the material efficiency in the removal of harmful substances from water is higher if greater is its surface area, there are tends of scientists to develop these materials with required and defined nanostructures. In addition to the specific surface area increasing, nanostructured materials exhibit a qualitatively new properties compared to the related structure at the micro or macro scale. In this manner, it is developed specific procedure for certain metal hydroxides and natural organic matter layering onto alumosilicate matrix as well as procedures of microalloying which both lead to significant changes of the surface acido-basic and electrical properties of the alumosilicate matrix. The nano-scale composites provide an opportunity to study the phase boundaries and phenomena occurring at the surface, interface boundaries and within intergranular area during composites synthesis or during their interaction with aqueous solutions.

AN OVERVIEW AND TRENDS IN USE OF COMPOSITES IN INDUSTRIAL PLANTS

Nanocomposites based on polymers represent an area of significant scientific interest and developing industrial practice. Despite the proven benefits of polymer based nano-composites in the scope of their mechanical properties, and some distinctive combination/synergism of improved structural features, the real application remains still relatively isolated and not well discussed. An insight in the historical (re)view on polymer nano-composites showed on the first type used based on the combination of natural fillers and polymers in the 90s [3-6] up to estimated 145 million USD spent at huge market of polymer based nano-composites in 2013 [7].

THE CONCEPTS OF INTERPHASE BOUNDARIES MODIFICATION, MICROALLOYING AND COATING/LAYERING IN THE COMPOSITE SYNTHESIS

Methods and techniques for managing properties of composite materials include the selection and modification of constituent materials as well as changing the interface boundaries within the composite. Some composites are most commonly fabricated by impregnation (infiltration) of the matrix or matrix precursor in the liquid state into the appropriate filler preform. The connection between the constituents depends on the microstructure and chemistry of the interface boundary. The matrix and filler are connected by chemical bonds, interdiffusion, van der Waals forces and mechanical interlocking [2]. The first three interactions require very close filler-matrix contact that can be achieved if the matrix or matrix precursor wetting the surface of filler during the infiltration of matrix or matrix precursors in the filler preform. Effective wetting means that the liquid is evenly distributed over the surface of filler, while a poor wetting means that the liquid drops formed on the surface. Wettability can be increased by applying the coatings, adding wetting agents or by chemical surface functionalization (the introduction of functional groups on the surface that increase wettability) thereby changing the surface energy. If the filler is carbon fiber, surface treatments involve oxidation treatments and the use of coupling agents, wetting agents, and/or coatings. Often, metals or ceramics are used as coatings for carbon fillers. Metallic coatings are usually formed by coating carbon fiber reinforcements with metals *i.e.* Ni, Cu and Ag. Examples of ceramic coatings are TiC, SiC, B_4C, TiB_2, TiN which are distributed by using Chemical Vapor Deposition (CVD) technique or by solution coating methods starting from organometalic compounds. Therefore, these are examples of application of coatings on carbon materials to illustrate the method of modification of surface properties.

In the case of metal-ceramic composites, certain liquid metals react with ceramic preform during infiltration. For instances, composites based on the Al–Al_2O_3 system can be obtained by Reactive Metal Penetration (RMP) method which is based on infiltration of ceramic preforms by a liquid metal, generally aluminium or aluminium alloys [8,9]. During the process, a liquid metal simultaneously reacts and penetrates the ceramic preform, usually silica or a silicate, resulting in a metal-ceramic composite characterized by two phases that are interpenetrated. Another example is the reaction between SiC and Al during the infiltration of molten aluminum in a preheated preform:

$$4Al + 3SiC \rightarrow Al_4C_3 + 3Si$$

(1)

From the equation it can be seen that Si is generated during the reaction which is then dissolved in molten aluminum, while Al_4C_3 occurs at the SiC-Al interfacial boundary. The degree of reaction increases with increasing temperature. On the contrary, there are metals that in liquid state difficult wet the surface of the ceramic resulting in metal infiltration hindering. The difficulty of wetting and bonding of liquid metals to ceramic surfaces is related to atomic bonding in the ceramic lattice and can be improved by application of coatings. Coated particles (composite particles) are composed of solid phase covered with thinner or thicker layer of another material [10.11]. These coatings - layers on the surface are important for several reasons. In such way, the surface characteristics of the initial solid phase are modified and sintering conditions as well as molten metal infiltration can be better controlled.

As can be seen from examples, the processing of composite materials often involves high temperature and pressure to cause the joining of constituent materials forming a cohesive material. Generally, the matrix dictates the required temperature, pressure and processing time during composite synthesis. Sintering is an important factor in achieving the desired microstructure of ceramic based composites and includes very complex processes. In addition to surface coatings, an important influence on sintering has been exhibited by an addition of microalloying components, which significantly determine a microstructure and properties of ceramics [12]. The presence of small amounts of impurities in the starting material can vastly influence their mechanical, optical, electrical, color, diffusivity, electrical conductivity, and dielectric properties of matrix. Microalloying, as a known modern procedure for changing the intrinsic semiconductor properties, by authors' original works (Purenovic et al.), get more and more important role in the control of some structurally sensitive properties of metals, alloys, ceramics, composites and other materials. It is known that the nature of matter is determined by its composition and structure. There are many structurally sensitive properties of materials, but among the most sensitive are the conductivity, electrode potential, magnetic, catalytic and mechanical properties. Microalloying means adding certain elements in small (ppm) quantities, thereby modified structure results in a significant change in the value of conductivity and the electrode potential. Conducted own investigation and the results obtained showed an excellent rational electrochemical behavior of composites such as microalloyed aluminum, microalloyed magnesium, as well as composite ceramics and quartz sand microalloyed with aluminum and magnesium, in contact with aqueous solutions of electrolytes or water which contain harmful ingredients in ionic, molecular and colloidal state. Microalloyed and structurally modified composite ceramics have high porosity (30%), with the macro-, meso-, micro- and submicropores. There is direct relationship between porosity and structure of these composite materials, especially when it comes to nanostructured fragmented crystals. It is worth to emphasize the domination of amorphous phases with crystalline substructure, which is impossible to be removed, and it would be inappropriate to be removed, because the contact of crystals with amorphous layer is responsible for numerous processes of electrons exchange. By certain processes and reactions in the solid phase, the amorphous microalloyed aluminum, microalloyed amorphous magnesium, amorphous-crystalline structure of composite microalloyed ceramics and amorphous-crystalline structure of microalloyed quartz sand could be obtained. Many metals, alloys and composite electrode materials manifested significant differences in the reversible thermodynamic potential and the steady corrosion potential.

The manufacturing processes used to make composite ceramics can cause the development of liquid phases during sintering, and their retention as remnant glass at triple junctions and along grain boundaries and interphase boundaries after cooling to room temperature. Formed thin intergranular films are relevant to creep behavior at high temperatures, and also responsible for the strength of the bonding at interfaces. However, the heat treatment at elevated temperatures which is used for joining constituent materials and establishing the cohesive forces shows a disadvantage because cooling can lead to disturbance of established bonds between phases. Namely, during the cooling, differences in coefficients of thermal expansion could result in unequal contraction by which established bonds are broken. This problem is particularly evident in metal-ceramic composites, where high temperatures are usually applied during synthesis.

PREPARATION OF MODERN NANO-COMPOSITES

Processing of nanocomposites based on layered silicates is rather challenging activity to achieve the full technical and engineering potential, which is the field with the largest growth forecast [13-16]. The modification of silicates by use of organic components is needed to allow intercalation, and also in order to improve compatibility/nano-distribution some additional ingredients have to be applied. The thermal treatment as step in processing sequence helps proper stabilisation of nanocomposites that has to take into consideration the oxidative stability of the polymer substrate, the influence of the nano-filler and the impact of modifiers and compatibilisers.

Montmorillonite of natural origin is among the most used nano-fillers. Traditional nano-fillers contain metal ions and other contaminants that may influence the thermooxidative stability and features of the nanocomposites. Organic modification of the (natural/traditional) clay is usually realized by cation exchange with a long-chain amines or quaternary ammonium salts. Content of such involved organic material content within the clay may be up to 40 mas.%. Therefore, the total thermal resistance of the composite material highly depends on the thermal stability of the organic ingredient. The thermal stability of the ammonium salts is limited at the processing temperatures applied (ex. extrusion, injection molding, etc.). Namely, thermal degradation of ammonium salts starts at 180°C and may be even tentatively reduced by catalytically active sites on the alumosilicate layer [17].

The compatibiliser applied as organically modified filler is often polypropylene-g-maleic anhydride in amount from 5 to 25% in the final composite formulation. The inferior stability of such low molecular weight filler comparing to the parent polymer affects the total stability of the final polymer based nanocomposites.

AN IMPROVEMENT OF COMPOSITES STABILITY

Nanocomposites may show higher stability due to increased barrier to oxygen, or lower stability because of undergone to hydrolysis through entrapped water [18,19]. In conventional practice stabilizer systems based on phenolic antioxidants and phosphites are applied, and in recent investigations new found components of filler degradation deactivators has been tested [20].

A traditional state-of-art polypropylene (PP) nanocomposite consisting of maleated PP and nano-clay is traditionally stabilized by a proven combination of phenolic antioxidant and phosphites. The polymer degradation may be completely prevented even after 5 extrusion cycles by using the patented stabilizer system AO-2 (based on oxazoline, oxazolone, oxirane, oxazine and isocyanate groups) [20], additionally improving mechanical properties of the resulting nano-composites and discoloration during processing and application.

The underlined thermal instability of the usual ammonium organic modifiers can be diminished by using the phosphonium, imidazolium, pyridinium, tropylium ions [21]. An alternative way to produce thermally stable nano-composites is the use of unmodified clays in combination with selected copolymers playing role of dispersants, intercalants, exfoliants and compatibilisers for PP nano-composites. In current processing of nanocomposites different structures are identified such as polyethyleneoxide based nonionic surfactants [22] and amphiphilic copolymers based on long-chain acrylates [23]. Recently, more specifically poly(octadecylacrylate-co-maleic anhydride) and poly(octadecylacrylate-co-N-vinylpyrrolidone) in the form of gradient copolymers are applied with unmodified montmorillonite for processing PP nano-composites. Such obtained nano-composites show partial exfoliation, the final visual appearance is similar to the classical ammonium modified systems, however better thermal and thermo-oxidative stability is proven [23]. The most important improvement is achieved in the mechanical vales comparing to the conventional polymer system.

NANOCOMPOSITES USE IN A COMPETITIVE ENVIRONMENT OF THE MATERIALS

Nanocomposites materials are very attractive from the scientific and practical point of view, although some other materials are also interesting, such as plastics, fillers, blends, and different additives fulfilling the specified product profile. In such competence, the lowest cost solution comprising acceptable material structure and properties/resistances would dominate. Even more, competitive (nano)composite materials would benefit from nanocomposites developments and keep their application fields with improved features. Most of nanocomposites materials applications are intended for long-term and outdoor use. This is important aspect on the need for relevant nanocomposites stability. Namely, it is known that inorganic fillers often show a negative effect on the oxidative stability to a varying extent. The interactions of the filler and the stabilizers over adsorption/desorption mechanisms are mainly responsible for the impact. The specific surface area of the filler and pore volumes, surface functionality, hydrophilicity, thermal and photo-sensation properties of the filler and transition metal content (ex. manganese, titanium, iron) have been found to be potential factors/elements of the interaction [24].

Polypropylene/montmorillonite nanocomposites, additionally stabilized with antioxidant, degrade much faster under photo-oxidative conditions than pure polypropylene [25,26].This phenomenon is attributed to active species/ sites in the clay generated by photolysis or photo-oxidation, and by consequence interaction between antioxidant, montmorillonite and maleic anhydride modified polypropylene. In natural clay present iron may additionally play an active role in the dramatic modification of material oxidation conditions [27], and nanoparticles also catalyze the decomposition process [28]. The use of so-called filler deactivators or coupling agents is potential solution for diminishing the negative influence of fillers on the (photo)oxidative stability by blocking active sites on the filler surface. Amphiphilic modifiers with reactive chemical groups in the form of polymers, olygomers or low molecular weight molecules such as bisstearylamide or dodecenylsuccinic anhydride have been proposed [29].Thus, stabilizer systems containing filler deactivators should have an affirmative effect in nano-composites for long-term stability.

NANO-COMPOSITES MATERIALS FOR WATER TREATMENTS: STATE-OF-THE-ART AND PERSPECTIVES

Clean drinking water is essential to human health, and also so-called technical water is a critical feedstock in a variety of key industries including electronics, pharmaceuticals and food processing industries. Taking into consideration that available supplies of fresh water are limited (due to population growth, extended deficiency, stringent health regulations, and competing demands from a variety of users/consumers) the world is facing with challenges to satisfy demands on high water quality standards and quantities (volumes). Benefits and trends in nano-scale science, chemistry and engineering impose that many of the current problems regarding green chemistry may be resolved using nano-sorbents, nano-catalysts, nanoparticles and nanostructured catalytic membranes. Nano-materials are characterized by a number of key physicochemical properties being particularly attractive for water purification treatments. Nanomaterials have much large specific surface area than bulk respect particles (mass to volume ratio), also they can be functionalized with reactive chemical groups specific in affinity to a given model compound. These materials may possess redox features and take part in shape- and structural-dependent catalyzed reactions of water purification. In aqueous solutions, they can serve as sorbents/catalysts for toxic metal ions, radionuclides, organic and inorganic solutes/anions [30]. Moreover, nano-materials can be used in selective targeting of biochemically constituents of aquatic bacteria and viruses. The nano-materials seems to be key components in future environmental friendly and cost-effective functional materials to desalinate public and polluted waters world-wide, for purification of water contaminated by pesticides, pharmaceuticals, phenol and other aromatics. The presence of heavy metals in water exhibits a variety of harmful effects on the living organisms in polluted ecosystems. The removal of heavy metals from water includes the following procedures: chemical precipitation, coagulation/flocculation, membrane processes, ion exchange, adsorption, electrochemical precipitation, etc. [31,32]. However, the application of composite materials in the controlling of pollutants in the environment and drinking water is significant [33,34], as described in further text.

The use of zeolites, natural or synthetic ones in waste water treatments is highly limited due to low adsorption capacity in the case of former and relatively small grain size in latter. Modification of natural or synthetic zeolites toward composite material which would satisfy both essential properties is a challenging task. Tailoring synthetic zeolite resulted in a composite porous host supporting microcrystalline active phase of vermiculite matrix [35]. The vermiculite-based composite showed the same hydraulic properties as natural clinoptilolite with similar grain size (2-5 mm), while the rate of adsorption and maximal adsorption capacity was improved four times. In other words, cation exchange capacity is increased when compared to natural zeolite with a comparative grain size, ion-exchange kinetics are substantially improved in comparison to natural zeolite, and hydraulic conductivity is considerably higher that synthetic powdered zeolite [35].

The development of new composite material based on use of inorganic polymeric flocculants as a combination of anionic and cationic poly-aluminium chloride (PACl) in one unique polyelectrolyte is proposed [36]. The incorporation of the anionic polyelectrolyte into PACl structure noticeably affects its initial properties (*i.e.* turbidity, Al species distribution, pH and conductivity). Interactions are taking place between Al species and polyelectrolytes molecules over hydrogen bonding (amino/amidic groups of the polyelectrolyte, and the –OH and –H groups of Al species are involved) and electrostatic forces/interactions. This resulted in new composite material. The main advantage of composite coagulants is lower residual aluminium concentration that remains in the treated sample, and more efficient treatments of waters (organic matter removal) can be realized [36]. Additional benefit is in cost effective process in the absence of specific equipment for handling the polyelectrolyte (ex. pumping system, etc.). Taking into account faster flocculation, increased efficiency and cost effectiveness, such new composite material seems to be promising one.

Porous ceramic composites can be prepared by silver nanoparticles-decoration using a silver nanoparticle colloidal solution and an aminosilane coupling agent [37]. The interaction between the nanoparticles and the ceramics comprises the coordination bonds between the $-NH_2$ group and the silver atoms on the surface of the nanoparticles. The composite can be stored for long periods without losing of nanoparticles, also being highly resistance to ultrasonic irradiation and washing. Such composite has shown high sterilization property as an antibacterial water filter [37]. This low cost composite, bearing in mind commonly available synthesis, simple preparation, the use of cheap and non-toxic reagents in the procedure, may be imposed as a potential solution for widespread use in water treatments.

Ultrafine AgO particles-decorated porous ceramic composites are prepared based on the main ingredient, cristoballite. The results on composite structure show that silver(II)oxide decorated diatomite-based porous ceramic composites possess crystal structure, and are composed of tetragonal cristoballite, monoclinic silver(II)oxide and cubic silver(I)oxide [38]. Such AgO-decorated porous ceramic composites show a strong antimicrobial activity and an algal-inhibition capacity. As the extension time is longer, the antibacterial effects are enhanced up to 99.9% [38].

Actual nanostructured composite materials based on multi-walled carbon nanotubes (MWCNT) and titania exhibited strong interphase structure between MWCNT and titania. This contact and interaction facilitated a homogeneous deposition/coverage of titania over MWCNT [39]. The photo-catalytic activity of the prepared composite materials was tested in the conversion of phenol from model watery solution under UV or visible light. The results showed higher photo-catalytic activity of the composite MWCNT and titania than over mechanical mixture proving an assumption on the existence of the interphase structure effect [39].

Nanocomposite membranes based on silica/titania nanotubes over porous alumina supports membranes were prepared [40]. An inserting of amorphous silica into nanophase titania caused the surpressed of phase transformation from anatase to rutile, and decreased the titania particle size. Good photo-catalytic activity of organic contaminants degradation, and wettability of composite membrane under UV-irradiation, helped to obtain high permeate flux across the composite membrane [40].

NEW ALUMOSILICATE BASED COMPOSITES CHEMI-CALLY MODIFIED BY COATINGS/THIN LAYERS – TESTED IN THE REMOVAL OF COLLOIDAL AND IONIC FORMS OF HARMFUL HEAVY METALS FROM WATER

Without new materials, there are no new technologies. Having in mind this fact, electrochemically active and structurally modified composites were obtained through microalloying and certain metals hydroxides layering, starting from bentonite as alumosilicate precursor. The composites have prognosed electrochemical, ion-exchanging and adsorption properties, as very sensitive structural and surface properties of materials. After the series of experiments, including composites interaction with synthetic waters, the obtained results are presented, analyzed and then systematized in the form of appropriate models of interactions.

Alumosilicate Composite Ceramic Microalloyed by SN for the Removal of Ionic and Colloidal Forms of MN

Usually, manganese does not present a health hazard in the household water supply. However, it can affect the flavor and color of water because it typically causes brownish-black staining of laundry, dishes and glassware [32]. Although manganese is one of the elements that are at least toxic, concentrations of manganese much higher than the maximum allowed concentration during long-term exposure can cause health damage. A number of known procedures for the manganese removal are not suitable for an elimination of its all chemical species due to reversible release of manganese into water systems. Therefore, some of these used procedures are at the edge of techno-economical viability. In order to remove ionic and colloidal forms of manganese, a new aluminosilicate-based ceramic composite with defined electrochemical activity was synthesized [41]. Synthesis procedure of the composite material consists of two phases. Firstly, composite particles were synthesized by applying Al/Sn oxide coating on the bentonite particles in an aqueous suspension. In the second phase, aluminium powder was added to the previously obtained plastic mass and after shaping in the form of spheres 1 cm in diameter and drying, sintering was performed at 900°C. Fig. 1 a), b) and c) presents the microstructure of composite by using different magnifications.

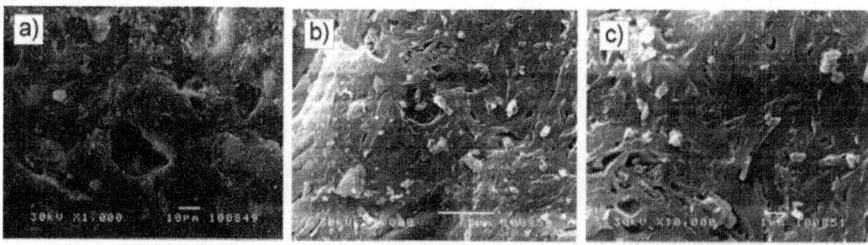

Figure 1: SEM images of the composite recorded at: a) low, b) medium and b) high magnifications.

During sintering a microalloying of composite by Sn occurred causing crystal grain surface layer amorphization and a creation of non-stoichiometric phases of Al_2O_3 with a metal excess [42,43]. In this way, microalloying causes electrochemical activity, which manifests itself in contact with the aqueous solutions of electrolytes and harmful substances in water. Therefore, the ceramics is unstable in contact with water and susceptible to corrosion because surface electrochemical processes taking place. The composite influence redox properties of water and electrochemically interacts with ionic and colloidal forms of manganese in synthetic water systems.

Alumosilicate matrix, whose particles are coated with Al/Sn oxides, was filled with a metal phase which is mostly aluminum with a small quantity of tin as a microalloying component. During the thermal treatment, liquid aluminium simultaneously reacts and penetrates the ceramics preform, resulting in metal/ceramic composite, where the all phases are interpenetrated forming a porous structure. In fact, the reduction of tin(II) occurred according to the following reaction:

$$2Al + 3SnO \rightarrow Al_2O_3 + 3Sn \tag{2}$$

The first reaction step is the reduction of Sn(II) to elemental Sn and its dispersion from the ceramics into the melt. Therefore, during the reaction, Sn is liberated into the liquid metal and diffuses towards the Al source. Moreover, oxygen partial pressure within the composite, at the Al-Al_2O_3 interface, can be estimated on the basis of thermodynamic parameters and calculated using the following equation [44]:

$$\Delta G° = RTlnPO_2 \tag{3}$$

The standard free energy of the reaction:

$$4/3Al + O_2 \rightarrow 2/3Al_2O_3 \tag{4}$$

at 900°C, given by Ellingham diagram [44] is -869 KJ/mol, and corresponding oxygen partial pressure: $PO_2 = 2.02 \cdot 10^{-39}$ Pa. Therefore, this low oxygen partial pressure during sintering provides reducing environment and the formation of nonstoichiometric oxide phases, with the metal excess, or with vacancies in oxygen sublattice. Nevertheless, Al_2O_3 belongs to the oxides of stoichiometric composition or with a negligible deviation from stoichiometry, it can occur as an amorphous and nonstoichiometric oxide with a metal excess during oxidation of aluminium. Common nonstoichiometric reactions occur at low oxygen partial pressures when one of the components (oxygen in this case) leaves the crystal [45,42]. A corresponding defect reaction is [45]:

(oxygen in this case) leaves the crystal [45,42]. A corresponding defect reaction is [45]:

$$O_O^x \rightleftarrows \frac{1}{2}O_2(g) + V_O^{\bullet\bullet} + 2e^{'} \tag{5}$$

As the oxygen atom escapes, an oxygen vacancy ($V_O^{\bullet\bullet}$) is created. Taking in mind that the oxygen is to be presented in neutral form, two resulting electrons would be easily excited into the conduction band.

Al–Sn alloys show a great activity compared to the thermodynamic Al^{3+}/Al potential of -1.66V vs. NHE, which stands for a pure aluminium. The activation is manifested by a shifting of the pitting potential in the negative direction and significant reducing of the passive potential region [43,46]. The addition of microalloying Sn to aluminium produced a considerable shift of the open circuit potential (OCP) in the negative direction [46].

During the process of composite ceramics sintering, significant changes in the structure of alumosilicate matrix were occurred. Namely, the polycrystaline alumosilicate matrix with amorphised grain and sub-grain boundary were obtained, where a main role possesses metallic aluminum itself, then a microalloyed tin and nonstoichiometric excess of these elements in ceramics, creating macro-, meso- and micro- pores with the reduced mobility of grain boundaries and termination of grain growth [47]. Aluminum and tin in conjunction with other admixtures present in composite ceramics cause drastic changes in the structure-sensitive properties and electrochemical activity. An active composite ceramics in contact with synthetic water containing manganese reduce and deposit the manganese in the macro-, meso- and micro- pores (eq. 6). Electrochemical activity is provided by electrochemical potential of Al atoms and free electrons that participate in redox processes.

$$2Al + 3Mn^{2+} \rightarrow 2Al^{3+} + 3Mn$$

(6)

The deposited manganese on microcathode parts of the structure can further form separate clusters and the adsorption layer [48,49]. Reduction processes take place until the Al^{3+} ions continue to solvate themselves in water. A part of Al^{3+} ions reacts with OH^- ions giving insoluble $Al(OH)_3$.

Interaction of Composite Material with Ionic and Colloidal Forms of MN in Synthetic Water

Interaction of the composite material with water manifests itself as decreasing in the redox potential of water, as shown in Fig. 2. This confirms the fact that the composite is electrochemically active in contact with water. During the interaction with water, aluminium from the composite is electrochemically dissolved into water providing electrons which can participate in the number of redox reactions of water yielding reduced species (molecules, ions and radicals) such as H_2, $OH^{\bullet}$, etc. [47].

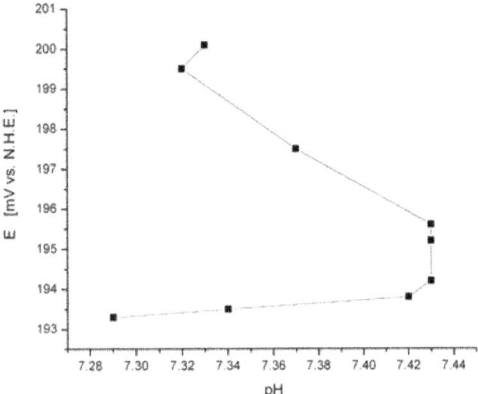

Figure 2: Redox potential of water dependence on pH during interaction of the composite with distilled water.

TDS value of distilled water immediately after contact with ceramics increases. It seems that increasing the TDS value is due to dissolution of Al^{3+}, Mg^{2+}, Na^+, SiO_3^{2-} from the bentonite based composite. Al^{3+} and SiO_3^{2-} ions are subjected to hydrolysis and polymerization reactions which are followed by spontaneous coagulation-flocculation processes and appearance of sludge after a prolonged period of time.

A reduction of manganese concentration in synthetic waters is shown in Fig. 3.

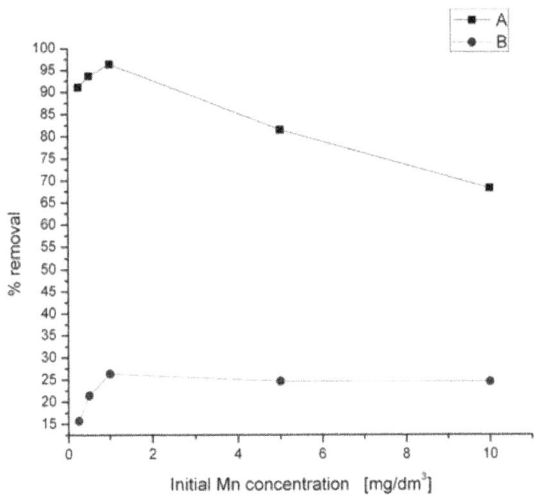

Figure 3: Percentage removal of Mn^{2+} (A) and colloidal MnO_2 (B) from synthetic wa-

ters (the composite dosage, 2 g/dm³; contacting time, 20 min; initial Mn concentrations in range $0.25 - 10$ mg/dm³; initial pH 5.75 ± 0.1; temperature, $20 \pm 0.5°C$).

Average initial pH of the synthetic waters was 5.75. After 20 min of contact with the composite material average pH was 6.70.

During the interactions of composite with synthetic waters, the colloidal MnO_2 was removed to a lesser degree than Mn^{2+}. The authors imposed that colloidal manganese possesses the following structure of micelles:

$$\{m[MnO_2]nSO_4^{2-} \ 2(n-x)K^+\}2xK^+ \tag{7}$$

Potential-determining ions in the structure of micelles are SO_4^{2-}. They are primarily adsorbed on MnO_2 and responsible for the stability of colloids. Therefore, it is clear that the reduction of manganese is more difficult and there is an electrostatic repulsion between colloidal particles and a composite with dominantly negatively charged surface sites. Thus, the removal efficiency of colloidal manganese is significantly lower compared with the ionic form of Mn^{+2}. During the electrochemical interactions of synthetic water containing Mn^{2+} and colloidal MnO_2 with the composite material, transferring of Al^{3+} ions in a solution increases the TDS value, as shown in Table 1.

Table 1: The results of synthetic waters analysis before and after treatment with composite material

$C_0(Mn)$ mg/dm3	TDS (mg/dm³)	pH	C(Mn) mg/d^{m}3	TDS (mg/dm³)	pH
Before Mn^{2+} synthetic water treatment			After Mn^{2+} synthetic water treatment		
0.25	3	5.75	0.0223	17	6.65
0.50	7	5.73	0.0318	21	6.71
1.0	10	5.71	0.0363	25	6.72
5.0	14	5.70	0.9271	29	6.70
10.0	28	5.76	3.9773	39	6.58
Before colloidal MnO_2 synthetic water treatment			After colloidal MnO_2 synthetic water treatment		
0.25	3	5.82	0.2108	18	6.73
0.50	6	5.75	0.3928	22	6.71
1.0	11	5.71	0.7366	25	6.72
5.0	14	5.72	3.768	29	6.75

10.0	28		5.75	7.549	39		6.67

The initial dissolution of the Al based alloys introduces both aluminium and alloying ions into the solution, and then the reposition of microalloying tin onto active sites at surface occurs [46], so it was not detected by ICP-OES analysis.

Aluminium ions generated during electrochemical processes of manganese removal may form monomeric species such as $Al(OH)^{2+}$, $Al(OH)_2^+$ and $Al(OH)_4^-$. During the time, these monomers have tendency to polymerize in the pH range 4–7 which results in oversaturation and formation of amorphous hydroxide precipitate according to complex precipitation kinetics. Many polymeric species such as $Al_6(OH)_{15}^{+3}$, $Al_7(OH)_{17}^{+4}$, $Al_8(OH)_{20}^{+4}$, $Al_{13}O_4(OH)_{24}^{+7}$, $Al_{13}(OH)_{34}^{+5}$ have been reported [50]. Average concentration of aluminium, immediately after 20 min of composite interaction with Mn^{2+} synthetic waters, was 0.2131 mg/dm³ and included all mentioned monomeric and polymeric species which were not coagulated. After a prolonged period of time concentration of aluminum has a tendency to decrease reaching values that are below 0.1 mg/dm³, due to precipitation of $Al(OH)_3$ sludge.

The increase in the pH during the experiments can be explained in terms of the electrochemical and the chemical reactions that take place in the system composite-synthetic water. Water reduction at cathodic parts of composite (eq. 8), the electrochemical dissolution of aluminum (eq. 9) and protolytic reactions (eq. 10-14) increase the pH value [51].

$$H_2O + e^- \leftrightarrows 1/2 H_2 + OH^- \tag{8}$$

$$2Al + 6H_2O \leftrightarrows 2Al^{3+} + 3H_2 + 6\,OH^- \tag{9}$$

$$Al(OH)_4^- + H^+ \leftrightarrows Al(OH)_3 + H_2O \tag{10}$$

$$Al(OH)_3 + H^+ \leftrightarrows Al(OH)_2^+ + H_2O \tag{11}$$

$$Al(OH)_2^+ + H^+ \leftrightarrows Al(OH)^{2+} + H_2O \tag{12}$$

$$Al(OH)^{2+} + H^+ \leftrightarrows Al^{3+} + H_2O \tag{13}$$

$$Al(OH)_3(s) \leftrightarrows Al^{3+} + 3OH^- \tag{14}$$

Bentonite Modified By Mixed Fe, Mg (Hydr) Oxides Coatings for the Removal of Ionic and Colloidal Forms of Pb(Ii)

Lead (Pb) is heavy metal which presents one of the major environmental pollutants due to its hazardous nature. It diffuses into water and the environment through effluents from lead smelters as well as from battery, paper, pulp and ammunition industries. Scientists established that lead is nonessential for plants and animals, while for humans it is a cumulative poison which can cause damage to the brain, red blood cells and kidneys [52].

In this subchapter, a cheap and effective composite material as a potentially attractive adsorbent for the treatment of Pb(II) contaminated water sources has been described. The procedure for obtaining a bentonite based composite involves the application of mixed Fe and Mg hydroxides coatings onto bentonite particles (0.375 mmol Fe and 0.125 mmol Mg per gram of bentonite) in aqueous suspension and subsequent thermal treatment of the solid phase at 498 K [53]. Bearing in mind layered structure of montmorillonite, the quite limited extent of isomorphous substitution of Mg for Fe in iron (hidr)oxides and significant differences in acid-base surface properties between these two (hydr)oxides, formation of heterogeneous coatings onto bentonite and specific structure of obtained composite have been achieved [54]. Different adsorption sites on such heterogeneous surface provide efficient removal of numerous chemical species of Pb(II) over a wide pH range.

The structural changes of montmorillonite during composite synthesis are mainly reflected in the reduction of d_{001} diffraction peak intensity in X-ray diffractograms and its shifting towards the higher values of 2θ. Moreover, it can be observed that the peak is broadened suggesting that the distance between the layers is non-uniform with disordered and partially delaminated structure. The crystallographic spacing d_{001} of montmorillonite in the native bentonite and the composite, computed by using Bragg's equation ($n\lambda = 2d \sin \theta$), is 1.54 nm and 1.28 nm, respectively. These changes in the structure took place because the d-spacing is very sensitive to the type of interlamellar cations, and the degree of their hydration [55].

The XRD patterns of the composite and starting (native) bentonite are presented in Fig. 4a and b, respectively.

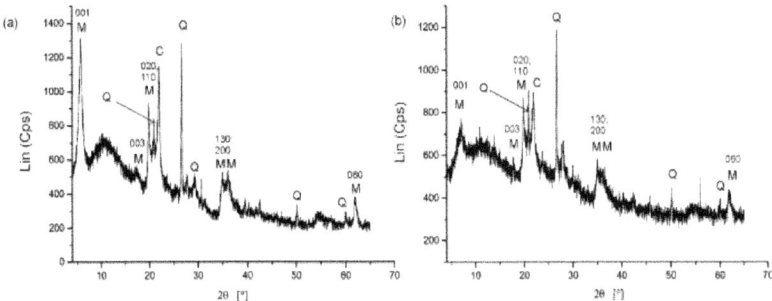

Figure 4: X-ray diffractograms of (a) composite and (b) native bentonite.

SEM micrographs (Fig. 5 a, b and c) show that bentonite and composite are composed of laminar particles arranged in layered manner, forming the aggregates with diameters up to 50 μm.

Figure 5: a) SEM of synthesized composite, (b) SEM of composite after interaction with Pb(II) solution and (c) surface morphology of the native bentonite.

No significant changes in the microstructure of composite occurred during the interaction with the aqueous solution of Pb(II).

Despite a thorough washing process, a large amount of NO_3^- is retained in the composite. A vibration mode at ca. 1389 cm^{-1} in FTIR spectrum confirms the NO_3^- stretching which indicates that some positive charged sites exist on the surface of composite and that they are counterbalanced by the NO_3^- which can be exchanged by other anions [53]. In addition, the formation of poorly crystallized magnesium hydroxonitrate in pH range 9-11 [56,57], where Fe/Mg coprecipitation was performed over bentonite particles, is very likely.

Specific Surface Area Determined By N_2 Adsorption/Desorption Using Bet Equation

The Fig. 6. shows the comparative nitrogen adsorption-desorption isotherms of native bentonite and composite.

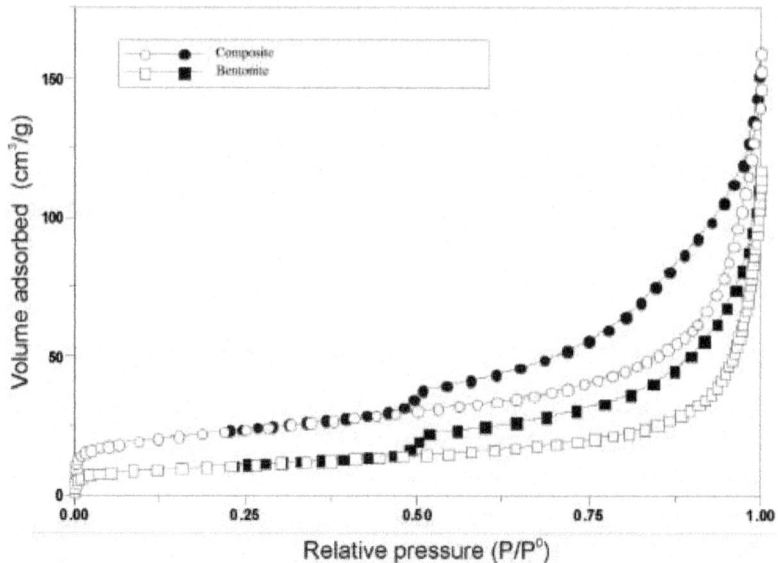

Figure 6: Nitrogen adsorption-desorption isotherms of native bentonite and composite.

The isotherms can be assigned to Type II isotherms, corresponding to nonporous or macroporous adsorbents. The hysteresis loops of Type H3 in the IUPAC classification occur at $p/p^0 > 0.5$, which is not inside the typical BET range. Furthermore, hysteresis loops of these isotherms indicate that they were given by either slit-shaped pores or, as in the present case, assemblages of platy particles of montmorillonite. Porous structure parameters are summarized in Table 2.

Table 2: Specific surface area and porosity of native bentonite and composite, determined by applying BET, BJH and D-R equation to N_2 adsorption at 77 K

Sample	$S_{BET}(m^2/g)$	Median mesopore diameter (nm)	Cumulative mesopore area (m^2/g)	Cumulative mesopore volume (cm^3/g)	Micropore volume (cm^3/g)
Bentonite	37.865	13.629	53.329	0.1202	0.0153
Composite	80.385	11.021	82.675	0.1716	0.0316

Compared to native bentonite, during the composite synthesis additional meso- and micropores were generated. Pore volumes (Gurvich) at p/p^0 0.999

for bentonite and composite are 0.180 cm^3/g and 0.243 cm^3/g, respectively. It was found that isotherms gave linear BET plots from p/p^0 0.03 to 0.21 for bentonite and from 0.03 to 0.19 for composite.

The composite has the specific surface area that is twice the size compared to the surface area of the native bentonite. This can be explained by the structural changes that occurred during the chemical and thermal modification of the native bentonite. The structural changes include delamination as well as the decrease of the distance between the layers of montmorillonite particles, because the interlayer water was lost under heating. The higher surface area of composite mainly results from the interparticle spaces generated by the three-dimensional co-aggregation of magnesium polyoxocations, iron oxide clusters and plate particles of montmorillonite. Macro- and mesopores arose from particle-to-particle interactions, while micropores were generated in the interlayer spaces of clay minerals due to irregular stacking of layers of different lateral dimensions [58].It is apparent that the changes of montmorillonite structure are responsible for the creation of new pore structure in the composite, which is then stabilized by the thermal treatment with the removal of H$_2$O molecules. The changes that involve partial dehydroxylation and cationic dehydration are brought about by thermal activation and they lead to various forms of cross-linking between oxides and smectite framework. As a result, composite does not swell and can be easily separated from water by filtration or centrifugation. There is a wide pore size distribution which supports disordered structure consisting of the delaminated parts with mesoporosity and the layered parts with microporosity.

The pH of the Pb(II) solution plays an important role in the adsorption process, influencing not only the surface charge of the adsorbent and the dissociation of functional groups on the active sites of the adsorbent but also the solution Pb(II) chemistry. The adsorption of Pb(II) on the composite decreased when pH decreased as shown in Fig. 7.

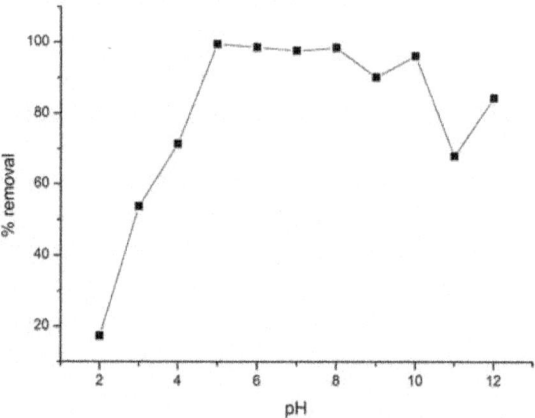

Figure 7: Effect of pH on adsorption of Pb(II) onto composite.

The adsorptive decrease at pH below 5 was caused by the competition between H^+ and Pb^{2+} for the negatively charged surface sites. Maximum retention is in the pH range 5-10. The main Pb(II) species in the pH range 6.5-10 are $Pb(OH)^+$ and $Pb(OH)_2$ which can easily form colloidal micelles characterized with the following imposed structure:

$$\{m[Pb(OH)_2]nPb(OH)^+ \cdot (n-x)NO_3^-\}xNO_3^- \tag{15}$$

The potential – determining ion is $Pb(OH)^+$ and that is the reason for the positive ZP of colloidal Pb(II) at the pH below 10 [59,60]. Therefore, colloidal micelles were easily attracted by the negatively charged composite surface. Particle size of colloidal Pb(II) at pH 7±0.1 was determined to be 268.7 ± 16.7 nm. At the pH range of 10-12 the predominant Pb(II) species are $Pb(OH)_2$ and $Pb(OH)_3^-$ which give rise to the formation of negatively charged colloidal micelles with the following structure:

$$\{m[Pb(OH)_2]nPb(OH)_3^- \cdot (n-x)Na^+\}xNa^+ \tag{16}$$

ZP values for Pb(II) colloidal solutions at pH 11.8 were - 50.7±3.6 mV with particle size of 252.7±28.2 nm. Having in mind surface heterogeneity of the composite and high point of zero charge value of $Mg(OH)_2$ (between pH 12 and pH 13) [61], negative ions and particles can be adsorbed on the positively charged surface sites at pH 10-12. Removal efficiency of $Pb(OH)_3^-$ was higher than negatively charged colloids, probably because the ionic species were involved in the process of ion exchange and chemisorption, while colloidal micelles could be bound to the surface dominantly by electrostatic forces.

Bentonite Based Composite Coated with Immobilized Thin Layer of Organic Matter

Synthesis of bentonite based composite material, described in this section, was carried out by applying thin coatings of natural organic matter, obtained by alkaline extraction from peat, mostly comprised of humic acids [62]. Humic acids have high complexing ability with various heavy metal ions, but it is difficult to use them as the sorbent because of their high solubility in water. However, they form stabile complexes with the inorganic ingredients of bentonite (montmorillonite, quartz, oxides, etc.) and can be additionally insolubilized and immobilized by heating at 350°C. After immobilization, humic acids represent an important sorbent for heavy metals, pesticides and other harmful ingredients from water. Humic acid are insolubilized by condensation of carboxylic and phenolic hydroxyl groups. Therefore, the aim was to remove manganese from aqueous solutions by treating it with synthesized composite as well as to study and explain the mechanism of composite interaction with manganese aqueous solutions. The composite does not release significant quantity of organic matter in water because it is tightly bonded to bentonite surface [63-65]. The degree of manganese removal was more than 94% at a range of initial manganese concentrations from 0.250 to 10 mg/l. The result of conductometric titration is given in Fig. 8. Equivalence point was located at the intercept of the first and second linear part of the titration curve. The value of the total acidic group content is calculated to be 215.18 μmol/g.

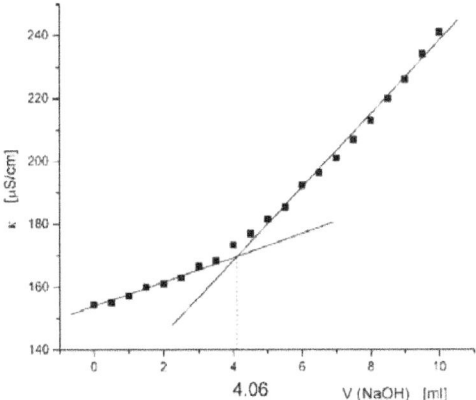

Figure 8: The conductometric titration of composite suspension (1 g in 250 ml of 1mM NaCl solution as background electrolyte) with 0.053 M NaOH.

experimental data of manganese adsorption onto composite are very well fitted by the Freundlich isotherm model (Fig. 9.) with a very high correlation coefficient value of 0.9948. The good agreement of experimental data with the Freundlich model indicates that there are several types of adsorption sites on the surface of the composite. The amount of adsorbed Mn(II) increases rapidly in the first region of adsorption isotherm and then the slope of isotherm gradually decreases in the second region. The adsorption capacity of composite is 11.86 mg/g, at an equilibrium manganese concentration of 16.28 mg/l.

After the treatment of model water with composite for the period of 20 min, the following results were obtained (Table 3).

Table 3: The results of water analysis before and after treatment with composite

Before water treatment			After water treatment			
C_0(Mn) mg/l	pH	Conductivity µS/cm	pH	Conductivity µS/cm	C(Mn) mg/l	%Mn Adsorption
0	6.43	8.01	6.67	11.43	0	0
0.250	6.37	9.57	7.11	13.76	0.0030	98.8
0.490	6.32	10.67	7.15	15.31	0.0039	99.2
1.0	6.30	14.67	7.12	31.10	0.0090	99.1
2.5	6.20	20.70	6.96	37.20	0.0187	99.25
5.0	6.19	32.80	6.83	49.40	0.0646	98.71
10.0	6.16	55.30	6.70	68.90	0.5314	94.69

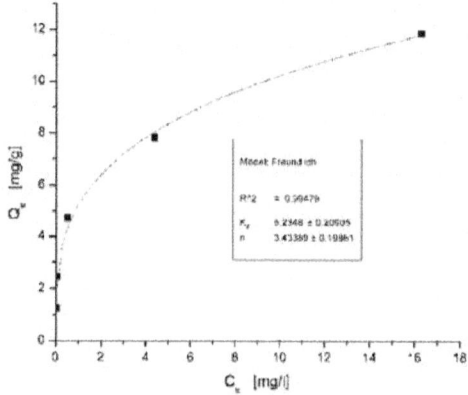

Figure 9: Freundlich adsorption isotherm for manganese adsorption onto composite.

During the thermal treatment in nitrogen atmosphere at 350 °C, the condensation of carboxyl and adjacent alcohol and phenol groups occurs. In this way the solubility of organic matter immobilized on bentonite matrix surface decreases [65]. Moreover, a part of carboxyl groups is decomposed by decarboxylation reaction, releasing CO_2 and CO. However, despite of this, a part of oxygen functional groups remains on the surface, and these groups act as sites that bind bivalent manganese forming inner-sphere complexes.

Besides organic functional groups, there are also Si-OH and Al-OH groups on the sites of crystal grain breaks, as well as permanent negative charge due to isomorphic substitution in clay minerals. They all contribute to the reduction of manganese concentration in the aqueous solution. Manganese retention by the formation of outer-sphere complexes, including ion exchange, can be showed by an Eq. (17)[66].

$$(\equiv S\text{-}O^-)_2...C^{n+}{}_{3\text{-}n} + Mn^{2+} \rightleftharpoons (\equiv S\text{-}O^-)_2...Mn^{2+} + (3\text{-}n)\ C^{n+}$$

$$(17)$$

in which C represents the cation that is exchanged.

The formation of inner-sphere complexes is represented by the Eqs. (18) and (19) and involves the release of hydrogen ions and the change of solution pH.

$$\equiv S\text{-}OH + Mn^{2+} \rightleftharpoons \equiv S\text{-}O\text{-}Mn^+ + H^+$$

$$(18)$$

$$\equiv 2S\text{-}OH + Mn^{2+} \rightleftharpoons (\equiv S\text{-}O)_2\text{-}Mn + 2H^+$$

$$(19)$$

According to these equations, it can be concluded that the pH value of the solutions decrease after the treatment. However, an opposite phenomenon can be experimentally observed (Table 3). The explanation for it is that hydrogen ions which are released during manganese retention participate in the protonation of surface groups:

$$\equiv S\text{-}OH + H^+ \rightleftharpoons \equiv S\text{-}OH_2^+$$

$$(20)$$

$$\equiv S\text{-}O^- + H^+ \rightleftharpoons \equiv S\text{-}OH$$

$$(21)$$

Therefore, the pH value of the Mn^{2+} aqueous solutions after treatment with composite had a higher value than the initial pH. This indicates that more hydrogen ions are bound to the surface than released by manganese binding. Namely, the composite exhibits amphoteric character due to the surface sites that act either as proton acceptors or as proton donors.

Organic matter decreases the PZC value of bentonite and neutralizes positive electric charge that comes from interlaminated cations, thus increasing composite affinity to manganese, even at lower pH values (67). Fig. 10. presents the pH dependence of residual Mn concentration, for the initial Mn concentration of 5 mg/l. The residual concentration of Mn decreases gradually with pH increasing in the range of 3.5-7 and then increases in the range of 7-10, with the apparent minimum at pH 7.

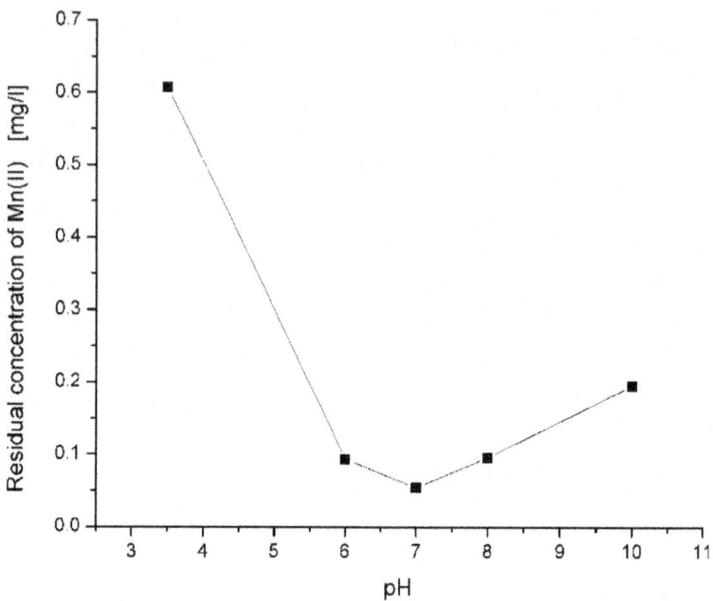

Figure 10: Residual concentration of Mn(II) as a function of model water pH.

The increase of pH value has dual effect on the removal of manganese. The increase of the pH value favours manganese removal due to increase of the number of deprotonated sites that are available for the binding of manganese. However, there is an increase in the solubility of organic matter which has been applied on the bentonite particles. The dissolved organic matter (humic acids) reacts with manganese forming complexes which bear a negative charge and have a weaker binding affinity for the composite surface than Mn^{2+}. Fig 10. indicates two opposite effects of the pH on manganese removal. The pH dependence of released organic matter (expressed as permanganate number) and turbidity (NTU) of solutions are shown in Fig. 11.

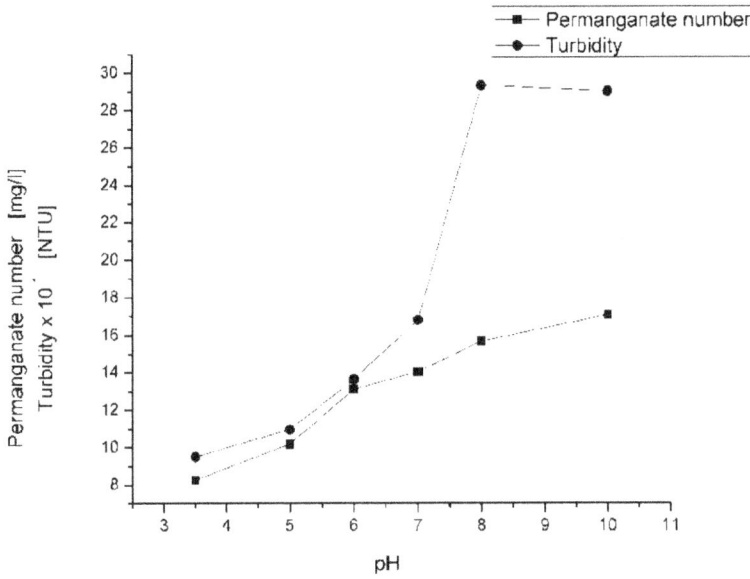

Figure 11: Premanganate number and turbidity of filtrate as function of pH (0.2 g of composite and 100 ml of 1mM Na_2SO_4 as background electrolyte).

The released organic matter contributes to the increased turbidity at higher pH values.

SUMMARY

The widespread industrial areas where nanocomposites can be applied are primary and conversion industry, modern coating technologies, constructional regions, and environmental (water, air) purification. In addition to the dominant use of composites as structural elements, important application of composite materials is in the water purification technologies. In this field of application, composites usually have the role of adsorbent, electrochemically active materials, catalysts, photocatalysts etc.

Bentonite is a natural and colloidal alumosilicate with particle size less than 10 μm, which is effectively used as sorbent for heavy metals and other inorganic and organic pollutants from water. Due to its positive textural properties and high specific surface area it can be used as low-cost matrix for synthesis of adsorbents or electrochemically active composite materials for the removal of pollutants in ionic and colloidal form from water. In this respect, three new/modified bentonite based composite materials have been synthetised and characterized.

Coated or composite particles are composed of solid phase covered with thinner or thicker layer of another material. These coatings - layers covering the surface of matrix are important for several reasons. In such way, the surface and textural characteristics of the initial solid phase are modified and sintering conditions can be better controlled. An important factor in achieving the desired microstructure of ceramics is sintering procedure that includes rather complex processes. A considerable influence on sintering has been exhibited by an addition of microalloying components, which significantly determined a microstructure and resulted properties of ceramics. The presence of small amounts of impurities in the starting material can vastly influence their mechanical, optical, electrical, color, diffusive, and dielectric properties of alumosilicate matrix. In summary, the process of diffusion mass transport in ceramic crystal regions are affected by temperature, oxygen partial pressure and concentration of impurities. A procedure for the removal of manganese in ionic (Mn^{2+}) and colloidal (MnO_2) forms from synthetic waters, by reduction and adsorption processes on electrochemically active alumosilicate ceramics based composite material has been described. Synthesis procedure of the composite material consists of two phases. Firstly, composite particles were synthesized by applying Al/Sn oxide coating onto the bentonite particles in an aqueous suspension. In the second phase, aluminium powder is added to the previously obtained plastic mass and after shaping in the form of spheres 1 cm in diameter and drying, sintering was performed at 900°C. Elemental tin, resulting from the reduction of Sn^{2+}-ion, comes into contact with liquid aluminum in the pores of the matrix performing aluminum microalloying and activation. Moreover, due to a low partial pressure of oxygen, nonstoichiometric oxides with metal excess are obtained, and they play an important role in the electrochemical activity of the composite material. In accordance with this, a redox potential of water is changed in contact with composite.

Another effective composite material as a potentially attractive adsorbent for the treatment of Pb(II) contaminated water sources has been synthesized by coating of bentonite with mixed iron and magnesium (hydr)oxides. The procedure for obtaining a bentonite based composite involves the application of mixed Fe and Mg hydroxides coatings onto bentonite particles in aqueous suspension and subsequent thermal treatment of the solid phase at 225°C. Formation of heterogeneous coatings on bentonite results in changes of bentonite acid-based properties, high specific surface area and positive adsorption characteristics. Different adsorption sites on such heterogeneous surface provide an efficient removal of numerous chemical species of Pb(II) (ionic and colloidal) over a wide pH range.

Third bentonite based composite material was obtained by applying thin coatings of natural organic matter, extracted from a peat, mostly based on humic acids. Humic acids are known due to high complexing ability to various heavy metal ions, but it is difficult to use them directly as the sorbent because of their high solubility in water. However, they form stabile complexes with the inorganic ingredients of bentonite (montmorillonite, quartz, oxides, etc.) and can be successfully insolubilized and immobilized by heating at 350°C. After immobilization, humic acids represent an important sorbent for heavy metals, pesticides and other harmful ingredients from water. Humic acid are insolubilized by condensation of carboxylic and phenolic hydroxyl groups. The composite such obtained can be effectively used as the sorbent for heavy metals.

ACKNOWLEDGEMENT

The authors acknowledge financial support from the Ministry of Education and Science of the Republic of Serbia.

REFERENCES

1. B. Ralph, H. C. Yuen, W. B. Lee, The processing of metal matrix composites- an overview. Journal of Materials Processing Technology 19976 63339

2. D. Chung, Materials. Composite, Science. . Springer, Media. B. Business, B.2010 2010

3. Usuki A, Kawasumi M, Kojima Y, Fukushima Y, Okada A, Kurauchi T. Swelling behavior of montmorillonite cation exchanged for ω-amino acids by ε-caprolactam. Journal of Materials Research 1993; 8 1179-1184.

4. Y. Kojima, A. Usuki, M. Kawasumi, Y. Fukushima, A. Okada, T. Kurauchi, properties. Mechanical, of, 6 clay hybrid. Journal of Materials Research 1993; 81185 .

5. E. Giannelis, Silicate. Polymer-layered, Advanced. Nanocomposites, Materials, 1996 829 .

6. R. Pfaendner, Industrial. Nanocomposites, of. opportunity, Polymer. challenge?, Stability. Degradatio, 20109 95369 .

7. Nanocomposites and nanotubes conference,1112 th March, 2008 www. nanosconference.com/home.asp;. accessed 30.04.2012.

8. 2010D. Manfredi, M. Pavese, S. Biamino, A. Antonini, P. Fino, C. Badini, Microstructure and mechanical properties of co-continuous metal/ceramic composites obtained from Reactive Metal Penetration of

commercial aluminium alloys into cordierite. Composites: Part A 2010; 41639 .

9.   1999E. Saiz, S. Foppiano, Chan. W. Moberly, A. P. Tomsia, Synthesis and processing of ceramic-metal composites by reactive metal penetration. Composites: Part A 1999; 30399 .

10. I. Haq, E. Matijevic, Preparation and properties of uniform coated inorganic colloidal particles. 12 Tin and its compounds on hematite. Progress in Colloid and Polymer Science 1998; 109185 .

11. F. Bergaya, B. K. G. Theng, G. Lagaly, Handbook of clay science. Elsevier; 2006

12. L. Olmos, C. L. Martin, D. Bouvard, Sintering of mixtures of powders: Experiments and modelling. Powder Technology 200919 190134 .

13. A. Michael, P. Dubois, silicate. Polymer-layered, preparation. nanocomposites, properties, of. a. uses, class. new, materials. of, Materials Science and Engineering 20002 281 .

14. S. S. Ray, M. Okamoto, Silicate. Polymer/layered, A. Nanocomposite, from. Review, to. Preparation, Processing, Progress in Polymer Science 20032 281539 .

15. F. Hussain, M. Hojjati, M. Okamoto, R. E. Gorga, article. Review, nanocomposites. polymer-matrix, manufacturing. processing, an. application, Journal. overview, Composite. of, Materials, 20084 401511 .

16. J. Moszo, B. Pukanszky, micro. Polymer, Structure. nanocomposites, properties. interactions, Journal of Industrial and Engineering Chemistry 20081 14535 .

17. W. Xie, Z. Gao, W. P. Pan, D. Hunter, A. Singh, R. Vaia, Thermal degradation chemistry of alkyl quaternary ammonium montmorillonite. Chemistry of Materials 20011 132979 .

18. A. Leszczynska, J. Njuguna, K. Pielochowski, J. R. Banerjee, (montmorillonite. Polymer/clay, with. nanocomposites, thermal. improved, Part. I. I. A. properties, study. review, thermal. on, of. stability, nanocomposites. montmorillonite, on. based, polymeric. different, matrixes, Thermochimica Acta 2007 454 1 122 .

19. Pandey J.K, Reddy K.R, Kumar A.P, Singh R.P, An overview on the degradability of polymer nanocomposites. Polymer Degradation and Stability 20058 88234 .

20. H. Wermter, R. Pfaender, patent. E. P. European, 1592741 assigned to Ciba Holding Inc., 2009.

21. A. Leszcynska, J. Njuguna, K. Pielichowski, J. R. Banerjee, Thermal

stability of polymer/montmorillonite nanocomposites: Part I: Factors influencing thermal stability and mechanisms of thermal stability improvement. Thermochimica Acta 200745 45375 .

22. G. Moad, K. Dean, L. Edmond, N. Kukaleva, G. Li, R. T. A. Mayadunne, Poly(ethylene. Non-Ionic, Surfactants. oxide)-based, Dispersants. Intercalants, for. Exfoliants, Nanocomposites. Polypropylene-Clay, Macromolecular Materials and Engineering 200629 29137 .

23. G. Moad, K. Dean, L. Edmond, N. Kukaleva, G. Li, R. T. A. Mayadaunne, R. Pfaendner, A. Schneider, G. Simon, Wermter, 2006 2006. Novel Copolymers as Dispersants/Intercalants/Exfoliants for Polypropylene-Clay Nanocomposites. Macromolecular Symposia 2006; 233170 .

24. N. S. Allen, M. Edge, T. Corrales, A. Childs, C. M. Liauw, F. Catalina, C. Peinado, A. Minihan, D. Aldcroft, Ageing and stabilisation of filled polymers: an overview. Polymer Degradation and Stability 19986 61183 .

25. H. Qin, C. Zhao, C. Zhang, G. Chen, M. Yang, Photo-oxidative degradation of polyethylene/montmorillonite nanocomposite. Polymer Degradation and Stability 20038 81497 .

26. S. Morlat-Therias, B. Mailhot, D. Gonzalez, J. L. Gardette, Photooxidation of Polypropylene/Montmorillonite nanocomposites Part II : Interactions with antioxidants. Chemistry of Materials 20051 171072 .

27. S. Morlat, B. Mailhot, D. Gonzalez, J. L. Gardette, of. Photooxidation, Montmorillonite. Polypropylene, Part. I. . nanocomposites, of. Influence, nanoclay, agent. compatibilising, of. Chemistry, Materials, 2004 16 3 377383 .

28. S. Chmela, A. Kleinova, A. Fiedlorova, E. Borsig, D. Kaempfer, R. Thormann, Photo-oxidation of sPP/organoclay nanocomposites. Applied Chemistry 2005 42 7 821829 .

29. B. Rotzinger, P. P. A. Talc-filled, concept. new, maintain. to, term. long, stability. heat, Polymer Degradation and Stability 20069 912884 .

30. N. Savage, M. S. Diallo, Nano materials and water purification: opportunities and challenges. Journal of Nanoparticle Reserch 2005 7331 .

31. Da Fonseca M.G, De Oliveira M.M, Arakaki L.N.H. Removal of cadmium, zinc, manganese and chromium cations from aqueous solution by a clay mineral. Journal of Hazardous Materials 2006 B137288 .

32. A. Dimirkou, M. K. Doula, Use of cl2optilolite and an Fe-overexchanged clinoptilolite in Zn2+ and Mn2+ removal from drinking water.

Desalination 2008; 224280 .

33.   Bladergroen B.J, Linkov V.M. Electrosorption ceramic based membranes for water treatment. Separation and Purification Technology 20012 25347 .

34.   T. Kanki, S. Hamasaki, N. Sano, A. Toyoda, K. Hirano, Water purification in a fluidized bed photocatalytic reactor using TiO2 -coated ceramic particles. Chemical Engineering Journal 200510 108 155-160.

35.   C. D. Johnson, F. Worrall, Novel granular materials with microcrystalline active surfaces-Waste water treatment applications of zeolite/vermiculite composites. Water Research 20074 412229 .

36.   Tzoupanos N.D, Zouboulis A.I. Preparation, characterisation and application of novel composite coagulants for surface water treatment Water Research 20114 453614 .

37.   Y. Lv, H. Liu, Z. Wang, S. Liu, L. Hao, Y. Sang, D. Liu, J. Wang, R. I. Boughton, Silver nanoparticle-decorated porous ceramic composite for water treatment. Journal of Membrane Science 200933 33150 .

38.   W. Shen, L. Feng, H. Feng, Z. Kong, M. Guo, silver. I. I. Ultrafine, particles. oxide, porous. decorated, composites. ceramic, water. for, treatment, Chemical Engineering Journal 201117 175592 .

39.   W. Wang, P. Serp, P. Kalck, Silva. C. Gomes, Faria. J. Luıs, Preparation and characterization of nanostructured MWCNT-TiO2 composite materials for photocatalytic water treatment applications. Materials Research Bulletin 20084 43958 .

40.   H. Zhang, X. Quan, S. Chen, H. Zhao, Fabrication and characterization of silica/titania nanotubes composite membrane with photocatalytic capability. Environmental Science and Technology 20064 406104 .

41.   M. Ranđelović, M. Purenović, A. Zarubica, J. Purenović, I. Mladenović, G. Nikolić, Alumosilicate ceramics based composite microalloyed by Sn: An interaction with ionic and colloidal forms of Mn in synthetic water. Desalination 2011 279(1-3) 353-358.

42.   Jeurgens L.P.H, Sloof W.G, Tichelaar F.D, Mittemeijer E.J. Composition and chemical state of the ions of aluminium-oxide films formed by thermal oxidation of aluminium. Surface Science 200250 506313 .

43.   Purenovic M.M. Influence of Some Alloying Elements and Admixtures on Electrochemical Behaviour of the System Aluminium- Oxide Layer- Electrolyte. PhD Thesis. Faculty of Technology and Metallurgy, University of Belgrade; 1978

44.   R. W. Chan, P. Haasen, Metallurgy. Physical, edition. fourth, Science. B.

V. Elsevier, Amsterdam, Netherlands; 1996

45. Rahaman M.N. Ceramic processing and sintering, second edition. Marcel Dekker, Inc. New York; 2003

46. S. Gudic, I. Smoljko, M. Kliskic, Electrochemical behaviour of aluminium alloys containing indium and tin in NaCl solution. Materials Chemistry and Physics 201012 121561 .

47. Cvetković V.S, Purenović J.M, Jovićević J.N. Change of water redox potential, pH and rH in contact with magnesium enriched kaolinite-bentonite ceramics. Applid Clay Science 20083 38268 .

48. Q. Wei, X. Ren, J. Du, S. Wei, S. Hu, Study of the electrodeposition conditions of metallic manganese in an electrolytic membrane reactor. Minerals Engineering 20102 23578 .

49. Cvetković V.S, Purenović J.M, Purenović M.M, Jovićević J.N. Interaction of Mg-enriched kaolinite-bentonite ceramics with arsenic aqueous solutions. Desalination 200924 249582 .

50. O. T. Can, M. Bayramoglu, M. Kobya, Decolorization of reactive dye solutions by electrocoagulation using aluminum electrodes. Industrial and Engineering Chemistry Research 20034 423391 .

51. P. Canizares, F. Martinez, M. Carmona, J. Lobato, M. A. Rodrigo, Continuous Electrocoagulation of Synthetic Colloid-Polluted Wastes. Industrial and Engineering Chemistry Research 20054 44 (8171-8177.

52. Lead in Drinking-water, Background document for development of WHO Guidelines for Drinking-water Quality, World Health Organization; 2003.

53. M. Ranđelović, M. Purenović, A. Zarubica, J. Purenović, B. Matović, M. Momčilović, Synthesis of composite by application of mixed Fe, Mg (hydr)oxides coatings onto bentonite- a use for the removal of Pb(II) from water. Journal of Hazardous Materials 2012 199-200 367-374.

54. R. M. Cornell, U. Schwertmann, Iron. The, Structure. Oxides, Reactions. Properties, Occurences, second. Uses, ed, W. I. L. E. Y. -V, C. H. Verlag, H. Gmb, K. Co, A. Ga, Weinheim; 2003

55. B. Caglar, B. Afsin, A. Tabak, E. Eren, Characterization of the cation-exchanged bentonites by XRPD, ATR, DTA/TG analyses and BET measurement, Chemical Engineering Journal 200914 149242 .

56. Krasnobaeva O.N, Belomestnykh I.P, Isagulyants G.V, Nosova T.A, Elizarova T.A, Teplyakova T.D, Kondakov D.F, Danilov V.P. Synthesis of Complex Hydroxo Salts of Magnesium, Nickel, Cobalt, Aluminum, and Bismuth and Oxide Catalysts on Their Base. Russian Journal of Inorganic Chemistry 2007 52 2 141146 .

57. Krasnobaeva O.N, Belomestnykh I.P, Isagulyants G.V, Nosova T.A, Elizarova T.A, Kondakov D.F, Danilov V.P. Chromium, Vanadium, Molybdenum, Tungsten, Magnesium,and Aluminum Hydrotalcite Hydroxo Salts and Oxide Catalysts on Their Base, Russian Journal of Inorganic Chemistry 2009 54 4 495499 .

58. J. Rouquerol, F. Rouquerol, K. S. W. Sing, Adsorption by Powders and Porous Solids: Principles, Methodology and Applications, Academic Press, San Diego USA; 1999

59. Q. Liu, Y. Liu, of. Distribution, I. I. Pb, in. species, solutions. aqueous, of. Journal, Colloid, Science. Interface, 2003268 .

60. M. Kosmulski, of. P. Z. C. Compilation, I. E. P. of, soluble. sparingly, oxides. metal, from. hydroxides, Advances. literature, Colloid. in, Science. Interface, 2009152 .

61. S. V. Krishnan, I. Iwasaki, Heterocoagulation, surface precipitation in a quartz-Mg(OH)2 system, Environmental Science and Technology 19862 201224 .

62. M. Ranđelović, M. Purenović, J. Purenović, M. Momčilović, Removal, 2 Mn2+ from water by bentonite coated with immobilized thin layers of natural organic matter. Journal of Water Supply: Research and Technology- AQUA 2011; 60 8 486493 .

63. F. Ayari, E. Srasra, M. Trabelsi-Ayadi, Characterization of bentonitic clay and their use as adsorbent. Desalination 200518 185391 .

64. C. A. Kolokassidou, I. Pashalidis, C. N. Costa, A. M. Efstathiou, G. Buckau, Thermal stability of solid and aqueous solutions of humic acid. Thermochimica Acta 200745 45478 .

65. S. Ghosh, Zhen-Yu, 1 Kang1 S, Bhowmik P. C, Xing B. S. Sorption and fractionation of a peat derived humic acid by kaolinite, montmorillonite, and goethite. Pedosphere 2009; 19 1 2130 .

66. Doula M.K. Removal 2 Mn2+ ions from drinking water by using clinoptilolite and a clinoptilolite Fe oxide system. Water Research 2006; 403167 .

67. J. Zhuang, G. R. Yu, Effects of surface coatings on electrochemical properties and contaminant sorption of clay minerals. Chemosphere 20024 49619 .

Chapter 4

TREATING OF WASTE WATER APPLYING BUBBLE FLOTATION

Ramiro Escudero, Francisco J. Tavera and Eunice Espinoza

[1]Instituto de Investigaciones Metalúrgicas, Universidad Michoacana de San Nicolás de Hidalgo, Morelia, Michoacán, México

INTRODUCTION

Recently, it might be exaggerated the impact of man activities as the solely cause of the tremendous changes in the global climate resulting from its activity effects on the environment nature; it has been suggested that it has started a new geological era, the "antroposoic age;" however, there are some clues, together with man influence in climate changes, that sun activity has a determining effect on the subject. Nevertheless, from the beginning of man species presence in the geological record it is observed an interaction of these species with the surroundings that increases with their degree of social and technological development.

The association between man and its environment presents a continuous growth because the development of its intellect and the complex nature of its thought; these features allowed man to adapt to a variety of geographical situations, spreading towards the entire planet, building technologies as a result of its social development needs.

In the historic record of civilization, in modern times, between the second half of the 1700's and the middle of the 1800's, there is an exponential increase in natural resources exploitation because the advance in the scientific knowledge and the technological development. The expansion trend of these features, the science and technology, has been maintained with no interruption, as well as their effect on the surroundings producing a deleterious consequence on the environs quality.

It is well understood that the equilibrium in the planet nature is very sensitive to man activities, therefore, it is necessary to intensify the carefulness to preserve the ecosystem quality in order to prevent catastrophic environmental disasters. A critic facet of natural environment is related to water pollution. Depletion of fresh water reserves makes it important to propose effective technologies to clean and recover water from polluted bionetwork. Therefore, advances in possible water treating improvements are issued in this chapter.

TREATING OF WASTE WATER APPLYING BUBBLE FLOTATION

From the time when natural resources started to exploit in a massive way, water is used as a vehicle to transport and process materials. Water is the most important solvent in nature, and the contact of water with a variety of substances makes it the origin of water pollution through the formation of dissolved solutes solutions which decrease the quality of the liquid.

Water is often encountered with high concentration of pollutants that could be in solution or dispersed as insoluble phases frequently forming emulsions and colloids. Among pollutants encountered in fresh water are heavy metals and organics. These materials regularly are difficult to process because to the enormous amounts of water that is produced by urban and industrial activities; therefore, the ideal process to clean polluted water should be operating in a continuous way to minimise fixed and operating costs.

Lead Carbonate Colloid in Residual Water: Continuous Ion Flotation

The ion flotation term was first introduced by Sebba [1], defined it as a technique used to collect a material which is in solution in an aqueous phase, and it may be even in a colloid structure. By adding modifiers of the chemical conditions of the aqueous media, and with the addition of a collector reagent, adequately electrically charged, the dissolved substance, or the colloid, is transformed in a product with hydrophobic sites which promotes its adsorption on the liquid/vapour "surface" of a gas bubble and, therefore, to float in the aqueous media up to its surface producing a foam that contains a concentrate of the original substance which can be collected.

By looking into the previous statement, it can be thought that there is a concentration relationship between the amount of dissolved substance and the required amount of collector reagent to create hydrophobic sites. Therefore, ion flotation technique is restricted to treat dilute aqueous solutions of such ion or colloid material, otherwise, the required amount of collector reagent may exceed the micelle critic concentration and the desired process would not be attained.

The goal idea in the application of ion or colloid flotation techniques may be to treat residual industrial or urban water, which means that it has to deal with enormous amounts of water and, therefore, a continuous process should be available to separate the solute impurity.

In the case of flotation as applied to mineral particles, it requires the gas bubbles to be large enough to produce a solid-bubble aggregate with a lower relative density than the mineral pulp density which is processed and, in this case, bubble shear must be controlled to avoid the detachment of solids from the bubble.

However, in ion flotation the material to be separated from the liquid presents a similar density as that of the liquid bulk, and the inertia of this material in front of the moving gas bubbles is such that it can be easily drawn by the liquid that flows surroundings the bubble. Consequently, it may be expected that the contact between the hydrophobic material and the gas bubble will be likely to occur by decreasing bubble size; in addition, when the formation of small bubbles takes place, it creates a larger gas surface area which improves the flotation process [2, 3].

Most of the experimental work reported on ion flotation has been carried out in flotation columns may be under the influence of a better performance of column flotation in mineral processing. In spite of this, the ascending bubbles in a liquid column produce axial mixing and turbulences that prevent an effective contact among the floatable material and the gas bubble.

At the present time, the ion flotation technique has not reached any commercial appliance; nevertheless, there are good expectations for different engineering applications. A possibility in commercial scale might be the treating of aqueous media containing heavy metals.

The presence of heavy metals in residual water is commonly observed as a product of mining and metallurgical works; if these waters are discharged directly this will cause a major environmental problem; therefore, it is necessary to treat these waters to remove impurities to acceptable levels to avoid any damage to the environment. However, still today, in some locations, the exigencies to keep a close control on the application of environmental regulations are poor and serious pollution problems exist [4].

To correct the problem, engineering work must be dedicated to produce a realistic procedure for its application in large scale as it is required in mining and metallurgical locations.

It has been proposed and validated in laboratory, and pilot plant tests, the use of froth flotation to treat lead polluted water in a continuous scheme operation. In that case, lead is present in the aqueous phase as a lead colloidal carbonate, with a lead concentration of 20 ppm. A series of spargered flotation cells are used to perform the separation of the lead specie [5]; Figure 1 shows a representation of the series of flotation cells.

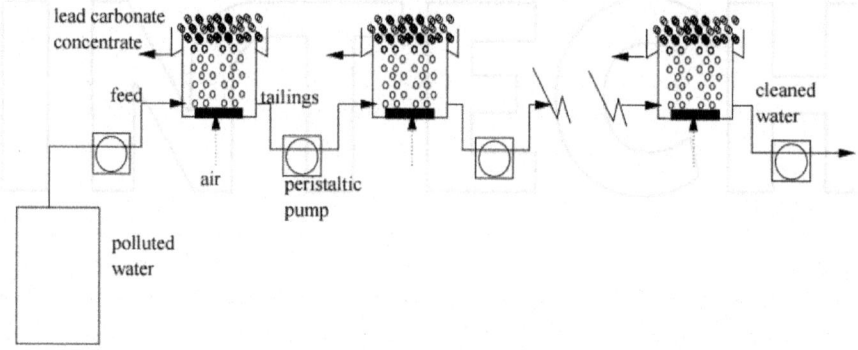

Figure 1: Illustration of a series of spargered flotation cells where porous gas spargers are installed to inject-disperse air in the form of small gas bubbles (< 1.5 mm).

In this flotation process, the water containing the lead colloidal carbonate is adequately conditioned by the addition of chemical reagent to fix an adequate pH to make possible to complex the lead specie with a suitable flotation reagent. Gas hold-up was measured below the froth region in the flotation cells by pressure measurements [3, 6], and locally at different points measuring electrical conductivity [7]; these measurements were related to the efficiency of producing large gas surface areas to perform the collection process of hydrophobic species.

In designing the chemical characteristics of the process, for example the pH of the aqueous phase, and the type of collector reagent, lead carbonate particles were separated and analysed in their Zeta potential (ζ) behaviour. Figure 2 shows the ζ performance of the lead carbonate.

The polluted water presented a pH of 7.4, therefore a cationic collector was used to produce a hydrophobic behaviour to the lead carbonate particles. The series of flotation cells were operated until the steady state conditions were achieved. At this point an apparent residence time was estimated ($\tau = [h_{cz}/J_l] \times [1 - \varepsilon_g]$; where h_{cz} is the collection zone height measured from the bottom to the froth interface, J_l is the superficial liquid velocity, and ε_g is the gas hold-up); as a first approximation this representation of residence time may be used to express the nature of mixing in the series of flotation cells.

The different streams (concentrate, tailings, and feed) were sampled to analyse their concentration of lead, and to determine the recovery of this material through the separation process. The recovery of lead may be related to the characteristics of the process in terms of the variables ε_g, bubble size, the superficial bubble surface flux, and the kinetics of the process.

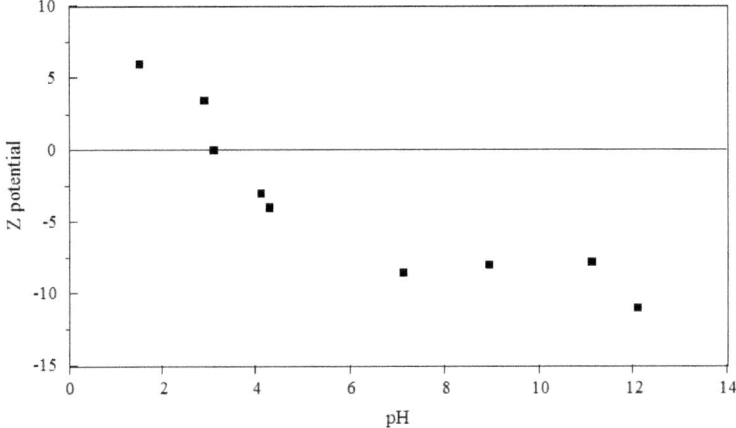

Figure 2: Measurements on the behaviour of Zeta potential (ζ) lead carbonate; measurements through the pH range between 1.4 and 12.

During the operation of the series of flotation cells, gas hold-up was monitored continuously in each flotation cell using measurements of pressure and electrical conductivity [6, 9]. Figure 3 presents the lead recovery in concentrate as a function of gas hold-up change in the first flotation cell.

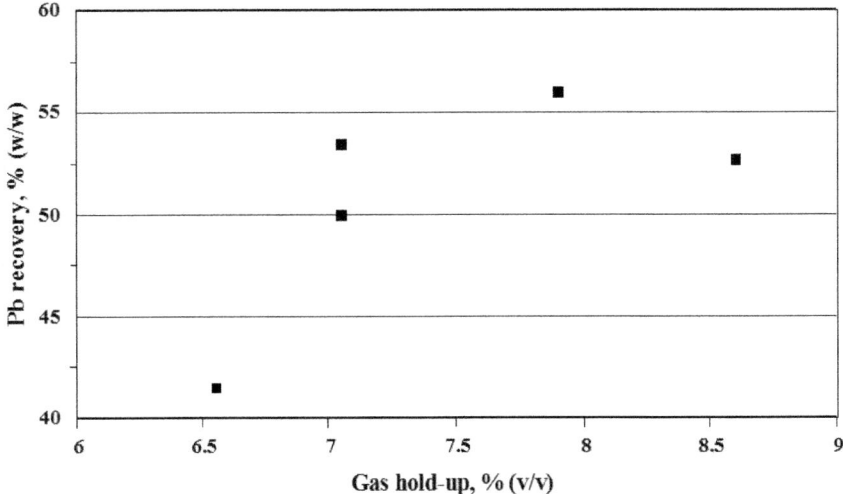

Figure 3: Typical graphic of experimental results on the flotation of lead represented as the lead recovery % and the corresponding gas hold-up.

The lead reported to the concentrate shows that its recovery increases as the amount of gas in the gas-water dispersion rises, as a result of the increase in the availability of bubble surface area to perform the collection process of hydrophobic species. The flotation of the lead specie presents a steady increase with increasing until the gas hold-up reaches 7.7% (v/v) above which there is a sudden reduction on that operation variable due to the formation of excessive turbulence produced by a considerable flow of gas in the cell.

Nevertheless, the magnitude of gas hold-up is related to bubble size, beside to the amount of gas which is fed into the flotation system. This may imply that the production of large gas hold-ups is related to formation of small gas bubbles. In this way, the recovery of lead from the aqueous media should be upgraded with the bubble size reduction. This effect is shown in Figure 4.

The bubble dimension was estimated by two methods: first a direct method using image analysis [10]; and the indirect method of drift flux analysis [3].

The experimental practice indicates that the decrease in the diameter of gas bubbles produces larger bubble surface area which "flows" through the flotation system collecting the hydrophobic species as they have a collision.

Production of large bubbles presents a hydrodynamic effect in the collection process, in addition to a smaller bubble surface area, which decreases the collection of lead carbonate colloid by minimizing the possible collision between the gas bubble and the hydrophobic lead carbonate surface. The contribution of these two effects of the bubble surface area is shown in fig. 4.

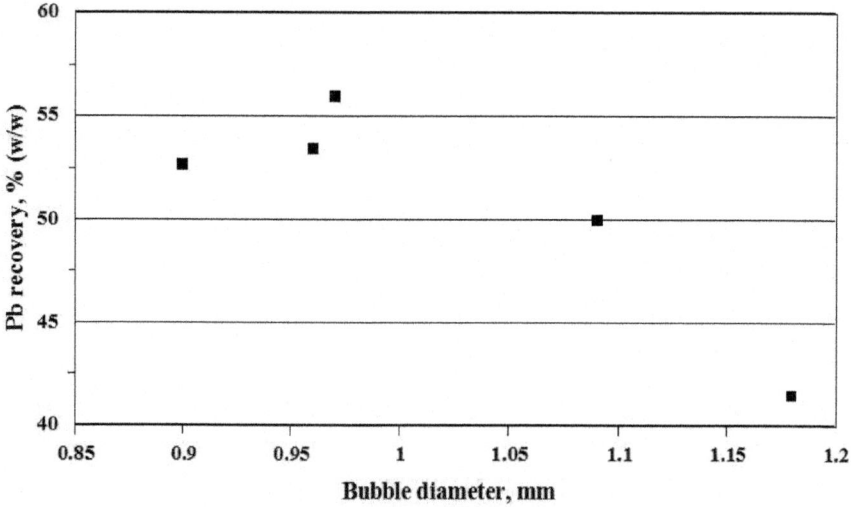

Figure 4: Effect of the bubble size on the separation of lead from the aqueous media.

In terms of the bubble surface area, the concept of "superficial bubble surface flux," Sb, has been suggested as a controlling ultimate flotation variable. Sb contains the chemical, hydrodynamics, and mechanical characteristics of the flotation system; Sb can be thought as the amount of bubble surfaces that flow through a unit of a cross sectional area normal to the bubbles motion, by unit of time, therefore with units of time^{-1}; therefore from geometry considerations, Sb = 6 (J_g/d_b), where J_g is the superficial gas velocity, and d_b is the bubble diameter [11]. It has been demonstrated that Sb is directly related to the flotation – separation kinetics [12]. Figure 5 shows the relationship between the recovery of floated lead and Sb.

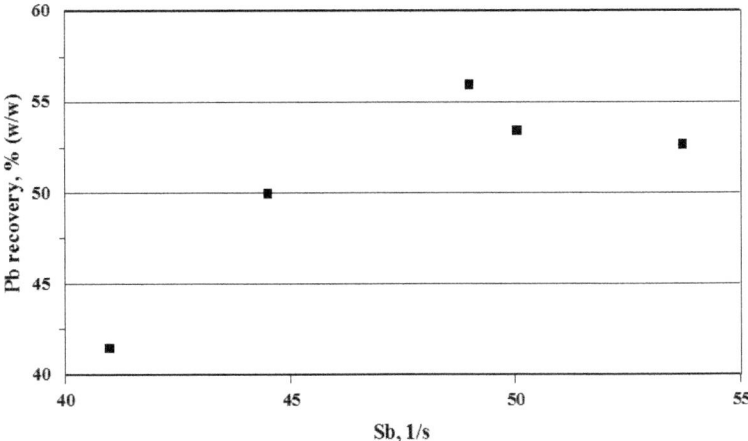

Figure 5: Relationship between the separation of lead carbonate from the aqueous phase by flotation, and the superficial bubble surface flux.

It can be noticed that the separation of lead from the water increases as the available bubble surface rises. Nevertheless, there is a maximum value of the lead recovery which corresponds to a given Sb value.

Above that Sb value, the relationship between Pb recovery and Sb changes due to an increase of mixing and circulation in the flotation system. This information indicates that the flotation of the lead colloid carbonate is very sensitive to the bubble size change, and therefore, to the superficial gas velocity and the superficial liquid velocity.

The behaviour of the flotation cells follow a model of the perfect mixer, therefore, flotation constant rate may be expressed as $\kappa = R/[\tau (1 - R)]$, where R is the lead recovery expressed as the weight fraction.[13]

The apparent rate constant for the lead colloid carbonate flotation is presented in Figure 6 as a function of the gas hold-up in the collection zone.

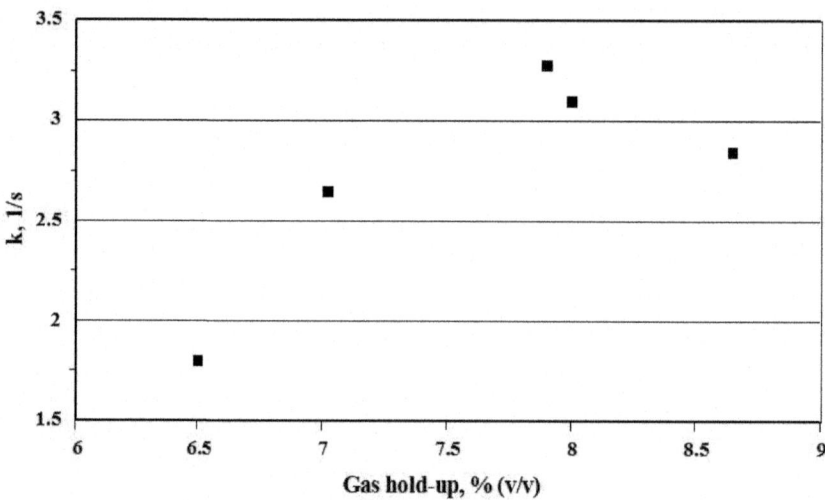

Figure 6: Behaviour of the apparent rate constant, as a function of the change on gas hold-up in the flotation of lead colloid carbonate.

It is observed that the flotation rate constant follows a linear relationship with respect to the change in the gas hold-up if the last is kept below 8%. Above this gas hold-up value the correlation between the flotation kinetics and the gas hold-up changes. This information suggests that the lead colloid carbonate flotation rate is strongly dependent from the hydrodynamic conditions of the system.

The previous statement indicates that the flotation rate constant must be related to the bubble size.Figure 7 presents the relationship between the flotation rate constant for the separation of the lead colloid carbonate, and the estimated bubble size.

This information describes that when the bubble size is between 0.97 mm and 1.2 mm presents a linear correlation with the apparent flotation rate constant. In this range of bubble size it can be seen that the kinetics of the flotation process is favoured with the decrease in the bubble diameter. But with the increase in bubble size above 1.2 mm the possibility to have a collision between the lead colloid carbonate and the bubble decreases, because the hydrodynamics of the bubble that moves faster in the liquid bulk.

When bubble size is smaller than 0.97 mm the linear trend between the apparent rate constant and bubble size is distorted because the effect of axial mixing. This axial mixing is formed by radial differences of gas distribution, becoming more evident as the gas hold-up increases.

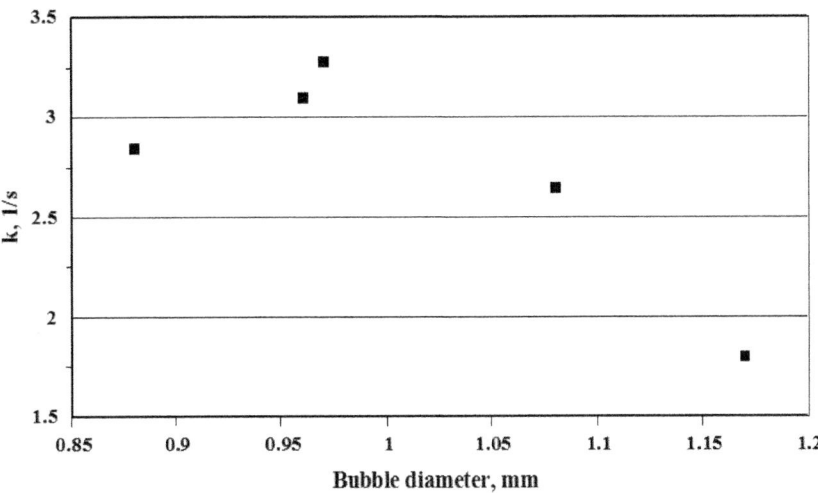

Figure 7: Relationship between the apparent rate constant and bubble size in the lead carbonate colloid separation.

In the series of flotation cells, the superficial gas velocity was maintained at 0.8 cm/s in each flotation cell, and the superficial liquid velocity was kept at 0.19 cm/s.

The lead carbonate colloid cumulative recovery, in the series of flotation cells, was estimated from the mass assaying in each cell.

Figure 8 represents the cumulative recovery of lead as a function of the number of flotation cells in the series of five flotation cells.

It is observed in fig. 8 that the cumulative recovery of lead through the series of 5 flotation cells is 93% (w/w). The total residence time of water in the flotation circuit is 13 minutes. This water processing time gives the impression that is short, as compared with the reported processing times in laboratory flotation columns testing (between 30 to several hundreds of minutes) [2].

This information indicates that the flotation cells instrumented with porous gas bubble generators can produce satisfactory bubble sizes to perform colloid flotation; it seems that differences in bubble size distribution are low enough to reduce axial mixing as compared with the natural circulation presented in flotation columns [13]. In this case lead content in water is reduced from 20 ppm to about 1 ppm.

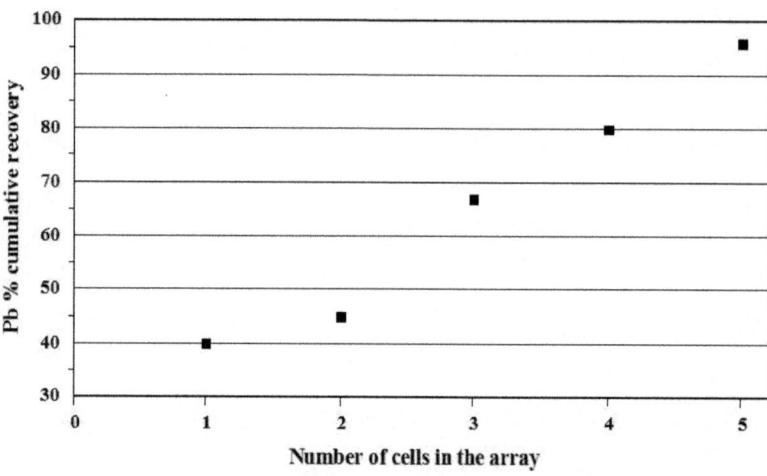

Figure 8: Cumulative recovery of floated lead as a function of the number of flotation cells series.

Water De-Oiling Using Column Flotation

Insoluble organic compounds in water are often encountered as pollutants, particularly in locations near urban centres or in regions associated with industrial and agricultural activities. Among these pollutants oily materials are frequently present in discharged water and should be removed from the aqueous phase before its emission.

In the production of food products such as vegetable cooking oil, residual water contains vestiges of oil, soap, and suspended solid particles from the processed oily seeds. Commonly, pollutants and water forms stable emulsions where the different phases are not easily coagulated and separated by differences of their densities.

In treating such emulsion products, different several approaches are used to eliminate the problem. Bacterial processing is often used to decompose the organic materials generating some non pollutant products; in these biological processes special care must be taken in providing adequate aeration and pH conditions of the polluted water in treating reactors where materials must stay for a while to be processed by living microorganisms. Some other techniques include water evaporation – condensation, but the cost tends to be high by the energy required to evaporate the water from the emulsion.

In searching for a short residence time process that allows to treat continuously large amounts of water, it was studied the possible separation of the organics content in water using flotation techniques. The use of flotation columns is reported as a possible solution to the problem [14], which implies a simple operation at a very low cost [3].

It was processed industrial residual water from a vegetable oil producer. The polluted water contained 40% v/v of organics consisted of residual oil, soap and solid particles, originated from the oil extraction and refining operations, in processing oily vegetable seeds.

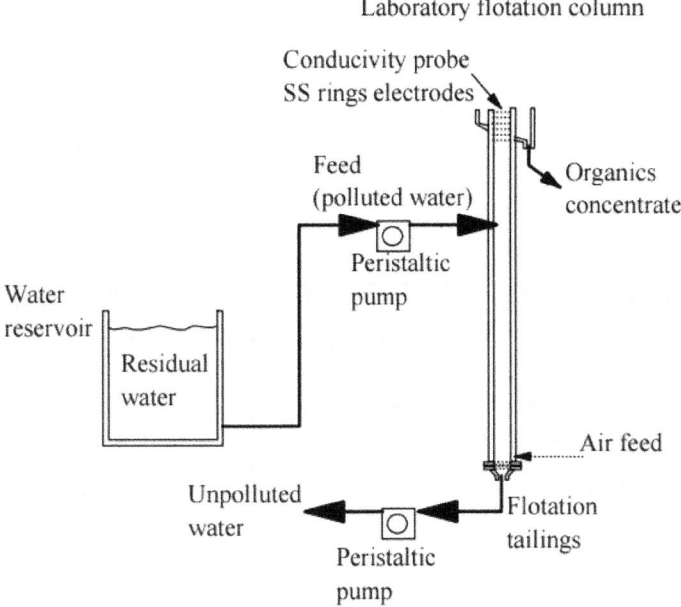

Figure 9: Representation of the laboratory flotation column assembly used to remove organics (oil, soap, solids) from water by froth flotation.

The flotation column was 0.1 m in diameter and 4 m height, and was fully instrumented to be operated automatically. The raw residual polluted water derived from the outlet water circuit at the industrial plant was placed in a 3 m³ reservoir from where the flotation column water feed was continuously taken. The flotation column feed and tailings flow rates were controlled to their desired values using automated peristaltic pumps. The gas hold-up was estimated by measurements of hydrostatic pressure at two heights in the column by means of electronic pressure transmitters. The froth depth was maintained at 10 cm (! 5 cm) from the column lip by varying the feed pump operation to rise or drop the froth zone-collection zone interface; the position of the froth level was monitored using an electrical conductivity probe which consisted of 6 steel ring electrodes placed flushed at the internal wall of the upper part of the column, within 5 cm separately. The superficial liquid velocity in the column was controlled by controlling to desired values the liquid flow rate controlling the tailings pump. The air was feed into the flotation column through a porous gas sparger made with filter cloth. The air flow rate in the column was measured and controlled in desired values by means of a mass air flow rate controller. The bubble diameter was estimated using drift flux analysis and validated against direct photographic image analysis. The flotation column was made of transparent acrylic, in such manner is possible to visually observe the presence of bubble coalescence, circulation and mixing, under different operating conditions.

In the first approximation a single laboratory flotation column was operated varying the superficial gas velocity under predetermined superficial liquid velocity conditions. The flotation measurements indicate that the flotation of organics is highly affected by the relative velocities between the liquid flow rate and the gas flow rate. The performance in the flotation column is reported to increase circulation and mixing as the liquid and gas flow rates increase in the column, this in turns will increase the bubble coalescence producing a rise in the collection zone turbulences, consequently decreasing the collision possibility between the bubbles and the oily globules. Nevertheless, when the superficial liquid velocity is in a small value range (0.1 cm/s – 0.15 cm/s) the organics recovery has an increase with the increase in the liquid velocity, perhaps due to an increase of the retention of the air bubbles by the liquid flow in the collection zone. This effect can be noticed in fig. 11 where the organics recovery is plotted as a function of the superficial liquid velocity.

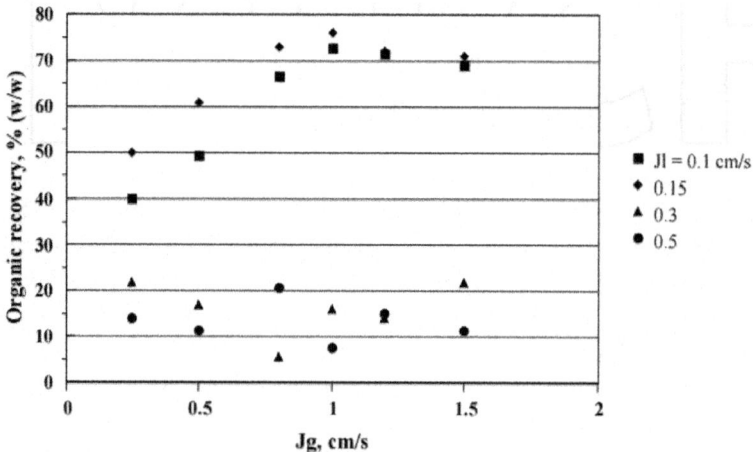

Figure 10: Organic recovery from water in the flotation column as the superficial gas velocity changes; the system is under different superficial liquid velocity.

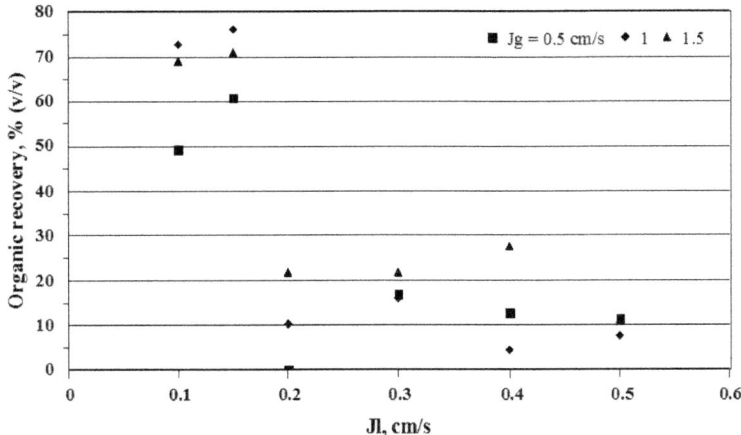

Figure 11: Organic recovery from water, reported in the flotation column concentrate, as a function of the superficial liquid velocity.

It can be observed that the de-oiling process efficiency is high when the superficial liquid velocity is kept below 0.2 cm/s, but above that value there is a drastic decline on the separation efficiency, this behaviour is observed in whole range of superficial gas velocities reported here. Also, it is thought that as the superficial liquid velocity increases in the collection zone of the flotation column, the possibility that small gas bubbles charged with hydrophobic organic materials are gone with the tailings flow out of the column, and for this reason, decreasing the organics recovery.

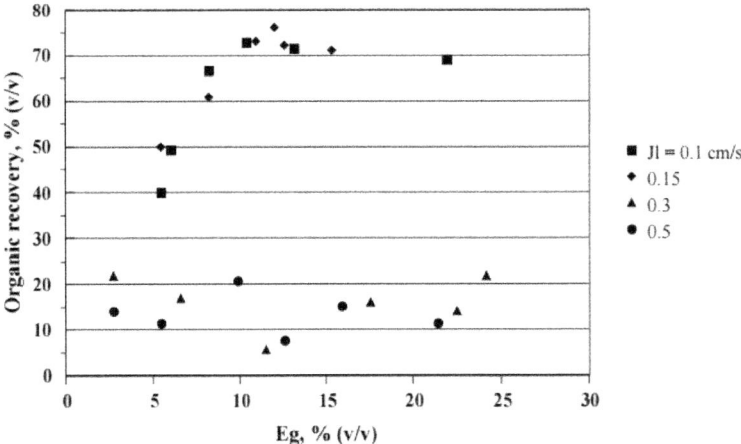

Figure 12: The effect of the superficial liquid velocity on the organic recovery/gas hold-up behaviour.

The experimental results plotted in fig. 12 suggest that large gas hold-up values are associated to small bubbles formation which will go down to the bottom of the flotation column because the effect of the downwards liquid stream; these experimental results support the impression that as the liquid flow rate increases the bubbles entrapment in the tailings stream also intensifies, producing low organic recoveries reported to the concentrate run. It can be noticed that under superficial liquid velocities above 0.15 cm/s the organic recovery in the concentrate ranges between 5% and 20%, which may be associated to large bubbles covered with the hydrophobic materials (which yields to such 5 to 20% organic recovery) and, therefore, most of the organic material collected by the gas bubbles is associated with the interactions with small bubble sizes. This idea is consistent with the hydrodynamic behaviour of gas bubbles moving in a liquid phase, since the oily globules present a lower density than the aqueous phase.

This explanation is supported by the gas bubble size estimates, from both image analysis and drift flux analysis, which are shown in Figure 13 in terms of the organic recovery in the concentrate.

The experimental bubble size estimates show the range of bubble dimensions is more constrained as the superficial liquid velocity is increased in the collection zone of the flotation column. This suggests that bubble-bubble coalescence is induced as the slip velocity among the phases (gas water counter-current flow) is increased.

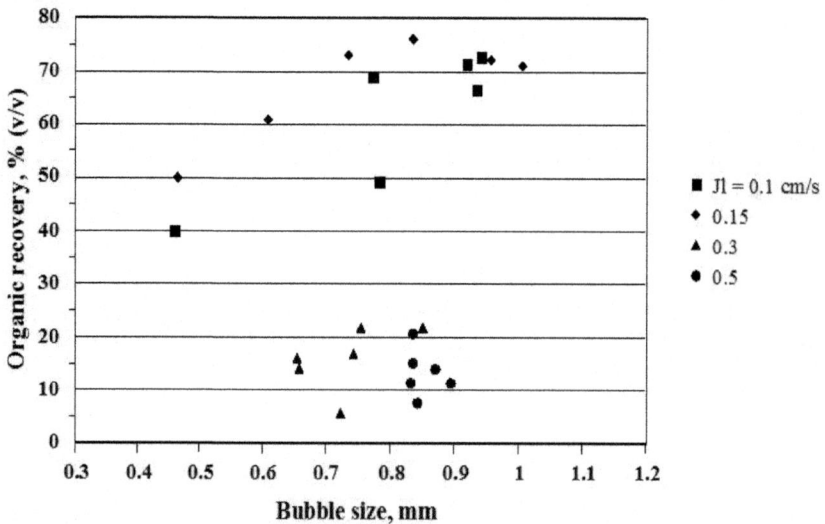

Figure 13: Effect of the bubble dimension on the recovery of organic floated.

Figure 14 presents the column flotation results on the water de-oiling process in terms of the organic recovery – superficial bubble surface flux behaviour. Under the lower values of the superficial liquid velocity in the collection zone of the column, it is very clear that the hydrophobic organic flotation increases when the gas surface area which flows in the flotation column increases.

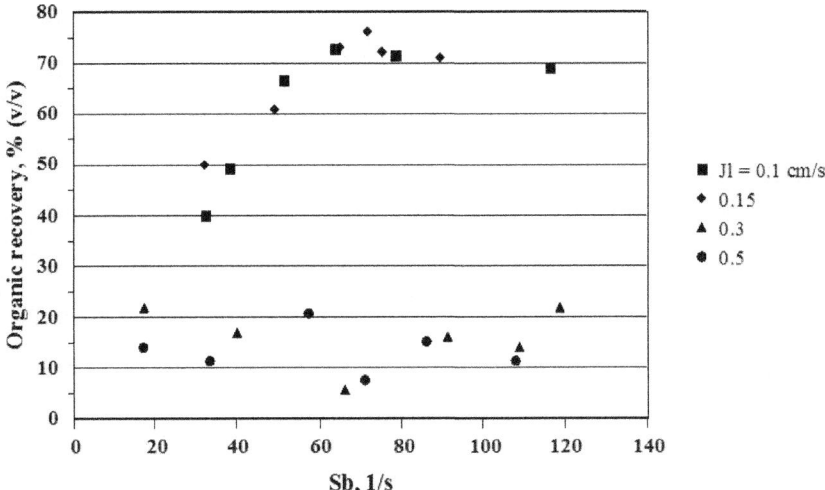

Figure 14: Behaviour of the organic recovery in concentrate with the change in the superficial bubble surface flux (Sb). The water de-oiling process in series stages of flotation columns. The water de-oiling process by column flotation was performed using a series of flotation stages. In this experience arrays of 2, 3 and 4 flotation column sequence were operated.

The experimental results and the estimates of the superficial bubble surface flux show that the hydrodynamic behaviour of the flotation column determines the separation efficiency and affects the relationship between the separation process and the available bubble surface area to perform the collection process. These experiences present that when the superficial liquid velocity is kept under the range between 0.1 and 0.15 cm/s, the change among the organic recovery and the superficial bubble surface flux follows a smooth pattern; however, when the superficial liquid velocity goes above 0.15 cm/s, the relationship between these process characteristics becomes erratic.

The experimental flotation columns array is schematically presented in fig. 15; this presentation shows that the residual water is fed into the first flotation column and the tailings from this column is used as the second column feed, and this arrangement is repeated up to four flotation columns.

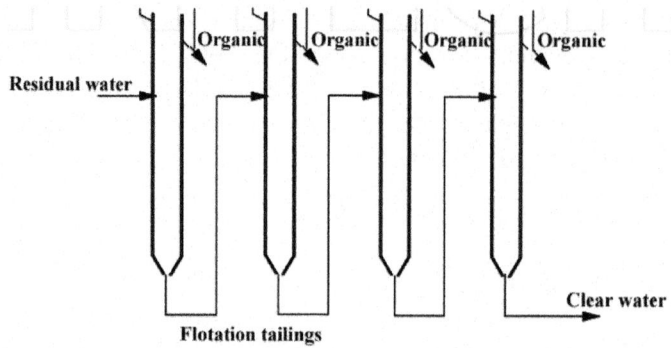

Figure 15: Experimental set up of flotation column series.

The experimental information shows that the de-oiling process can be performed effectively using a series of flotation columns. In this experience about 99% organic recovery is achieved using a series of four columns. The minimum residence time reported in order to achieve 99% organic recovery from the very stable water-oil (~40% oil) suspension is in the order of 100 minutes in a continuous four flotation columns operation [14].

Waste Water Treatment by Crystallization. Cleaning of Water Contaminated with Copper, Lead, and Nickel

It is well known the fact that although three quarters of this planet is occupied by water, not all is available for human consumption. 97% of this water is in oceans and seas whereas only 3% is called fresh water or sweet water. Of that 3%, the 77.6% is frozen polar ice, the 21.8% is groundwater, and only the 0.6% is available surface water for human consumption [15].

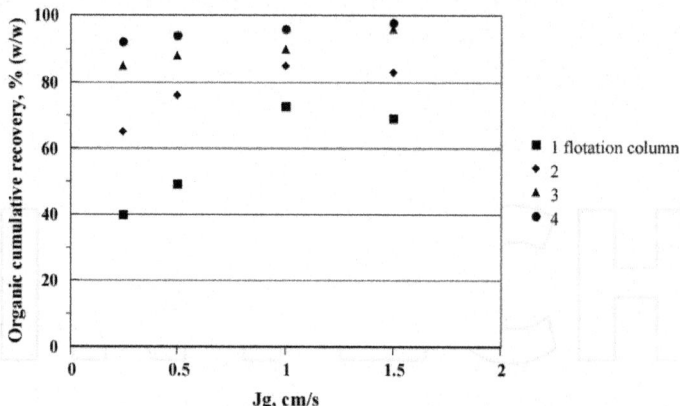

Figure 16: Experimental results of water de-oiling in one flotation column and a series of 2, 3 and 4 flotation columns.

According to the United Nations Industrial Development Organization (UNIDO), it is likely that in the year 2025 the industrial activity will consume twice as much water than at present; on the other hand, pollution may be multiplied by four due as well to this industrial activity [16]. Researchers at the Swiss Institute of Environmental Science and Technology recently reported that the rainwater that fall on the European Continent contains so many toxic pesticides dangerous for human consumption. In addition to acid rain, many other pollutants are deposited in water sources or streams, as in the third world, where 90% of the waste water is ending up in local rivers and streams [17]. The consumption of this contaminated water by the population leads to serious health problems such as gastrointestinal and cardiovascular disorders, malformations in newborns, among others. The World Health Organization (WHO) establishes the recommended levels of trace elements in drinking water (set for an average daily consumption), for both that help the body´s metabolism as representing toxicity to humans. For instance, in the case of lead the recommended values are from 0.01 to 0.05 mg/l, whereas for copper is among 0.5 and 2.0 mg/l, depending of the government of each country [18, 19].

To solve the problem of contaminated effluents, are applied some processes in order to leave the water clear, colorless, odorless, and disinfected. The suspended solids, organic compounds, and pathogenic organisms, in most of the waste water treatment plants are removed; nevertheless, these processes do not eliminate the heavy metals dissolved in the effluent [20].

If the heavy metals are in solid state they may be extracted through a primary treatment (i.e., sedimentation or flotation) [21], or secondary (i.e., activated sludge) [22]. On the other hand, if heavy metals are in solution, the use of lagoons, hydroponic bed, activated zeolite and membranes, phytoremediation, bird feathers, nanostructures, bioremediation, chemical cristallyzation, among others, have been proposed [23, 24, 25, 26, 27, 28].

Most of the processes mentioned above observe a certain limit as regards efficiency since the physicochemical aspects affecting such processes are not well understood, and the pollution problem is partially solved since the new "aggregate" with the removed contaminant remains as such.

This part of the chapter establishes from a thermodynamic point of view (pH, electrochemical potential, ionic strength, activity coefficient), the conditions to predict the formation of certain species (precipitated or dissolved) in distilled water contaminated with lead, copper, and nickel, and open to the atmosphere. The knowledge of the mechanisms of crystallization or chemical precipitation of the ionic species contribute to selectively agglomerate and separate the copper, nickel, and lead species in order to clean water contaminated with these heavy metals.

The role of heavy metals in reactions involving liquid-liquid and liquid-solid relationships is not enough studied yet. In the case of waste water treatment several research has been done although the mechanisms of precipitation of certain species and the selectivity of their capture in not clear [26, 27,28]. For liquid-liquid, and solid-liquid systems, the interaction between species is ruled by the following expressions:

$$I = \frac{1}{2}\sum m_i z_i^2$$

(1)

Where I is the ionic strength, which is a measure of the intensity of the electric field in the system [29,30], m is the molality of i; and z is the charge of the corresponding ion i (i represents every specie involved in a given reaction). In this work the involved species are the salts: $Pb(NO_3)_2$, $CuSO_4$, and $NiSO_4 \cdot 6H_2O$, the water and the pH modifiers, H_2SO_4, and KOH. The chemical activity is a corrected concentration [31], and physically is the actual amount of reagent that takes part during the reaction; in this case is the concentration of metallic ions in the media that affectively react. The average activity coefficient is calculated as follows:

$$\gamma\pm = 10^{\left(A|z+z-|\sqrt{I}\right)}$$

(2)

Being A the constant value from Debye –Hückel equation for liquid media and pressure of 1 atmosphere, $|z + z -|$ is the absolute value of the sum of the electric charge of the dissolved ions. The activity of given i species can be calculated according to:

$$a[i] = \gamma_i m_i$$

(3)

Where m is the molarity of the i species. In order to calculate the electrochemical potential, Eh, the equation proposed by Garrels [29] was applied:

$$Eh = E° - \left(\frac{0.05916}{Z}\right)\log Q$$

(4)

Being $E°$ the standard potential, Z the number of electrons participating during the reaction, and Q is the reaction quotient. $E°$ can be calculated through the following expression:

$$E° = \frac{-\Delta G°}{ZF}$$

$$(5)$$

Where $\Delta G°$ is the Gibbs free energy of the corresponding reaction, and F is the Faraday constant (96487 C/mol = 23 060.9 Cal/Vol•mol). By applying the former equations there is possible to build $Eh - pH$ diagrams and to use them as tools to understand the conditions under which given ionic or precipitated species are chemically stable.

In this part of the chapter lead, nickel, and copper salts were dissolved in distilled water and physicochemical parameters such as ionic strength, activity coefficient, activity, and electrochemical potential were calculated, in a pH ranging from 3 to 13. Lead, nickel, and copper precipitates were identified and the corresponding formation reactions were established in order to build a Pourbaix diagram.

The obtained information makes possible at first to design a procedure to clean water contaminated with the heavy metals mentioned here through the route sedimentation-flotation or filtering-flotation. The experimental results also provide information regards deposited species on mineral surfaces during milling which affect the behavior of collectors during flotation decreasing its metallurgical performance.

Preparation of Diluted Solutions of Cu, Ni, and Pb in Distilled Water

Lead nitrate ($Pb(NO_3)_2$), copper sulphate ($CuSO_4$), and hexahydrated nickel ($NiSO_4 \cdot 6H_2O$) were dissolved separately and simultaneously in distilled water. The pH of the media was varied in 3,5,7,9,11, and 13. After 24 hours the precipitated solids were separated and analyzed through X-ray diffraction (XRD), and scanning electron microscopy (SEM) techniques. The remaining lead, copper, and nickel in every solution were quantified by atomic absorption spectroscopy (AAS) analysis. The pH was modified by adding sulfuric acid (H_2SO_4), and potassium hydroxide (KOH). The initial metal concentration in each solution was 40 ppm.

The chemical analysis of precipitates was carried out by X-ray diffraction (XRD), and scanning electron microscopy (SEM). On the other hand, the quantitative chemical analysis from liquids, were carried out by atomic absorption spectroscopy (AAS).

With the quantitative and qualitative chemical analysis data, the values of activity, activity coefficient, ionic strength, and electrochemical potential were calculated. The former information was used to calculate the corresponding transformation lines as function of the pH. The resulting equilibrium diagrams are shown below.

Precipitation of Lead Species

Visually the formation of lead precipitates starts at pH 3, although these solids are re-dissolved at pH 5. Lead crystals are formed again at pH 7, and finally the precipitates are dissolved once more at pH 11. From XRD analysis at pH 3 the detected species are the lead sulfate ($PbSO_4$), and hydrated lead nitrite ($Pb(NO_2)_2 \cdot (H_2O)$), which indicates the decomposition and hydration of the salt originally dissolved. The precipitated solids at pH from 7 to 11 correspond to an hydroxicarbonate $Pb_3(CO_3)_2 \cdot (OH)_2$), also known as hydrocerusite.

Taking into consideration that hydrocerusite forms under alkaline conditions, and in absence of ionic sulfate, the following reaction is suggested:

$$Pb_3(CO_3)_2 \bullet (OH)_2 + 6H^+ \quad \rightarrow \quad 3Pb^{2+} + 4H_2O + 2CO_2$$

(6)

The re-dissolution of hydrocerusite is carried out according to the following reaction proposed by Pankow [31], and thermodynamically it occurs at pH 11.2:

$$Pb_3(CO_3)_2 \bullet (OH)_2 + 5H_2O \quad \rightarrow \quad 3Pb(OH)_3^- + 2CO_2 + 3H^+$$

(7)

On the other hand, the formation of carbonated species is explained by considering the replacement of sulfates or sulfites to carbonates in an open system to air, according the following reaction proposed by Azareño [32]:

$$PbSO_4 + CO_3^- \quad \rightarrow \quad PbCO_3 + SO_4^-$$

(8)

In the case of the decomposition of lead sulfate:

$$PbSO_4 + 2H^+ \rightarrow Pb^{2+} + H_2SO_4$$

(9)

From the above reactions it is possible to observe that the dissolution or precipitation of lead species just depends on pH. The Pourbaix diagram built according the calculated variables is shown in Figure 17. As known, in the following equilibrium diagrams the dashed lines represent the zone where aqueous species are stable; within these lines both the aqueous and precipitated species co-exist. For the copper case, the vertical lines correspond to the transformations shown in reactions (6), (7), and (9).

In light of the above, from pH up to 3.9, both ($PbSO_4$), and Pb^{2+} co-exist; whereas from pH 3.9 to 6 the all lead is dissolved. In the pH range from 6.0 to 11.2 the steady species are the Pb^{2+} and the $Pb_3(CO_3)_2 \bullet (OH)_2$. For pH larger than 11.2 the hydrocerusite is dissolved again and a lead hydroxide is formed.

According to the results from quantitative chemical analysis, the concentration of Pb^{2+} in the liquid media decreases because of the presence of lead precipitates at pH higher than 5, and the ionic lead increases again for pH higher than 11.

Table 1 shows the calculated values of activity for equations (6), and (9), as well as their Gibbs free energy, and the equilibrium pH. Thermodynamically the lead precipitation starts at pH 3.9 ($PbSO_4$), although visually it is noticed at pH 3. In the case of the $Pb_3(CO_3)_2 \cdot (OH)_2$, this visually starts at pH 7, whereas according to thermodynamics the precipitation of such specie would initiate at pH 6. Differences between observations and calculations are due to human errors and it suggests the use of another technique (i.e., conductivity measurements) to detect accurately the moment at which the precipitation phenomena take place [33].

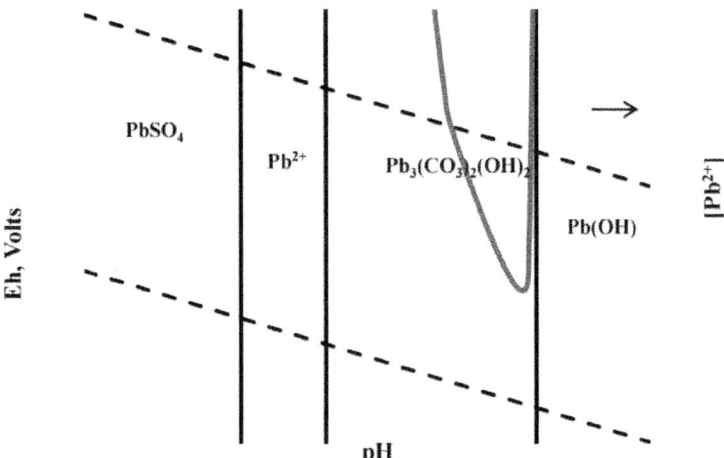

Figure 17: Transformation lines with changes on the metal concentration in the bulk solution. Pb – S – H_2O system.

Table 1: Calculated values of activity of Pb^{2+}, Gibbs free energy, and the equilibrium pH for equations (6), and (9)

Reaction	aPb²⁺	ΔG° Reaction (Kcal/mol)	Equilibrium pH
$PbSO_4 + 2H^+ \rightarrow Pb^{2+} + H_2SO_4$	3.84E-04	-5975.14	3.9
$Pb_3(CO_3)_2(OH)_2 + 6H^+ \rightarrow 3Pb^{2+} + 4H_2O + 2C$	2.14E-04	-24617.8	6.0

Precipitation of Copper Species

From qualitative DRX analysis the detected crystalline species are the hydrated copper hydroxisulphate ($Cu_4SO_4(OH)_6 \cdot H_2O$), cupric hydroxide ($Cu(OH)_2$), and cupric oxide ($CuO$). In the pH range from 3 to 5.5 there is not precipitation of any copper specie, whereas among pH 7.5 and 10.5 the three former species co-exist. From pH 5.5 to 7.5, and 10.5 to 13, the precipitated species are the hydrated copper hydroxisulphate, and the cupric oxide, respectively.

The solid, $Cu_4(SO_4)(OH)_6 \cdot H_2O$, in a given pH transforms itself and co-exist with both the cupric hydroxide and the cupric oxide, and finally the cupric hydroxide transforms to cupric oxide at pH 10.5.

In the case of formation of the hexahydrated copper hydroxisulphate :

$$Cu_4SO_4(OH)_6 \cdot H_2O + 6H^+ \rightarrow 4Cu^{2+} + SO_4^{2-} + 7H_2O \tag{10}$$

For the precipitation of the cupric hydroxide:

$$4Cu(OH)_2 + 2H^+ + CuSO_4 \rightarrow CuSO_4(OH)_6 \cdot H_2O + Cu^{2+} + H_2O \tag{11}$$

Being the formation of the cupric oxide defined by the following reaction:

$$4CuO + CuSO_4 + 3H_2O + 2H^+ \rightarrow Cu_4SO_4(OH)_6 \cdot H_2O + Cu^{2+} \tag{12}$$

The Pourvaix diagram for the $Cu - S - H_2O$ system is shown in Figure 18, and Table 2: it shows the Gibbs free energy values for reactions (10), (11), and (12).

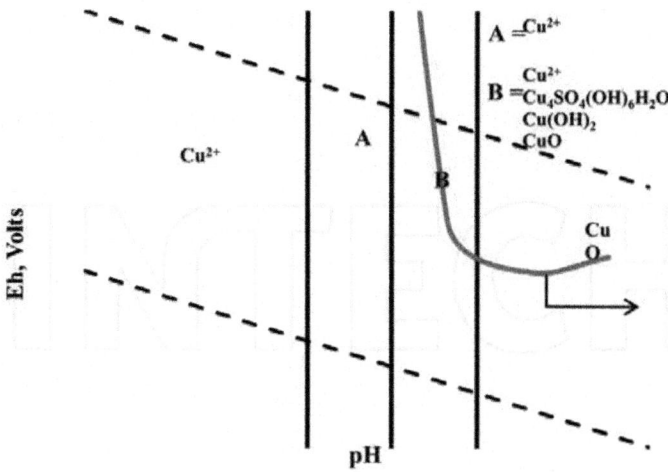

Figure 18: Pourvaix diagram for the $Cu-SO_4-H_2O$ system. A and B means the liquid and solid species that co-exist in a respective region.

Table 2: Activity, Gibbs free energy, and equilibrium pH values for reactions (10), 11), and (12)

Reaction	aCu^{2+}	ΔG° Reaction (Kcal/mol)	Equilibrium pH
$Cu_4SO_4(OH)_6 \cdot H_2O + 6H^+ \rightarrow 4Cu^{2+} + SO_4^{2-} + 7H_2O$	8.00E-04	193.6	5.6
$4Cu(OH)_2 + 2H^+ + CuSO_4 \rightarrow CuSO_4(OH)_6 \cdot H_2O + Cu^{2+} + H_2O$	4.08E-05	176.37	7.6
$4CuO + CuSO_4 + 3H_2O + 2H^+ \rightarrow Cu_4SO_4(OH)_6 \cdot H_2O + Cu^{2+}$	4.08E-05	143.4	9.7

Precipitation of Nickel Species

In this case and according the DRX analysis the only detected specie was the nickel hydroxide (Ni(OH)$_2$). Visually the precipitation of such specie is detected at pH 9; although, thermodynamically the nickel hydroxide starts forming at pH 7.6. Figure 19 shows the Eh-pH diagram for the Ni-SO$_4$-H$_2$O system.

From the above information the proposed reaction is as follows:

$$Ni(OH)_2 + 2H^+ \rightarrow Ni^{2+} + H_2O \tag{13}$$

The corresponding thermodynamic values are shown in Table 3. Sean et al., [20], Chanturiya et. al., [21], Liu [22], Shigehito et. al., [23], and Guo-riu Fu [24], reported the precipitation of two nickel phases named α – nickel hydroxide, and β – nickel hydroxide, within the pH range established in this work. These species were detected by using both: XRD and IR analysis techniques.

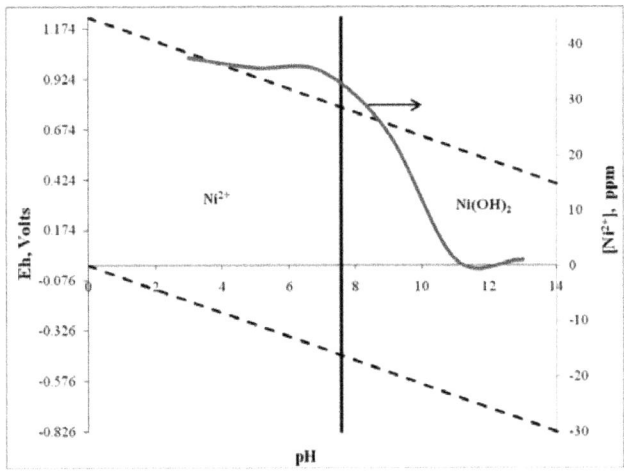

Figure 19: Pourbaix diagram for the Ni – SO$_4$ – H$_2$O system. Changes on concentration of the ionic nickel are included.

Table 3: Activity coefficient, Gibbs free energy, and equilibrium pH for reaction (13)

Reaction	aCu²⁺	ΔG° Reaction (Kcal/mol)	Equilibrium pH
$Ni(OH)_2 + 2H^+ \rightarrow Ni^{2+} + H_2O$	4.47E-03	-17447.1	7.57

Selective Precipitation of Copper, Nickel, and Lead

With all the above information it is possible to design a cleaning route for water contaminated with lead, copper and nickel. By adjusting the pH of the media selectively any of the metals can be precipitated and then separated by filtration or sedimentation.

Lead, copper, and nickel salts were simultaneously dissolved in distilled water. The amount of every salt was adjusted to fix the initial concentration of every metal at 160 g/l.

According to the experimental data and the thermodynamic calculations, each metal can be selectively precipitated and removed from the contaminated effluent. After each pH adjustment of the liquid, the solids were separated by centrifugation. The proposed procedure is as follows:

a.   At pH values up to 4 the only precipitated specie is the lead sulfate (PbSO₄), remaining in the liquid media both the nickel and the copper. Figure 20 shows the XRD analysis of the solids crystallized under pH 3.

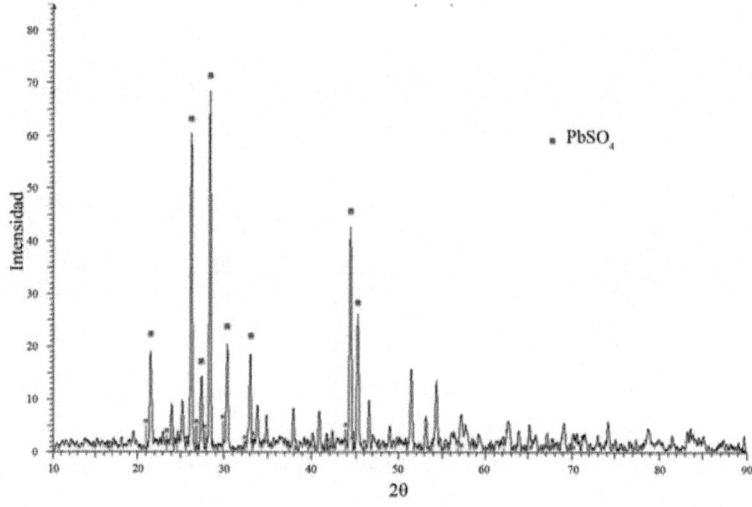

Figure 20: XRD analysis for solids precipitated at pH 3. Experimental results for the

case of lead, copper, and nickel salts simultaneously dissolved in distilled water.

1. By keeping the pH at 7.5, the copper species will precipitate as hydroxysulphate, and only the nickel is expected to remain in solution. Figure 21 shows the XRD analysis of the solids precipitated at this stage. Considering that the pH of equilibrium for the Ni-H$_2$O system is 7.6, it seems normal to observe the onset of precipitation of some nickel species.

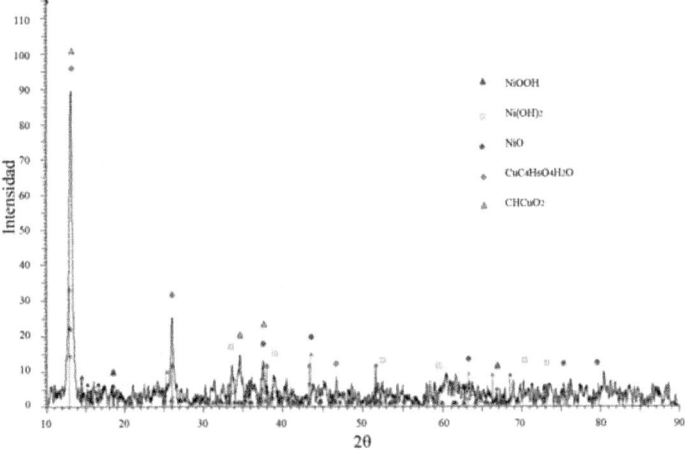

Figure 21: XRD analysis for solids precipitated at pH 7.5. Experimental results for the case of lead, copper, and nickel salts simultaneously dissolved in distilled water.

1. Finally, if the pH of the liquid is fixed at values larger than 10, the nickel will precipitate as hydroxide. Figure 22 shows the XRD result after fixing the pH at 11.

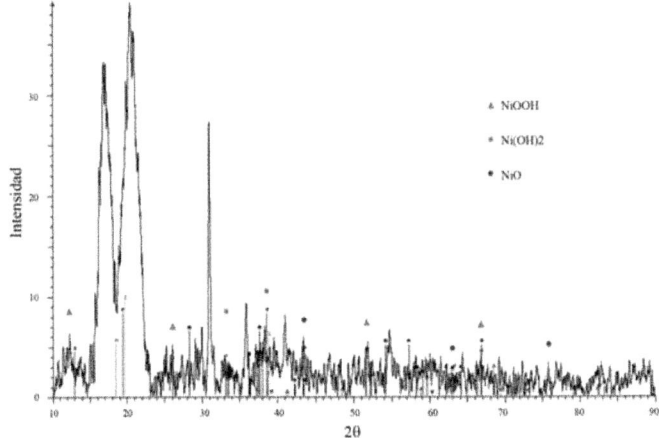

Figure 22: XRD analysis for solids precipitated at pH 11. Experimental results for the

case of lead, copper, and nickel salts simultaneously dissolved in distilled water.

Waste Water Treatment De-Oiling By Using Microorganisms

The biological treatments are procedures for cleaning waste water effluents; they have in common the use of microorganisms (i.e., bacteria), to carry out the removal of undesirable components of the water, using the metabolic activity of such microorganisms. The traditional application is the removal of biodegradable organic matter, either soluble or colloidal, and the elimination of compounds containing nutrients (nitrogen and phosphorous). The following describes an experimental procedure, carried out at laboratory level, to eliminate vegetable oils, first from a synthetic emulsion and then, from an industrial effluent. The microorganisms tested here, are both aerobic and nonpathogenic. The bioreactor used was a flotation cell made of transparent acrylic material.

Determining the Lipolytic Capacity of Microorganisms

With the appropriate conditions for the growth of the bacteria colony (pH, temperature, dissolved oxygen, surface tension, and nutrients), it was possible to quantify the lipolytic capacity of the microorganisms: *Candida Kefir*, *Candida Parapsilosis*, *Pseudomonas Fluorescens*, *Bacillus Cereus*, and *Bacillus Coagulans*, individually and in bulk. Since the purpose of this stage of the work was to establish the lipolytic character of the above mentioned microorganisms, the measurements of residual oil in water were carried out at the beginning and at the end of the experimental digestion. The experimental results are shown in Table 4.

Table 4: Residual oil in the synthetic emulsion after microbial action

SAMPLES	RESIDUAL OIL (g/l)	RESIDUAL OIL (%)
Undigested sample	0.241	100
Candida Kefir	0.233	96.5
Candida parapsilosis	0.023	9.78
Bacillus cereus	0.059	24.5
Pseudomonas fluorescens	0.005	2.3
Bacillus coagulans	0.019	7.9
Bulk	0.023	94.7

From table 1 the microorganism with higher lipolytic capacity is the *Pseudomonas fluorescence*, leaving only 2.3 % of residual oil in the emulsion. The same experimental results expressed in terms of the biomass developed (growth of colony) is shown in Table 5. As can be observed, the amount of biomass increases with the oil removed from the emulsion, due to the amount of nutrients (lipids) ingested by the microorganisms.

The microorganism with more biomass developed during the ingestion of lipids was the *Pseudomonas fluorescens* (253.5%), followed by the *Bacillus coagulans* (197.9%), and at the end appears the *Candida parapsilosis* (178%). Table 6 shows the variations in pH, temperature, and dissolved oxygen, monitored during this first stage of experimentation and corresponding to the particular case of the *Pseudomonas fluorescens*. The entire experiment finished after 132 hours.

Table 5: Development of biomass in the degradation of a synthetic emulsion

SAMPLES	**BIOMASS (g/l)**	**BIOMASS (%)**
Undigested sample	0.247	100
Candida Kefir	0.328	132.7
Candida parapsilosis	0.344	139.2
Bacillus cereus	0.440	178.0
Pseudomonas fluorescens	0.626	253.5
Bacillus coagulans	0.489	197.9
Bulk	0.329	158.5

Table 6: Experimental results after digestion of the synthetic emulsion by the microorganism *Pseudomonas fluorescens*

Sampling (hrs)	pH	Temperature (°C)	Surface Tension (dyn/cm)	Dissolved Oxygen (mg/l)	Residual Oil (%)	Biomass (%)
0	5.9	17	36.8	6.4	100	100
12	5.7	21	38.1	6.6	95	104
24	5.1	18	36.6	7	87	100
36	4.7	20	36.5	6.8	81	102
48	4.6	18	36.6	7	71	105
60	4.4	20	36.9	6	67	108
72	4.3	17	38.9	6.5	58	101
84	4.3	19	36.9	6.1	55	111
96	4.1	17	41.3	5.9	47	112

108	4.1	20	41.2	6	41	116
120	4.1	18	41.5	6	33	127
132	4.1	21	42.1	6	23	146

From table 6 the pH of the system decreased during the first 100 hours of the experiment. After this time, the pH of the emulsion remained constant at pH 4.1. This acidification is caused by the rupture of triglycerides, releasing fatty acids and glycerol, together with the carbonic acid formed when water combines with the carbon dioxide from the bacterial metabolism. Regarding the surface tension of the emulsion, this increases with the amount of lipids ingested by the microorganisms, because of the ionic species eliminated from the emulsion, increasing the cohesive forces of the fluid [39]. This increase in the surface tension value is more evident after 96 hours of starting the cleaning process of the synthetic emulsion.

The increase in the surface tension coincides with the increase in the biomass after the 96 hours above referred. As shown in table 6, inoculation lasted approximately 80 hours, and the bacterial colony growth remained. Then after 52 hours, the exponential growth phase of the colony is observed, increasing 45% compared with the initial biomass. It is not observed a decrease in the biomass because there are still nutrients (oil) to degrade by the microorganisms.

The behavior of the dissolved oxygen (DO) in the emulsion also has a direct relationship with the colony growth; during the inoculation phase, the DO practically remains constant ($\sim$ 6.6 mg/l), then there is a decrease in the amount of dissolved oxygen by the growth of the colony of microorganisms.

De-Oiling Of an Industrial Effluent

A sample of the effluent from an edible vegetable oils factory was collected. The amount of oil, quantified according the Soxhlet technique [40], was of 41.45 mg/l, which is above the maximum permissible value for the environmental standards. The initial pH of the sample was 5.6 and this value does not affect the growth of the *Pseudomonas fluorescens* bacteria, since this microorganism is naturally acidophilus. Table 7 shows the measured variables to the original sample.

Table 7: Characteristics of an effluent from an edible vegetable oils factory

Characteristic	Result
pH	5.8

Surface Tension (dyn/cm)	37.9
Dissolved Oxygen (mg/l)	0.35
Oil content (mg/l)	41.5

A set of experiments were run by using a transparent acrylic plexiglass cell as biological reactor (30 cm by side and 40 cm in height), with 4 taps properly located to adjust 4 sensors for measuring continuously: pH, temperature, dissolved oxygen, and surface tension of the system. A bubble generator (made of filter cloth) was installed at the bottom of the reactor. The air feed to the cell was measured and controlled through a mass flow controller. The system was operated under stationary regimen, and open to the atmosphere. Table 4.5 shows the experimental variables applied in this particular research.

Table 8: Experimental conditions for de-oiling of an industrial effluent using a laboratory biological reactor

Variable	Value
Volume of emulsion fed to the cell	7.0 liters
pH	5.6
Air fed to the cell	25 l/min
Microorganism	*Pseudomonas fluorescens*e
Sampling time	Every 48 hours
Volume of each sample	20 ml
Test duration	8 days

Figure 23 shows changes in pH during the bio-treatment of the industrial sample. As can be observed the pH of the liquid media increases with the time (from 5.6 to 8.5). This is associated with the stressful conditions in the system: the special nutrients are depleted, waste products accumulate, bacterial overpopulation occurs (causing high demand of oxygen for aerobic metabolism), conditions that limit the growth colony and cause cell death.

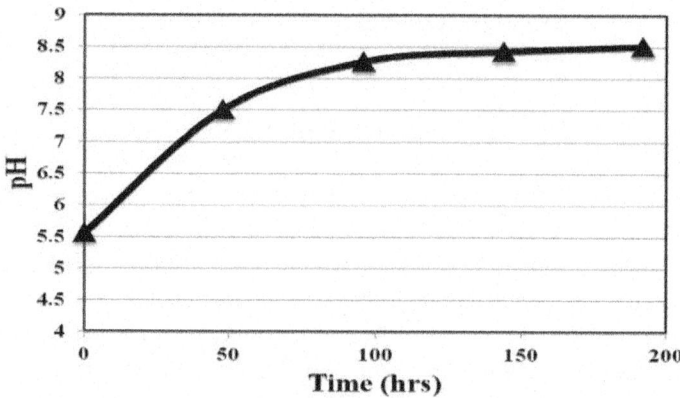

Figure 23: Changes in pH during the bio-treatment of the industrial effluent. Pseudomonas fluorescens bacteria were used to degrade oil.

Even the pH of the media begins to be unsuitable for cell growth. At this time there are deader than alive bacteria, then the scarcity of nutrient sources (sugars and lipids), the protein degradation process of the bacteria begins, and generates ammonia as degradation product. This process provides organic nitrogen sources as nutrient which transforms the system into alkaline, increasing the pH value [41].

On the other hand, as in the case of the synthetic emulsion, the surface tension of the effluent increases with the time, because of the amount of lipids ingested by the microorganisms. As mentioned above, the cohesive forces of the fluid increase because of the ionic species eliminated from the emulsion (see Figure 24).

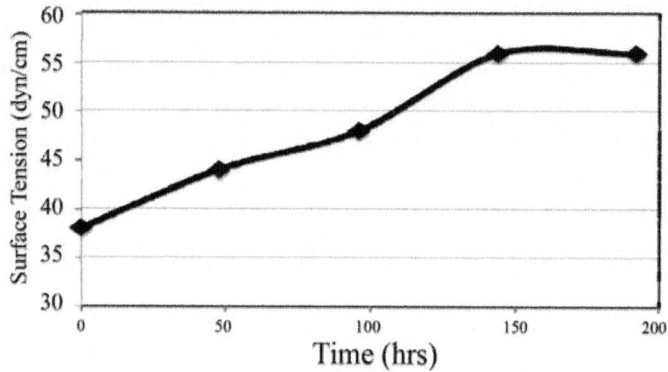

Figure 24: Changes in the surface tension of the effluent with the time, due to the increment in the amount of lipids digested by the microorganisms.

The lipolytic activity of the microorganisms is presented in Figure 25. The initial concentration of fat in the effluent was equal to 41.5 mg/l (taken as 100%); at 48 hours after initiation of the experiment the oil content in the emulsion decreased to 34.2 mg/l (82.4%); at 96 hours of incubation, was reached a concentration below 10 mg/l (24%), and finally, after 144 hours of the vegetable oil content in the effluent was 4.6 mg/l (11%) and remained constant from this point.

The constant trend in the last two experimental points, shown in figure 3, coincides with the trends of both: pH and surface tension, shown in figures 1 and 2, respectively, which also agrees with the behavior of the biomass in the last two points of sampling, as can be seen in Figure 26.

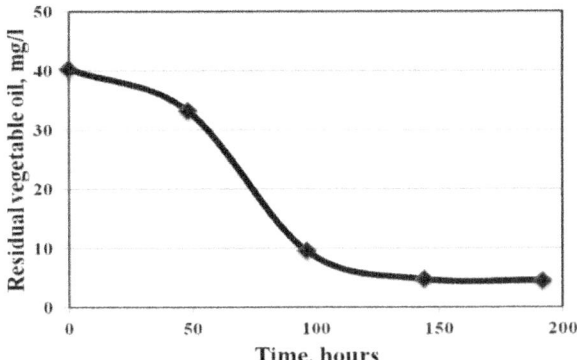

Figure 25: Variations of the vegetable oil content in the industrial effluent treated by the microorganism *Pseudomonas fluorescens*.

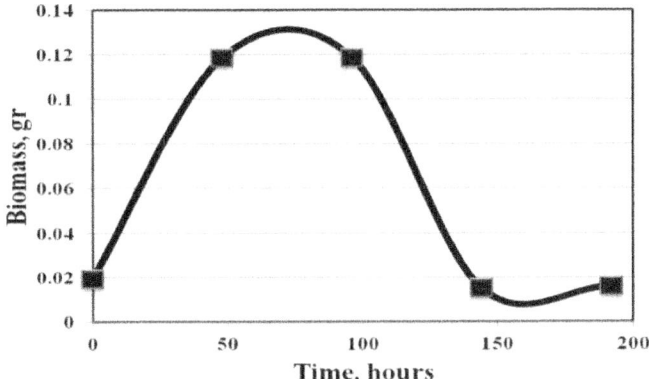

Figure 26: Biomass development (Pseudomonas fluorescens) during the degradation of an industrial effluent.

Is observed in figure 26 that biomass increased five times their initial concentration in a period of 48 hours of incubation, this development continued until approximately 100 hours of testing; this period matches with the maximum lipolytic degradation in the effluent. At this time the fat concentration of the effluent reached the lowest value (4.6 mg/l); the bacterial growth was limited to the shortage of nutrients and metabolic waste buildup in the effluent.

CONCLUSIONS

The information derived from the set of experiments conducted to establish the feasibility of using flotation devices in non-mineral systems, such as waste water treatment, can be used to set the following conclusions:

The experimental information shows that the de-oiling process can be performed effectively using a series of flotation columns. In this experience about 99% organic recovery is achieved using a series of four columns. The minimum residence time reported in order to achieve 99% organic recovery from the very stable water-oil (~40% oil) suspension is in the order of 100 minutes in a continuous four flotation columns operation.

Aereated flotation cells can be used as oxygen suppliers to aerobic nonpathogenic microorganisms (i.e., *pseudomonas fluorescens* bacteria), used during the biological treatments for cleaning waste water effluents; these microorganisms degrade the fat components of the water (either ionic or colloidal), using their metabolic activity. The experimental results shows that through these bacteria, around the 90% of the vegetable oil content in the effluent can be removed.

On the other hand, by precipitating heavy metal ions, it is possible to design a process for cleaning waste water contaminated with heavy metals (i.e., lead, copper, and nickel), through a sedimentation-flotation route.

The thermodynamic al analysis of the metal-water systems (activity, activity coefficient, ionic strenght, electrochemical potential), allows accurately predict the crystallization of the ionic species under controlled conditions of temperature and pH.

The experimental results shows that these data converge well enough with the thermodynamic calculations for dissolved or precipitated lead species, being possible the selective precipitation of the three metals tested here, and leaving the water without any kind of metal, ionic or solid.

ACKNOWLEDGEMENT

The authors deeply appreciate the support given to this investigation to the

Universidad Michoacana de San Nicolas de Hidalgo, and the Mexican Government through the National Council for Science and Technology.

REFERENCES

1. F Sebba, 1962Ion Flotation. Elsevier Publishing Co., New York.

2. F. M Doyle, 2003Ion flotation- its potential for hydrometallurgical operations. Int. J. Miner. Process., 72, 387399

3. J. A Finch, G. S Dobby, 1990Column Flotation. Pergamon Press, Oxford.

4. R. H. C Lara, R. G Cruz, M. G. F Monroy, 2005Mecanismos de la alteración de la galena asociados al efecto ambiental en medios calcáreos. Artículo presentado en el XL Congreso Mexicano de Química, Sociedad Química de México, 2529de Septiembre, Morelia, México.

5. F. J Tavera, M Reyes, R Escudero, F Patino, 2010On the treating of lead polluted wáter: flotation of lead coloidal carbonate. J. Mex. Chem. Soc., 54204208

6. F. J Tavera, R Escudero, A Uribe, J. A Finch, 2000Ni- DETA flotation in aqueous media: application of flotation columns. AFINIDAD, LVII (190), 415423

7. F. J Tavera, C. O Gomez, J. A Finch, 1996A novel gas hold-up probe and application in flotation columns. Trans. Instn. Min. Met., 105 (C), 99104

8. J. C Maxwell, 1892A Treatise of Electricity and Magnetism. 3rd Edition, 1 (II), Chapter IX, Oxford University Press, London, 435449

9. F. J Tavera, C. O Gomez, J. A Finch, 1998Conductivity flow cells for measurements on dispersions. Can. Met. Quart., 37 (1), 1925

10. R Escudero, 2000Characterisation of rigid porous spargers by permeability and its relevance to scaling-up. Ph. D. thesis, McGill University, Montreal, Canada.

11. F. J Tavera, R Escudero, J. A Finch, 2001Gas hold-up in a flotation column: laboratory measurements. Int. J. Miner. Process., 61, 2340

12. B. K Gorain, 1997The effect of bubble surface area flux on the kinetics of flotation and its relevance to scale- up. Ph. D. thesis, University of Queensland, Australia.

13. O Levenspiel, 1972Chemical reaction engineering. 2nd ed., McGraw Hill, Inc., New York.

14. F. J Tavera, R Escudero, A Sánchez, 2004Treating of water polluted with organics in flotation columns. Proceedings of the International Symposium on Environment; 9-70703-293-6Mexico, 95105

15. V. I Grover, 2006Water: global common and global problems. Science Publishers.

16. [wwwunido.org]

17. M Barlow, T Clarke, 2004Oro azul: las multinacionales y el robo organizado de agua en el mundo. Ed. Paidós.

18. Z. C Alvez, 1995Qualidade da água. Meio Ambiente por inteiro. Jornal da Secretaria Municipal do Meio Ambiente (1), 7.

19. wwwwho.int/

20. R Crites, and G Tchbanoglous, 2004Tratamiento de Aguas Residuales en Pequeñas Poblaciones. (M. Camargo, L. P. Pardo, & G. Mejía, Trads.) Mc Graw Hill.

21. Metcalfand Eddy, INC. (1994Ingeniería Sanitaria, tratamiento, evacuación y reutilización de aguas residuales (Tercera Edición ed.). (J. d. Montsoriu, Trad.) Ed. Labor.

22. Celis HidalgoJ., Junod Montano, J.,and Sandoval Estrada, M. (2005Recientes Aplicaciones de Depuración de Aguas Residuales con Plantas Acuáticas, Teoría (14

23. Alba EliasF., Vergara Gonzalez, E. P., Ordieres, J. B., Martinez de Piston, F., González Marcos, A., and Ortiz Marcos, I. (s.f.). Tratamiento de aguas contaminadas con metales pesados empleando compost usado de champiñon como agente bioregenerador. 16.

24. López MartinezA., Cuapio Ortiz, L., Cárdenas Puebla, S., Balcazar, M., Jauregui, V., and Bonilla Petriciolet, A. (2007Determinación de Concentración de Cr y Cd Adsorbido por Plumas de Pollo. Tesis de Maestría. (I. T. Aguascalientes, Ed.)

25. E Soto, R. D Miranda, C. A Sosa, and J. A Loredo, 2006Optimización del Proceso de Remoción de Metales Pesados de Agua Residual de la Industria Galvánica por Precipitación Química. Información Tecnológica, 17 (2), 33-42.

26. S Aksu, and F. M Doyle, Potential-pH Diagrams for Copper in Aqueous solutions of various organic complexing agents, Electrochemical Society Proceedings 2000

27. Reyes Pérez M2005Tratamiento continuo, de aguas contaminadas con Cu y Pb, por flotación iónica en celdas con dispersores porosos; efecto de las propiedades de la dispersión aire-líquido en la separación, Tesis de maestría, IIM, UMSNH, (2005).

28. M. A Barakat, Removal of Cu (II), Ni (III) and Cr(III) Ions from Wastewater Using Complexation- Ultrafiltration Technique, Journal of

Environmental Science and Technology, 12008

29. R. M Garrels, and C. L Christ, Minerals, Solutions, and Equilibria, Harper & Rowe, N. Y. 4501965

30. L. A Cisternas, Diagramas de fases y su aplicación, Reverte, (2009

31. J. F Pankow, Aquatic chemistry concepts, CRC Press, (1991

32. O. A Azareno, J. P Núnez, L. A Figueroa, D. E León, S. S Fernández, S. R Orihuela, R. M Caballero, R. R Bazán, and Yi Choy A. S. Flotación de Minerales Oxidados de Plomo. Revista del Instituto de Investigación de la Facultad de Ingeniería Geológica, Minera, Metalúrgica y Geográfica. 52002

33. F. J Tavera, D Colwell, R Escudero, and J Finch, 2000Estimation of Gas Holdup in Froths by Electrical Conductivity: Aplication of the Standard Addition Method. Revista de Química Teórica y Aplicada AFINIDAD, Barcelona 57España, Abril,.

34. R. S Sean, and S. D Thomas, 2002Raman Spectroscopy of Co(OH)2 at High Pressures: Implications for Amorphization and Hidrogen Repulsion. Physical Review. 66B, 134301134301

35. V. A Chanturiya, T. N Matveeva, and L. B Lantsoba, 2003Investigation into Products of Dimethyl Dithiocarbamate and Xantate Sorption of Sulfide Minerals of Copper-Nickel Ores. Journal of Mining Science. 39281286

36. Z Liu, F. M Doyle, Modeling Metal Ion Removal in Alkylsulfate Ion Flotation Systems. Minerals and Metallurgical Processing, 182001

37. D Shigehito, H Akinobu, B. A Bienvenu, and M Mizuhata, 2009O. H Ni, Films Fabricated by Liquid Phase Deposition Method. Thin Solids Films, 51715461554

38. F Guo-riu, H Zhong-ai, X Li-jing, J Xiao-qing, X Yu-long, W Yao-xian, Z Zi-yu, Y Yu-ying, and W Hong-ying, 2009Electrodeposition of Nickel Hydroxide films on Nickel Foil and its Electrochemical Performances for Supercapacitor. International Journal of Electrochemical Science. 410521062

39. AtkinsPeter. Jones, Loretta. "Principios de Química: los caminos del descubrimiento". 3ra edición, 1ra reimpresión 2006Ed. Médica-Panamericana. B.A. Arg. 17

40. S. M Marta, and S. C Francisca, 2000The Chemical Composition of "Multimistura" as a Food Supplement. Food Chemistry. 6814144

41. wwwbiologia.edu.ar

Chapter 5

WASTE WATER TREATMENT METHODS

Adina Elena Segneanu[1], Carmen Lazau[1], Paula Sfirloaga[1], Paulina Vlazan[1], Cornelia Bandas[1], Ioan Grozescu[1] and Cristina Orbeci[2]

[1]National Institute for Research and development in Electrochemistry and Condensed Matter –INCEMC Timisoara, Romania

[2]Politehnica University Bucuresti, Romania

INTRODUCTION

The last decades have shown a reevaluation of the issue of environmental pollution, under all aspects, both at regional and at international level. The progressive accumulation of more and more organic compounds in natural waters is mostly due to the development and extension of chemical technologies for organic synthesis and processing.

Population explosion, expansion of urban areas increased adverse impacts on water resources, particularly in regions in which natural resources are still limited. Currently, water use and reuse has become a major concern. Population growth leads to significant increases in default volumes of waste water, which makes it an urgent imperative to develop effective and affordable technologies for wastewater treatment.

The physico-chemical processes common treatment (coagulation and flocculation) using various chemical reagents (aluminum chloride or ferric chloride, polyelectrolytes, etc.) and generates large amounts of sludge. Increasing demands for water quality indicators and drastic change regulations on wastewater disposal require the emergence and development of processes more efficient and more effective (ion exchange, ultrafiltration, reverse osmosis and chemical precipitation, electrochemical technologies). Each of these treatment methods has advantages and disadvantages.

Water resources management exercises ever more pressing demands on wastewater treatment technologies to reduce industrial negative impact on natural water sources. Thus, the new regulations and emission limits are imposed and industrial activities are required to seek new methods and technologies capable of effective removal of heavy metal pollution loads and reduction of wastewater volume, closing the water cycle, or by reusing and recycling water waste.

Advanced technologies for wastewater treatment are required to eliminate pollution and may also increase pollutant destruction or separation processes, such as advanced oxidation methods (catalytic and photocatalytic oxidation), chemical precipitation, adsorption on various media, etc.. These technologies can be applied successfully to remove pollutants that are partially removed by conventional methods, e.g. biodegradable organic compounds, suspended solids, colloidal substances, phosphorus and nitrogen compounds, heavy metals, dissolved compounds, microorganisms that thus enabling recycling of residual water. (Zhou, 2002) Special attention was paid to electrochemical technologies, because they have advantages: versatility, safety, selectivity, possibility of automation, environmentally friendly and requires low investment costs (Chen, 2004; Hansen et. al., 2007).

The technologies for treating wastewater containing organic compounds fall within one of the following categories:

- Non-destructive procedures – based on physical processes of adsorption, removal, stripping etc.
- Biological destructive procedures – based on biological processes using active mud.
- Oxidative destructive processes – based on oxidative chemical processes which, in their turn, can fall within one of the following categories:
- Incineration;
- WO - "Wet Oxidation", operating in conditions of high temperature and pressure, with the versions:
- WAO - "Wet Air Oxidation" (wet oxidation with O_2 air oxidative agent);
- CWAO - "Catalytic Wet Air Oxidation" (catalytic wet oxidation with O_2 air oxidative agent);
- SWA - "Supercritical Water Oxidation" (oxidation with O_2 air oxidative agent in supercritical conditions).
- Liquid oxidation: AOPs - "Advanced Oxidation Processes", operating in conditions of temperature and pressure and use as oxidative agents O_3, H_2O_2 and even O_2, catalysts and/ or UV radiations.

ADVANCED OXIDATION PROCESSES

Advanced oxidation processes (AOPs) are widely used for the removal of recalcitrant organic constituents from industrial and municipal wastewater.

In this sense, AOP type procedures can become very promising technologies for treating wastewater containing non-biodegradable or hardly biodegradable organic compounds with high toxicity. These procedures are based on generating highly oxidative HO radicals in the reaction medium.

- H_2O_2
  H_2O_2 + UV (direct photolysis)
  H_2O_2 + $Fe^{2+/3+}$ (classic, homogeneous Fenton)
  H_2O_2 + Fe/support (heterogeneous Fenton)
  H_2O_2 + $Fe^{2+/3+}$ + UV (VIS) (Photo-Fenton)
- O_3
  O_3 (direct ozone feeding)
  O_3 + UV (photo-ozone feeding)
  O_3 + catalysts (catalytic ozone feeding)
- H_2O_2 + O_3
  TiO_2 (heterogeneous catalysis)
  TiO_2 + UV (photo-catalysis)

The preferential use of H_2O_2 as oxidative agent and HO radicals generator is justified by the fact that the hydrogen peroxide is easy to store, transported and used, and the procedure is safe and efficient.

The technologies developed so far indicate the use of zeolites, active coal, structured clay, silica textures, Nafion membranes or Fe under the form of goethit (α-FeOOH), as support materials for the catalytic component.

The AOPs (Advanced Oxidation Processes) can be successfully used in wastewater treatment to degrade the persistent organic pollutants, the oxidation process being determined by the very high oxidative potential of the HO radicals generated into the reaction medium by different mechanisms(Pera-Titus et al., 2004).

AOPs can be applied to fully or partially oxidize pollutants, usually using a combination of oxidants. Photo-chemical and photocatalytic advanced oxidation processes including UV/H_2O_2, UV/O_3, $UV/H_2O_2/O_3$, $UV/H_2O_2/Fe^{2+}(Fe^{3+})$, UV/TiO_2 and $UV/H_2O_2/TiO_2$ can be used for oxidative degradation of organic contaminants. A complete mineralization of the organic pollutants is not necessary, being more worthwhile to transform them into biodegradable aliphatic carboxylic acids followed by a biological process (Wang and Wang, 2007).

The oxidation process is determined by the very high oxidative potential of the HO radicals generated into reaction medium by different mechanisms.

In the case of the AOPs Fenton-type procedure (hydrogen peroxide and Fe^{2+} as catalyst), the generation of hydroxyl radicals takes place through a catalytic mechanism in which the iron ions play a very important role (Andreozzi et al., 1999; Esplugas, et al., 2002) the main reactions involved being presented in equations (1) – (4):

$$Fe^{2+} + H_2O_2 \rightarrow Fe^{3+} + HO^- + HO \qquad (1)$$

$$Fe^{3+} + H_2O_2 \leftrightarrow H^+ + [Fe(OOH)]^{2+} \qquad (2)$$

$$[Fe(OOH)]^{2+} \rightarrow Fe^{2+} + HO_2 \qquad (3)$$

$$HO_2 + Fe^{3+} \rightarrow Fe^{2+} + H^+ + O_2 \qquad (4)$$

In the presence of UV radiations (photo-Fenton process), an additional number of HO radicals are produced both through direct H_2O_2 photolysis and through UV radiations interaction with the iron species in aqueous solutions (eq. 5-7) (Spacek et al., 1995; Pignatello, J.J., 1992):

$$H_2O_2 + UV \rightarrow 2HO \qquad (5)$$

$$Fe^{3+} + H_2O + UV \rightarrow Fe^{2+} + H^+ + HO \cdot \qquad (6)$$

$$[Fe(OH)]^{2+} + UV \rightarrow Fe^{2+} + HO \cdot \qquad (7)$$

The main parameters which determine the efficiency of the oxidation process are: the structure of the organic compounds, the hydrogen peroxide and the catalyst concentrations, the wave length and intensity of UV radiations, the initial solution ph and the reaction contact time.

As recalcitrant organic pollutants continue to increase in air and wastewater streams, environmental laws and regulations become more stringent (Gayaa et al, 2008). The main causes of surface and groundwater contamination are industrial effluents (even in small amounts), excessive use of pesticides, fertilizers (agrochemicals) and domestic waste landfills. Wastewater treatment is usually based on physical and biological processes. After elimination of particles in suspension, the usual process is biological treatment (natural decontamination), but unfortunately, some organic pollutants, classified as bio-recalcitrant, are not biodegradable. In this way advanced oxidation processes (AOPs) may become the most widely used water treatment technologies for organic pollutants not treatable by conventional techniques due to their high chemical stability and/or low biodegradability (Munoz et al.2005). Advanced oxidation processes are indicated for removal of organic contaminants such as halogenated hydrocarbons (trichloroethane, trichlorethylene), aromatics (benzene, toluene, and xylene), pentachlorophenol (PCP), nitrophenol, detergents, pesticides, etc. These processes can also be applied to oxidation of inorganic contaminants such as cyanides, sulfides and nitrites (Munter, 2001). A general classification of advanced oxidation processes based on source allowing radicals. This classification is presented in Figure 1.

Heterogeneous photocatalysis has proved to be of real interest as efficient tool for degrading both aquatic and atmospheric organic contaminants because this technique involved the acceleration of photoreaction in presence of semiconductor photocatalyst (Guillard, 1999). Thus these processes can be classified in: advanced oxidation processes based on ozone based advanced oxidation processes H_2O_2, photocatalysis, POA "hot" technologies based on ultrasound, electrochemical oxidation process, oxidation processes with electron beam. These processes involve generation and subsequent reaction of hydroxyl radicals (_OH), which are one of the most powerful oxidizing species. Photocatalytic reaction is initiated when a photoexcited electron is promoted from the filled valence band of semiconductor photocatalyst to the empty conduction band as the absorbed photon energy, h_, equals or exceeds the band gap of the semiconductor photocatalyst leaving behind a hole in the valence band. Thus in concert, electron and hole pair (e−–h+) is generated (Horvath, 2003). An ideal photocatalyst for photocatalytic oxidation is characterized by the following attributes: photo-stability, chemically and biologically inert nature, availability and low cost (Carp et. al., 2004). Many semiconductors such as TiO_2 (Lazau, 2011), ZnO (Daneshvar et al., 2007), ZrO_2 (Lopez et al. 2007), CdS (Yingchun, 2011), MoS_2 (Kun Hong, 2011), Fe_2O_3 (Seiji, 2009) and WO_3 (Yuji, 2011) have been examined and used as photocatalysts for the degradation of organic contaminants. TiO_2 is most preferred one due to its chemical and biological inertness, high photocatalytic activity, photodurability, mechanical robustness and cheapness. Thus, these materials were used in the degradation of phenol, 1,4-dichlorobenzene (Papp et al., 1993),methanol (Nobuaki et. al., 2007), azo dye (Daneshvar, 2003), trichloromethane, hexachloro cyclohexane (Byrappa et. al., 2002), trichloroethylene and dichloropropionic acid (Nikola, 2001). To avoid the problem of filtration, many methods were proposed to immobilize the photocatalysts, but in these conditions the photocatalyst is expected to be used for a relatively long time, especially for industrial applications (Venkata, 2004). Various substrates have been used as a catalyst support for the photocatalytic degradation of polluted water. For example glass materials: glass mesh, glass fabric, glass wool, glass beads and glass reactors were very commonly used as a support for titania. Other uncommon materials such as microporous cellulosic membranes, alumina clays, ceramic membranes, monoliths, zeolites, and even stainless steel were also experimented as a support for TiO2 (Gianluca, 2008).

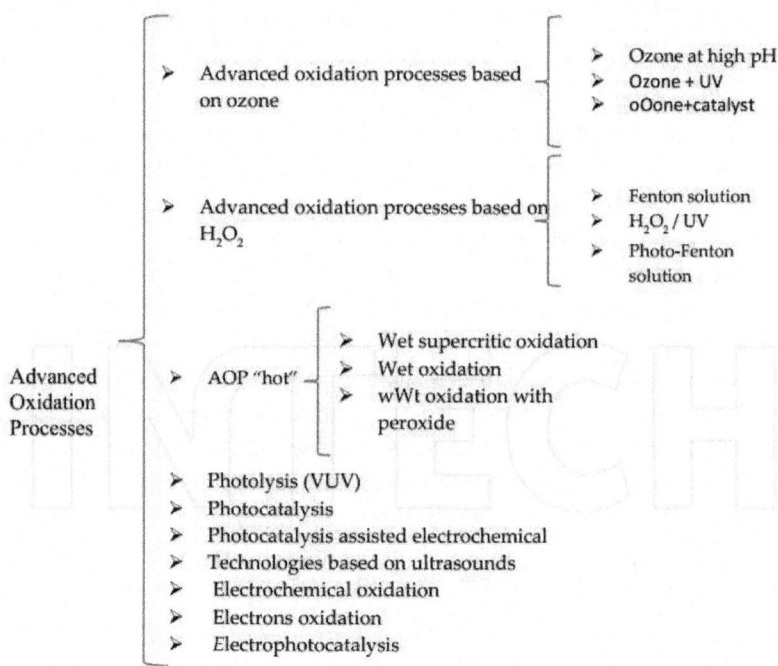

Figure 1: Classification of advanced oxidation processes

Advanced oxidation processes (AOPs) and electrochemical oxidation is based on the in-situ generation of OH radicals, which allow its non-selective reaction with organics allowing organics mineralization by its conversion into CO_2. The electrochemical methods are very promising alternatives for organics degradation because of their environmental compatibility, versatility, simplicity, and easy possibility of automation. The electrochemical oxidation performance depends strongly on the electrode material. To generate OH radicals by electrooxidation, several types of anodes with high overpotential for oxygen potential are suitable, i.e., DSA-type, PbO_2, boron-doped diamond (BDD) electrodes etc. Recently, electrochemical oxidation with a boron-doped diamond electrode is one of the most promising technologies in the treatment of the industrial effluents containing organics. BDD electrode exhibited a very good chemical stability and its application in the electrooxidation of organics led to complete mineralization into CO_2 in relation with applied potential or current density. A major drawback of the electrochemical oxidation consists of the high energy consumption to the mineralization. The presence of a catalyst in the electrical field or combined and direct photoelectrochemical application can enhance the treatment efficiency with lower energy consumption (Ratiu et. al., 2010).

Electrochemical and photochemical technologies may offer an efficient means of controlling pollution. Their effectiveness is based on the generation of highly reactive and non-selective hydroxyl radicals, which are able to degrade many organic pollutants. Electrolysis, heterogeneous photocatalysis, or photo-assisted electrolysis may be regarded as advanced oxidation processes (AOPs) and used in the supplementary treatment of wastewaters. The efficiency of the electrochemical oxidation depends on the anode material and the operating conditions, e.g., current density or potential. In general, in most applications of photoelectrocatalysis in the degradation of organics, the applied anodic bias potential is lower than the oxidation potential of organics on the electrode, due to direct electro-oxidation does not complicate the photocatalytic mechanism (Ratiu et. al.,2011).

The efficiency of photoelectrochemical degradation for organic pollutants depends not only on the selection of a suitable supporting electrolyte and pH values, but also on the electrode potential and preparation conditions of the semiconductors involved. In a photoelectrochemical system, photoelectrons and photoholes can be separated under the influence of an applied electric field. The problem of the separation of semiconductor particles from the treated solution, so persistent in heterogeneous photolysis, is not an issue in photoelectrochemical systems. There are numerous semiconductors which can be used as photoelectrocatalytic materials, such as TiO_2, WO_3, SnO_2, ZnO, CdS, diamond, and others (Hepel 2005).

Particular Aspects

In the case of the chlorinated phenols, the number and the position on aromatic ring of the chlorine atoms modifies the oxidation efficiency (Pera-Titus et al., 2004).

The oxidation rate constant decreases linearly with increasing number of chlorine content on the aromatic ring. Also, the increase of chlorine content will block some favorable positions susceptible to hydroxyl radical attack.

The oxidation process is also controlled by the presence of another species in reaction medium (intermediate products) in the sense that they interact with the catalyst component in a different manner. The species of reductive character accelerate the oxidation process because they reduce Fe^{3+}(inactive) to Fe^{2+} (active) and thus the generation of OH radicals intensifies (Du et al. 2006; Riga, A., et al, 2007). The acid type species lower the pH of the reaction medium and can form stable complexes with Fe^{3+} or Fe^{2+} ions, strongly slowing the oxidation process. The presence into reaction medium of the inorganic ionic species (Cl^-, ClO_4^-, NO_3^-, HCO_3^-, CO_3^{2-}, SO_4^{2-}, $H_2PO_4^-$) modifies the rate of the oxidation of the organic compounds as function of their nature and concentration. The inorganic anions can change the overall efficiency of the system by different ways. The influence of ClO_4^- and NO_3^- ions is less pronounced than another anions because they do not form complexes with Fe(II) and Fe(III) and do not react with HO (Lu et al., 1997; Siedlecka, E. M., 2007). Cl⁻, SO_4^{2-} and $H_2PO_4^-$ anions decrease the rate of decomposition of H_2O_2 by forming ferric unreactive complexes and react with hydroxyl radicals forming Cl_2^-, SO_4^- and H_2PO_4 radicals who are less or much less reactive than HO (Siedlecka, E. M., 2007; De Laat et al., 2004). The influence of Cl⁻ is in correlation with the solution pH and its concentration, being insignificant at low concentration (<5 mM) but becomes very important at higher concentration values (>28 mM) (Kwon et al., 1999). As function of the nature of the inorganic anions, at higher concentration of 0.1 M, the inhibition order of the oxidation rate is the following: $H_2PO_4^- > Cl^- > HCO_3^- > CO_3^{2-} > SO_4^{2-} > NO_3^-$ (Riga, A., et al., 2007).

The presence of the inorganic species inside the reaction medium influences the rate of the oxidation process as function of their nature and concentration. The inorganic anionic species reduce the 4-CP oxidation efficiency by Fe(II) and Fe(III) complexes forming, HO radicals scavenging or iron precipitate forming.

NO_3^- induces a small influence on 4-CP oxidation efficiency. This may be explained by the absence of the interactions between NO_3^- and the catalyst ($Fe^{2+/3+}$) and hydroxyl radicals. The anions Cl⁻, SO_4^{2-} and PO_4^{3-} modify drastically the 4-CP oxidation efficiency, especially at high concentration into reaction medium. They interact with Fe^{2+} and Fe^{3+} forming chloro-, sulfato- and phosphate-iron complexes which are inactive in HO generation mechanism. Also, Cl⁻, SO_4^{2-} and $H_2PO_4^-$ anions interact with hydroxyl radicals (scavenging process), forming less reactive species (Cl_2^-, SO_4^- and H_2PO_4) into reaction medium.

The decrease of the 4-CP oxidation degree by the photo-Fenton procedure is correlated with the nature of the anions as following: $Cl^- > PO_4^{3-} > SO_4^{2-} >> NO_3^-$ (Orbeci et al., 2008).

The presence of the insoluble inorganic species (bentonite) modifies the 4-CP oxidation efficiency in different manner. Into reaction medium, 4-CP can be adsorbed by the bentonite substratum or can be destruct by oxidation, both processes increasing the 4-CP removal degree from the solution. The presence of the insoluble inorganic species (bentonite) modifies the oxidation efficiency by additional 4-CP and UV sorption processes, especially at high solution turbidity values (Orbeci et al., 2008).

The efficiency of the various AOPs depends both on the rate of generating the free radicals and the extent of contact between the radicals and the organic compound. Also, the pH has a significant role in determining the efficiency of Fenton and photo-Fenton oxidation processes (Gogate and Pandit, 2004). Limitations due to the use of homogeneous catalysts, such as limited pH range, production of Fe containing sludge, and deactivation could be overcome by heterogeneous catalysts.

The optimum pH range in the case of homogeneous photo-Fenton process is 2.5-4, a correction of solution pH being necessary. Also, at the end of the oxidation process, iron precipitation and catalyst separation and recovery are necessary. These disadvantages can be avoided using the heterogeneous photo-Fenton procedure by immobilization of active iron species on small particulate solid supports. In this case, different iron-containing catalysts can be used, such as the iron bulk catalysts (iron oxy-hydroxyl compounds: hematite, goethite, magnetite) or iron supported catalysts (zeolites, clays, bentonite, glass, active carbon, polymers etc.) (Duarte and Madeira, 2010; Feng et al., 2005; He et al., 2005; Leland and Bard, 1987; Nie et al., 2008; Ortiz de la Plata et al., 2010; Vinita et al. 2010).

The use of the heterogeneous photo-Fenton procedure in the catalytic component version (Fe in various oxidation states) precipitated on solid support presents several drawbacks:

- catalyst's relatively high cost – associated with the cost of the so-called support, with the cost of the Fe compounds and with the operations necessary for Fe compounds to fix on the support;
- decrease in the efficiency of the UV radiations due to their partial adoption on the solid support;
- progressive solubility of the catalytic component (Fe) during oxidative processes and as a result of a progressive loss of catalytic activity.

In the case of the heterogeneous photo-Fenton process, a relevant fraction of the incident UV radiation can be lost via scattering, due to particulate solid support suspended into the reaction medium. As a consequence, the photo-Fenton process may be seriously affected (Herney-Ramirez et al., 2010). Also, the solution pH affects the iron leaching from the support, at pH values less than 3 a higher amount of iron being released into the solution (Duarte and Madeira, 2010). By using a zero-valent iron with iron oxide composite catalysts, the oxidation process proceeds via hydroxyl radicals generated from Fe^{2+}(surf) species and H_2O_2 in a Fenton like mechanism. The Fe^{2+}(surf) species are formed by electron transfer from Fe^0 to Fe^{3+} at the interface metal/oxide (Moura et al., 2005, 2006). The experimental data (Nie et al., 2008) indicate that the hydrogen peroxide provides a driving force in the electron transfer from Fe^{2+} to Fe^{3+}, while the degradation of organic pollutants increases the electron transfer at the interface of Fe^0/iron oxide due to their reaction with hydroxyl radicals.

The degradation of organic pollutants using photo-Fenton processes occurs by intermediate oxidation products formation. In the case of phenol oxidation by Fenton reagent, a series of intermediates were identified, corresponding mainly to ring compounds and short-chain organic acids *(Zazo et al., 2005)*.Most significant among the former were catechol, hydroquinone, and *p*-benzoquinone; the main organic acids were maleic, acetic, oxalic, and formic, with substantially lower amounts of muconic, fumaric, and malonic acids. Oxalic and acetic acid appeared to be fairly refractory to the Fenton oxidation process. In the Fenton process, carboxylic acids like acetic and oxalic acid may be formed as end products during the degradation of phenol while in photo-Fenton process, both these acids were identified during the early stages of phenol degradation and were oxidized almost completely at the end of the process *(Kavitha and Palanivelu, 2004)*. The chlorophenols are common persistent organic contaminants, which show low biodegradability, posing serious risks to the environment once discharged into natural water *(Du et al., 2006)*.

Studying the degradation of 4-chlorophenol by an electrochemical advanced oxidation process, several authors (Wang and Wang, 2007*)* have proposed the following possible pathways: (a) 4-chlorophenol dechlorination to phenol; (b) hydroxylation of phenol to hydroquinone; (c) dehydrogenation of hydroquinone to benzoquinone; (d) oxidation of benzoquinone (with aromatic ring cleavage) to aliphatic carboxylic acids such as maleic acid, fumaric acid, malonic acid; (e) oxidation of maleic and fumaric acids to oxalic acid, formic acid and finally, to carbon dioxide and water.

The main intermediate products detected by HPLC analyses were chlorocatechol and benzoquinone after 60 min reaction time and aliphatic carboxylic acids after 120 min reaction time. Benzoquinone and hydroquinone-like intermediates such as catechol, hydroquinone and 4-chlorocatechol can reduce the ferric ion to ferrous ion and the oxidation process becomes faster *(Du et al., 2006)*. It is not necessary to degrade 4-chlorophenol to the final products of CO_2 and H_2O, being more worthwhile to treat to the biodegradable stage-aliphatic carboxylic acids followed by a biological process *(Wang and Wang, 2007)*. The photocatalytic processes may be used as a pre-treatment of toxic chemicals, including chlorophenols, in order to convert them into fully biodegradable compounds.

Recently, a series of pharmaceuticals such as analgesics, antibiotics, steroids etc. have been detected in the water feeding systems of several countries in Europe, the USA and Australia *(Bound and Voulvoulis, 2004; Kaniou et* al., 2005). Unless antibiotics are removed from wastewaters through specific purification processes, they can affect the microbial communities in filtering systems using active sludge and, in general, the bacteria found in water, and, as a result, they can disturb the natural elementary cycles. The accumulation of antibiotics in surface waters represents a potential danger in the case in which they are used as sources of drinking water. Photocatalytic oxidation of antibiotics in aqueous solution is based on the oxidative potential of the HO radicals (2.80 V) generated in the reaction medium though photocatalytic mechanisms, in the presence of H_2O_2 and UV radiations. Through the photo-Fenton procedure, the efficiency of the oxidation is controlled by the nature and the structure of the organic substrate, the initial pH of the solution, the concentration of H_2O_2 and of the catalytic component (Fe^{2+}) as well as by the time the reaction medium stands in the area where UV radiations act.

The kinetic assessment of the oxidative degradation process applied to antibiotics of the type amoxicillin, ampicillin and streptomycin (pseudo 1st degree Lagergren kinetic model) suggests that the oxidative process occurs in two successive steps, with the formation of reaction intermediates. The ratio of the 1st degree kinetic constant values corresponding to the two oxidation stages depends on the structure of the antibiotics and indicates a marked decrease in the oxidation rate in the second stage. This decrease can be attributed to the formation of reaction intermediates such as inferior organic acids with a high stability in regard to oxidation and/or blocking active catalytic centers through the formation of compounds of the $Fe^{2+/3+}$ species with the reaction intermediates, compounds which are inactive in the process of generating HO radicals (Orbeci et al., 2010).

Advanced oxidation processes of Fenton and photo-Fenton type can be used for antibiotics degradation from wastewater (Orbeci et al., 2010) or for increasing their biodegradability in biological wastewater treatment (Elmolla and Chaudhuri, 2009). Unlike complete amoxicillin degradation, the mineralization of the organic compounds from solution is not complete in the Fenton oxidation process due to formation of refractory intermediates (Ay and Kargi, 2010).

The photo-Fenton process degradation of amoxicillin by using iron species as catalyst ($FeSO_4$ and potassium ferrioxalate complex) and solar radiation reduces the bactericide effect of amoxicillin but the toxicity may persist due to intermediates formed during the oxidation process. The toxicity decreases significantly when these intermediates are converted into short chain carboxylic acids, allowing further conventional treatment (Trovó et al., 2011). The homogeneous photo-Fenton process is limited by the narrow working pH range (2.5-4) and requires the correction of solution pH for iron precipitation and catalyst separation and recovery. Otherwise, high amounts of metal-containing sludge can be formed and the catalytic metals are lost in these sludge. Because of these disadvantages, several attempts have been made to develop heterogeneous photo-Fenton procedure by immobilization of active iron species on solid supports. Since iron is relatively inexpensive and nontoxic, it has been widely used in different environmental treatment processes (Herney-Ramirez et al., 2010; Nie et al., 2008). In the heterogeneous photo-Fenton process, different iron-containing catalysts can be used, such as the iron bulk catalysts (iron oxy-hydroxyl compounds: hematite, goethite, magnetite) or iron supported catalysts (zeolites, clays, bentonite, glass, active carbon, polymers etc.) (Feng et al., 2005;He et al., 2005; Leland and Bard, 1987; Nie et al., 2008; Ortiz de la Plata et al., 2010; Vinita et al. 2010).

Antibiotics can be more or less extensively metabolized by humans and animals. Depending on the quantities used and their rate of excretion, they are released in effluents and reach sewage treatment plants (Alexy et al., 2004; Bound and Voulvoulis, 2004; Kümmerer, 2009).

Available data on antibiotics (ampicillin, erythromycin, tetracycline and penicilloyl groups) indicate their capability to exert toxic effects to living organisms (bacteria, algae etc.), even at very low concentration. These antibiotics are practically non-biodegradable having the potential to survive sewage treatment, leading to a persistence of these compounds in the environment and a potential for bio-accumulation (Arslan-Alaton et al., 2004). The presence of antibiotics in the environment has favored the emergence of antibiotic-resistant bacteria, increasing the likelihood of infections as well as the need to find new and more powerful antibiotics. As expected, antibiotic-contaminated water is incompatible with conventional biological water treatment technologies (Rozas et al., 2010). Antibiotics have the potential to affect the microbial community in sewage systems and can affect bacteria in the environment and thus disturb natural elementary cycles (Kümmerer, 2009). If they are not eliminated during the purification process, they pass through the sewage system and may end up in the environment, mainly in the surface water.

This is of special importance, since surface water is a possible source of drinking water (Kaniou et al., 2005). The antibiotics degradation by advanced oxidation processes has proven to be reasonably suited and quite feasible for application as a pre-treatment method by combining with biological treatment (Arslan-Alaton et al., 2004). The pre-treatment of penicillin formulation effluent by advanced oxidation processes based on O_3 and H_2O_2/O_3 did not completely remove the toxicity of procaine penicillin G from the effluents, leading to serious inhibition of the treatment of activated sludge (Arslan-

Alaton et al., 2006). One of the novel technologies for treating polluted sources of industrial wastewater and drinking water is the photo-Fenton process by which hydroxyl radicals are generated in the presence of H_2O_2, Fe^{2+} catalyst and UV radiation.

The advanced oxidation processes or even the hybrid methods may not be useful in degrading large quantity of the effluent with economic efficiency and hence it is advisable to use these methods for reducing the toxicity of the pollutant stream to a certain level beyond which biological oxidation can ensure the complete mineralization of the biodegradable products (Gogate and Pandit, 2004).

Removal of organic compounds in wastewater is a very important subject of research in the field of environmental chemistry. In this sense, photocatalysis is a handy promising technology, very attractive for wastewater treatment and water potabilization (Nikolaki et al., 2006; Lim et al., 2009). Using titanium dioxide for water splitting after UV irradiation, it has been shown that this can encompass a wide range of reactions, especially the oxidation of organic compounds. The study of the photodegradation for a large series of substances such as halogenated hydrocarbons, aromatics, nitrogenated heterocycles, hydrogen sulfide, surfactants and herbicides, and toxic metallicions, among others has clearly shown that the majority of organic pollutants present in waters can be mineralized or at least partially destroyed. The photocatalytic treatment of many organic compounds has been successfully achieved. The photocatalytic activity is dependent on the surface and structural properties of the semiconductor such as crystal composition, surface area, particle size distribution, porosity, band gap and surface hydroxyl density (Ahmed et al., 2010).

A variety of semiconductor powders (oxides, sulfides etc.) acting as photocatalysts have been used. Most attention has been given to TiO_2 because of its high photocatalytic activity having a maximum quantum yields, its resistance to photo-corrosion, its biological immunity and low cost. There are two types of reactors: reactors with TiO_2 suspended in the reaction medium and reactors with TiO_2 fixed on a carrier material (Lim et al.; 2009, Mozia, 2010; Li et al., 2009). A very promising method for solving problems concerning the separation of the photocatalyst from the reaction medium is the application of photocatalytic membrane reactors (PMRs), having other advantages such as the realization of a continuous process and the control of a residence time of molecules in the reactor (Mozia, 2010). In case of polymer membranes, there is a danger for the membrane structure to be destroyed by UV light or hydroxyl radicals. The investigations described (Chin et al., 2006; Molinari et al., 2000) show that the lowest resistance exhibit membranes prepared from polypropylene, polyacrylonitrile, cellulose acetate and polyethersulfone, UV light leading to a breakage in the chemical bonds of the methyl group. The least effect of the UV/oxidative environment on the membrane stability was observed in case of polytetrafluoroethylene and polyvinylidene fluoride membranes (Chin et al., 2006).

The self-assembly of TiO2 nanoparticles was established through coordinance bonds with –OH functional groups on the membrane surface, improving reversible deposition, hydrophilicity and flow and diminishing the irreversible fouling (Mansourpanah et al., 2009). TiO_2-functionalized membranes may be obtained by several methods, but the sol-gel process is ubiquitous because it has many advantages such as purity, homogeneity, control over the microstructure, ease of processing, low temperature and low cost (Alphonse et al., 2010).

The advanced oxidation processes based on the photo-activity of semiconductor-type materials can be successfully used in wastewater treatment for destroying the persistent organic pollutants, resistant to biological degradation processes. TiO_2 is the most attractive semiconductor because of its higher photocatalytic activity and can be used suspended into the reaction medium (slurry reactors) or immobilized as a film on solid material. A very promising method for solving problems concerning the photocatalyst separation from the reaction medium is to use the photocatalytic reactors in which TiO_2 is immobilized on support. The immobilization of TiO_2 onto various supporting materials has largely been carried out via physical or chemical route.

The application of photocatalysis in water and wastewater treatment has been well established, particularly in the degradation of organic compounds into simple mineral acids, carbon dioxide and water (Pera-Titus et al., 2004; Cassano and Alfano, 2000). Titanium dioxide (TiO_2), particularly in the anatase form is a photocatalyst under ultraviolet (UV) light. A reactor refers to TiO_2 powder which is suspended in the water to be treated, while the immobilized catalyst reactor has TiO_2 powder attached to a substrate which is immersed in the water to be treated. Immobilised TiO_2 has become more popular due to the complications in the TiO_2 suspension systems (Hoffmann et al., 1995).

The TiO_2 immobilisation procedure developed not long ago can be used in determining a suitable immobilization procedure, particularly if economical and simple equipment is necessary. The overall performance of the TiO_2 coating can be affected by various factors depending on the coating methods. In addition, it is also difficult to evaluate the photocatalytic efficiency of the coatings through photocatalytic activity (Augugliaro et al., 2008). Invariably, the TiO_2 was immobilized using the sol–gel technique (Addamo et al., 2008). In this case, some problems are noted: decrease of surface area; potential loss of TiO_2; decreased adsorption of organic substances on the TiO_2 surface; mass transfer limitations. No polymeric support was considered to immobilise TiO_2 since the polymeric material can undergo photo-oxidative degradation by illuminated TiO_2 (Cassano and Alfano, 2000; Augugliaro et al., 2008).

Researchers have used photocatalytic oxidation to remove and destroy many types of organic pollutants. After photocatalysis was recognized to be a great oxidation mechanism, researchers began testing it on many different compounds, and in many different processes (Cassano and Alfano, 2000;Hoffmann et al., 1995; Addamo et al., 2008).

Phenolic compounds, a kind of priority pollutants, often occur in the aqueous environment due to their widespread use in many industrial processes such as the manufacture of plastics, dyes, drugs, antioxidants, and pesticides. Phenols, even at concentrations below 1 lg/L, can affect the taste and odor of the water (Pera-Titus et al., 2004). Therefore, the identification and monitoring of these compounds at trace level in drinking water and surface waters are imperative. Chlorophenols represent an important class of very common water pollutants. 4-chlorophenol is a toxic and non-biodegradable organic compound and can often be found in high quantity in the waste waters from various industrial sectors (Pera-Titus et al., 2004; Augugliaro et al., 2006). A severe toxicity of 4-chlorophenol requires the development of a simple, sensitive and reliable analytical method.

Among the advanced oxidation processes investigated in thelast decades, photocatalysis in the presence of an irradiated semiconductor has proven to be very effective in the field of environment remediation.The use of irradiation to initiate chemical reactions is the principle on which heterogeneous photocatalysis is based; infact, when a semiconductor oxide is irradiated with suitable light, excited electron–hole pairs result that can be applied in chemical processes to modify specific compounds.The main advantage of heterogeneous photocatalysis, when compared with the chemical methods, is that in most cases it is possible to obtain a complete mineralization of the toxic substrate even in the absence of added reagents.The radical mechanism of photocatalytic reactions, which involve fast attacks of strongly oxidant hydroxyl radicals, determines their un selective features.

Environmental applications of photocatalysis using TiO_2 have attracted an enormous amount of research interest over the last three decades (Hoffmann et al., 1995; Linsebigler et al., 1995; Mills and Le Hunte, 1997; Stafford et al., 1996). It is well established that slurries of TiO_2 illuminated with UV light can degrade to the point of mineralization almost any dissolved organic pollutant. Nevertheless, photocatalysis, particularly in aqueous media, has still not found widespread commercial implementation for environmental remediation. The main hurdle appears to be the cost, which is high enough to prevent the displacement of existing and competing technologies by photocatalysis. TiO_2 research has progressed on multiple tiers, whereby the study of fundamental processes promotes material development, allowing no repeated uses that promise to play a larger and larger role in engineering sustainable technologies.

TiO_2 is the most used semiconductor because of its higher photocatalytic activity, resistance to the photocorrosion process, absence of toxicity, biological immunity and the relatively low cost (Nikolaki et al., 2006; Han et al., 2009). TiO_2 crystalline powder can be used suspended into a reaction medium (slurry reactors) or immobilized as a film on a carrier material. The anatase form of TiO_2 is reported to give the best combination of photoactivity and photostability. An ideal photocatalyst for oxidation is characterized by the following attributes: photostability; chemically and biologically inert nature; availability and low cost; capability to adsorb reactants under efficient photonic activation (Bideau et al., 1995, Pozzo et al., 1997, Balasubramanian et al., 2004, Gaya and Abdullah, 2008, Siew-Teng Ong et al., 2009, Hanel et al., 2010). The support must have the following characteristics: (a) transparent to irradiation; (b) strong surface bonding with the TiO_2 catalyst without negatively affecting the reactivity; (c) high specific surface area; (d) good absorption capability for organic compounds; (e) separability; (f) facilitating mass transfer processes and (g) chemically inert (Pozzo et al., 1997). As solid support, different materials were investigated: natural or synthetic fabrics, polymer membranes, activated carbon, quartz, glass, glass fiber, optical fibers, pumice stone, zeolites, aluminum, stainless steel, titanium metal or alloy, ceramics (including alumina, silica, zirconia, titania), red brick, white cement etc. (Bideau et al., 1995, Rachel et al., 2002, Balasubramanian et al., 2004, Kemmitt et al. 2004, Gunlazuardi and Lindu, 2005, Hunoh et al. 2005, Chen et al., 2006, Medina-Valtierra et al., 2006, Gaya and Abdullah, 2008, Lim et al., 2009, Siew-Teng Ong et al., 2009, Hanel et al., 2010, Zita et al., 2011).

The immobilization of TiO_2 on different supporting materials has largely been carried out via a physical or chemical route: dip coating, porous material impregnation, sol-gel method, reactive thermal deposition, chemical vapor deposition, electron beam evaporation, spray pyrolysis, electrophoresis, electro-deposition (Rachel et al., 2002, Gunlazuardi and Lindu, 2005, Medina-Valtierra et al., 2006, Hanel et al., 2010, Zita et al., 2011).

The methods most commonly used for deposition of TiO_2 on supports are sputtering and sol-gel dip-coating or spin-coating. Often the fixation of TiO_2 on solid supports reduces its efficiency due to various reasons such as reduction of the active surface, a more difficult exchange with solution, introduction of ionic species etc. A degree of 60–70% reduction in performance is reported in aqueous systems for immobilized TiO_2 as compared to the unsupported catalyst (Gaya and Abdullah, 2008), the best results being obtained by the immobilization of TiO_2 on glass fiber (Rachel et al., 2002, Lim et al., 2009). The separation and/or removal technologies based on membrane and photocatalytic processes have a great potential for application in advanced wastewater treatment. Separation membranes have become essential parts of the human life because of their growing industrial applications in high technology such as biotechnology, nanotechnology and membrane based separation and purification processes. Available technologies to deal with chlorophenolic compounds include the advanced oxidation processes (AOPs) based on the formation of hydroxyl radicals with high oxidation potential.

The hybrid method consists in a photocatalytic procedure using a reactor equipped with TiO_2-functionalized membrane (cylindrical shape) and high pressure mercury lamp for UV radiations generation, centrally and coaxially positioned. The TiO_2-functionalized membranes have been obtained by sol-gel method synthesis of TiO_2 (from tetrabutylortotitanate) as nanoparticles, formed directly in porous membrane regenerated cellulose type. Solutions of methyl, ethyl and propyl alcohols have been used as reaction medium. The experiments were performed at 30 ± 2^0C using synthetic solutions of 4-CP (analytical grade reagent). The amount of hydrogen peroxide (30% w/w) used was calculated at 1.5 time H2O2/4-CP stoichiometric ratio. The degradation process was studied by monitoring the organic substrate concentration changes function of reaction time using chemical oxygen demand analysis (COD).

The experimental data show the catalytic role of TiO_2-functionalized membranes in the oxidation process. The oxidation is preceded by an adsorption process and the transfer of 4-chlorophenol from the solution to the photocatalytic reaction zone through the functionalized membrane. Titanium dioxide, deposited on the membrane, acts as a photocatalyst in the presence of UV radiations leading to a higher efficiency of the oxidation process in a short reaction time. The catalytic activity of TiO_2-functionalized membranes is influenced by the nature of the alcohol used in obtaining them. This can be explained by the crystallite size of TiO_2 and their dispersion on membrane. However, at a higher reaction time, the determined solution COD values tend to increase, indicating that the TiO_2-functionalized membranes become unstable. This can be attributed to a partial solubilization process of membrane into reaction medium with a strong oxidizing potential.

The presence of phenol and phenolic derivatives in water induces toxicity, persistence and bioaccumulation in plant and animal organisms and is a risk factor for human health. The technologies of separation and/or removal of phenolic derivatives based on membrane and photocatalytic processes play an important role.

Available technologies to deal with phenolic compounds include the advanced oxidation processes, based on the formation of very active hydroxyl radicals, which react quickly with the organic contaminant. Among the AOPs, the photocatalytic process is one of the most attractive methods because the reagent components are easy to handle and environmentally benign.

WASTEWATER DECONTAMINATION BY PROCESSES OF ABSORPTION

Rapid development of industry and society led to serious environmental problems such as contamination of groundwater and surface chemical treatment with organic compounds coming from agriculture (pesticides, herbicides, other.) or inorganic compounds in industry (pigments, heavy metals, etc.)

A method used for the wastewater decontamination is the contaminants absorption on the catalyst surface. The most known adsorbent substances cleansing practice are: activated carbon, silica gel, discolored soils, molecular sieves, cotton fibers etc.

After the manner in which the contact is realized between wastewater and adsorbent is distinguished static and dynamic adsorption. In the first case finely adsorbent divided is stirred with water and after a time it's separated by decantation or filtration. In the dynamic adsorption case, wastewater passes through a fixed, mobile or a fluidized absorbent layer with a continuous flow.

Another alternative to the wastewater treatment is the use of technologies based on magnetic nanomaterials AB_2O_4 type (A = Co, Ni, Cu, Zn, B = Fe^{3+}) used as catalysts for degradation of organic compounds or absorbents to retain the surface pollutants heavy metals (mercury, arsenic, lead and others). Their importance and complexity led to research programs development on magnetic materials, with new or improved properties in last decades. These properties are dependent on chemical composition and microstructural characteristics, which can be controlled in the fabrication and synthesis processes.

These materials must have a relatively high surface area, a smaller particle size, and porous structure. In particular, the magnetic properties of the powders makes them to be easily recovered by magnetic separation technology after adsorption or regeneration, which overcomes the disadvantage of separation difficulty of common powdered adsorbents (Qu, et.al, 2008).

Challenges in synthesis of nanostructured catalysts are that many reactions employ mixed catalysts consisting of different oxide metals, and that the function of active centers is not only determined by the constituent atoms but also by the surrounding crystal or surface structures; it is thus necessary to accurately control the synthesis of nanostructured catalysts (Rickerby, et.al, 2007).

Magnetite nanoparticles have highest saturation magnetization of 90 emu/g among iron oxides. Therefore, magnetite nanoparticles can be used to adsorb arsenic ions followed by magnetic decantation. Other iron oxides and hydroxides have been reported to have arsenic ability (Hai, et. al, 2009). Oxidation resistance is an important factor for arsenic removal under atmospheric conditions.

By diverse synthesis methods (hydrothermal, ultrasonic hydrothermal, sol-gel, coprecipitation and other), was obtained ferrite nanomaterials derived from magnetite (FeO, Fe_2O_3) substituting the Fe^{2+} ion in different concentrations (0.5, 0.8, 1, 1.2, 1.5) with Co^{2+}, Cu^{2+}, Ni^{2+}, Zn^{2+} ions (Vlazan, 2010; Fannin, et.al, 2011).

MEMBRANE OXIDATIVE PROCESSES IN WATER TREATMENT

This topic gives an overview of the hybrid photocatalysis-membrane processes and their possible applications in water and wastewater treatment. Different configurations of photocatalytic membrane reactors (PMRs) are described and characterized. They include PMRs with photocatalyst immobilized on/in the membrane and reactors with catalyst in suspension. The advantages and disadvantages of the hybrid photocatalysis-membrane processes in terms of permeate flow, membrane fouling and permeate quality are discussed. Moreover, a short introduction to the heterogeneous photocatalysis and membrane processes as unit operations is given.

The detailed mechanism of the photocatalytic oxidation of organic compounds in water has been discussed widely in the literature and will be presented here in brief only. The overall process can be divided into the following steps.

- Diffusion of reactants from the bulk liquid through a boundary layer to the solution-catalyst interface (external mass transfer).
- Inter-and/or intra-particle diffusion of reactants to the active surface sites of the catalyst (internal mass transfer).
- Adsorption of at least one of the reactants.
- Reactions in the adsorbed phase.
- Desorption of the product(s).

- Removal of the products from the interface of the bulk solution.

A photocatalyst should be characterized by: (I) high activity, (II) resistance to poisoning and stability in prolonged at elevated temperatures, (III) mechanical stability and resistance to attrition, (IV) non-selectivity in most cases, and (V) physical and chemical stability under various conditions. Moreover, it is desirable for the photocatalyst to be able to use not only UV, but also visible light and to be inexpensive. Different semiconducting materials, such as oxides (TiO_2, ZnO, CeO_2, ZrO_2, WO_3, V_2O_5, Fe_2O_3, etc.) and sulfides (CdS, ZnS, etc.) have been used as photocatalysts.

Recently, numerous investigations have been focused on different modifications of TiO_2 in order to improve its activity under UV irradiation or to reduce the band gap energy so that it is able to utilize the visible light. The best photocatalytic performances with maximum quantum yields have been always obtained with TiO_2. Anatase is the most active allotropic form of TiO_2 among the various ones available. Unfortunately, due to a wide band gap (about 3.2 eV), TiO_2 is inactive under visible light.

The most important operating parameters which affect the efficiency of the photocatalytic oxidation process can be summarized as follows: reactor design; light wavelength and intensity; loading of the photocatalyst; initial concentration of the reactant; temperature; pH of reaction medium; oxygen content; the presence of inorganic ions (Mozia, 2005; Gogate and Pandit, 2004; Herrmann, 2006).The photocatalytic reactors can be divided into two main groups:

- Reactors with TiO_2 suspended in the reaction mixture: in the case of reactors with TiO_2 suspended in the reaction mixture (I), the photocatalyst particles have to be separated from the treated water after the oxidation process.

- Reactors with TiO_2 fixed on a carrier material (glass, quartz, stainless steel, pumice stone, titanium metal, zeolites, pillared clays, membranes etc.).

A very promising method for solving problems concerning separation of the photocatalyst as well as products and by-products of photo-oxidation process from the reaction mixture is application of photocatalytic membrane reactors (PMRs). PMRs are hybrid reactors in which photo-catalysis is coupled with a membrane process. The membrane would play both the role of a simple barrier for the photocatalyst and of a selective barrier for the molecules to be degraded. Membrane processes are separation techniques which are widely applied in various sectors of industry including food, chemical and petrochemical, pharmaceutical, cosmetics and electronic industries, water desalination, water and wastewater treatment and many others. The main advantages of membrane processes are: low energy consumption; low chemicals consumption; production of water of stable quality almost independent on the quality of the treated water; automatic control and steady operation allowing performance of a continuous operation; low maintenance costs; easy scale up by simple connecting of additional membrane modules

Synthetic membranes may be: organic or inorganic materials; homogeneous or heterogeneous; symmetrical or asymmetrical; porous or dense; electrically neutral or charged

In this sense, the driving forces are: pressure difference; concentration difference; partial pressure difference or electrical potential difference.

Most of the PMRs described in the literature combine photo-catalysis with pressure driven membrane processes such as:

- Micro-Filtration (MF)
- Ultra-Filtration (UF)
- Nano-Filtration (NF)
- Dialysis
- Pervaporation (PV)
- Direct Contact Membrane Distillation (DCMD)

Hybrid photocatalysis-membrane processes are conducted in the installations often called "photocatalytic membrane reactors". However, in the literature, other names for these configuration scan be also found, including "membrane chemical reactor"(MCR), "membrane reactor", "membrane photoreactor", or, more specific, "submerged membrane photocatalysis reactor" and "photocatalysis–ultrafiltration reactor"(PUR).nFor the hybridization of photocatalysis with membrane process it will be useful to apply a general term of "photocatalytic membrane reactor".

Photocatalytic membrane reactors design show in figure 2 (a, b).

Photocatalytic membranes for the PMRs can be prepared from different materials and indifferent ways.Figure 3 presents two possible types of asymmetric photocatalytic membranes (Bosc et al., 2005). In the first case, photoactive separation layer is deposited on a non-photoactive porous support (Fig. 3a)the photoactive layer, being also the separation layer (skin) is formed on a porous non-photoactive support. In the second case, a non-photoactive separation layer is deposited on a photoactive porous support (Fig. 3b) the separation layer is non-photoactive and is deposited on a porous active support.

The main advantage of PMRs with photocatalytic membranes is that this configuration allows one to minimize the mass transfer resistances between the bulk of the fluid and the semiconductor surface.

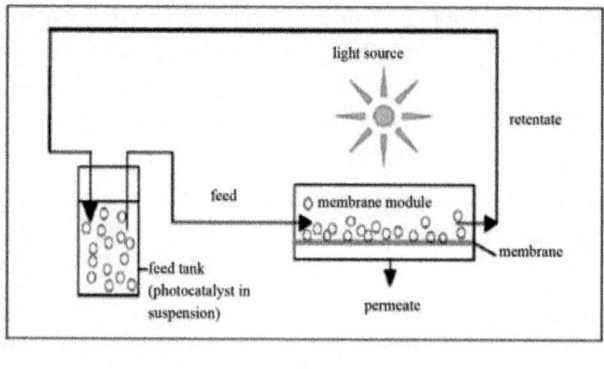

(a)

(b)

Figure 2: a. PMR utilizing photo-catalyst in suspension: irradiation of the membrane module (Mozia S., 2010), b. PMR utilizing photo-catalyst in suspension: irradiation of the feed tank (Mozia S., 2010).

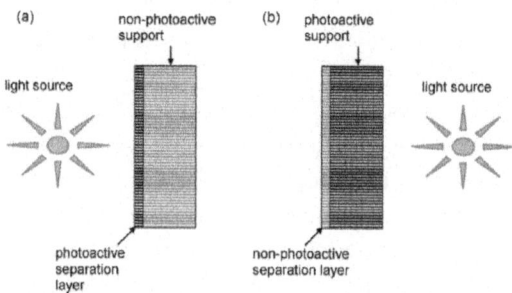

Figure 3: Asymmetric photocatalytic membranes (Bosc et al., 2005).

CONCLUSION

Choosing the most suitable method of wastewater treatment studies require both increasing the effectiveness and economic efficiency (operating and investment costs).

Advanced oxidation processes (AOPs) are good alternatives for removal the toxic compunds from wastewater. The AOPs can be successfully used in wastewater treatment to degrade the persistent organic pollutants, the oxidation process being determined by the very high oxidative potential of the HO radicals generated into the reaction medium by different mechanisms. AOPs can be applied to fully or partially oxidize pollutants, usually using a combination of oxidants. Photo-chemical and photo-catalytic advanced oxidation processes including UV/H_2O_2, UV/O_3, $UV/H_2O_2/O_3$, $UV/H_2O_2/Fe^{2+}(Fe^{3+})$, UV/TiO_2 and $UV/H_2O_2/TiO_2$ can be used for oxidative degradation of organic contaminants. A complete mineralization of the organic pollutants is not necessary, being more worthwhile to transform them into biodegradable aliphatic carboxylic acids followed by a biological process. The efficiency of the various AOPs depend both on the rate of generation of the free radicals and the extent of contact between the radicals and the organic compound.

Photocatalytic oxidation in water treatment has proved its efficiency at many pilot-scale applications. However, wide marketing of commercially available solar detoxification systems is obstructed by the general market situation: a new water treatment procedure has an opportunity to be implemented only when its cost is at least two-fold lower than the cost of a procedure currently in use. Photocatalysis, also called the "green" technology, represents one of the main challenges in the field of treatment and decontamination systems, especially for water and air. Its operating principle is based on the simultaneous action of the light and a catalyst (semi-conductor), which allows for pollutant molecules to be destroyed without damaging the surrounding environment.

In recent years, applications to environmental remediation have been one of the most active subjects in photocatalysis.

REFERENCES

1.   M Addamo, V Augugliaro, Di Paola, A., García-López, E., Loddo, V., Marcì, G., Palmisano, L., (2008Photocatalytic thin films of TiO2 formed by a sol-gel process using titaniumtetraisopropoxide as the precursor, Thin Solid Films 516, 38023807

2.   S Ahmed, M. G Rasul, W. N Martens, R Brown, M. A Hashib, 2010Heterogeneous photocatalytic degradation of phenols in wastewater:

A review on current status and developments", Desalination, 261(1-2), 318

3.   R Alexy, T Kümpel, K Kümmerer, 2004Assessment of degradation of 18 antibiotics in the Closed Bottle Test. Chemosphere 57505512

4.   P Alphonse, A Varghese, C Tendero, 2010Stable hydrosols for TiO2 coatings", Journal of Sol-Gel Science and Technology, 56(3), 250-263.

5.   R Andreozzi, V Caprio, A Insola, R Marotta, 1999Advanced oxidation processes (AOP) for water purification and recovery, Catalysis Today, 53 (1), 51-59

6.   Anita RachelMachiraju Subrahmanyam, Pierre Boule, (2002Comparison of photocatalytic efficiencies of TiO2 in suspended and immobilised form for the photocatalytic degradation of nitrobenzenesulfonic acids, Applied Catalysis B: Environmental 37(4), 301-308.

7.   I Arslan-alaton, A. E Caglayan, 2006Toxicity and biodegradability assessment of raw and ozonated procaine penicillin G formulation effluent. Ecotox. Environ. Safe. 63131140

8.   I Arslan-alaton, S Dogruel, E Baykal, G Gerone, 2004Combined chemical and biological oxidation of penicillin formulation effluent. J. Environ. Manag. 73155163

9.   V Augugliaro, M Litter, L Palmisano, J Soria, 2006The combination of heterogeneous photocatalysis with chemical and physical operations: A tool for improving the photoprocess performance, Journal of Photochemistry and Photobiology C: Photochemistry Reviews 7127144

10. F Ay, F Kargi, 2010Advanced oxidation of amoxicillin by Fenton' reagent treatment. J.Hazard.Mater. 179622627

11. G Balasubramanian, D. D Dionysiou, M. T Suidan, I Baudin, Laîné J.M., (2004Evaluating the activities of immobilized TiO2 powder films for the photocatalytic degradation of organic contaminants in water. Applied Catalysis B: Environmental, 47(2), 73-84.

12. M Bideau, B Claudel, C Dubien, L Faure, H Kazouan, 1995On the "immobilization" of titanium dioxide in the photocatalytic oxidation of spent waters, Journal of Photochemistry & Photobiology A: Chemistry, 91(2), 137-144.

13. F Bosc, A Ayral, C Guizard, 2005Mesoporous Anatase Coatings for Coupling Membrane Separation and Photocatalyzed Reactions" J. Membr. Sci. 2651319

14. J. P Bound, N Voulvoulis, 2004Pharmaceuticals in the aquatic environment- a comparison of risk assessment strategies. Chemosphere 5611431155

15. K Byrappa, K. M. L Rai, M Yoshimura, 2002Hydrothermal preparation of TiO2 and photocatalytic degradation of hexachloro cyclohexane and dichlorodiphenyl trichloromethane", Journal of Environmental Science and Technology, 21, 10851090

16. A. E Cassano, O. M Alfano, 2000Reaction Engineering of Suspended Solid Heterogeneous Photocatalytic Reactors", Catalysis Today, 58(2-3), 167197

17. O Carp, C. L Huisman, A Reller, 2004Photoinduced reactivity of titanium dioxide, Prog. Solid State Chem. 32(1-2), 33-177.

18. A. G Chakinala, P. R Gogate, A. E Burgess, D. H Bremner, 2009Industrial wastewater treatment using hydrodynamic cavitation and heterogeneous advanced Fenton processing. Chemical Engineering Journal, 152498502

19. S. Z Chen, P. Y Zhang, W. P Zhu, L Chen, Sheng-Ming Xu, (2006Deactivation of TiO2 photocatalytic films loaded on aluminium: XPS and AFM analyses, Appl. Surf. Sci. 252 (20), 7532-7538.

20. S. S Chin, K Chiang, A. G Fane, 2006The stability of polymeric membranes in a TiO2 photocatalysis process", Journal of Membrane Science, 275(1-2), 202211

21. N Daneshvar, D Salari, A. R Khataee, 2003Photocatalytic degradation of azo dye acid red 14 in water: investigation of the effect of operational parameters, Journal of Photochemistry and Photobiology A: Chemistry, 157111

22. N Daneshvar, S Aber, Seyed Dorraji M. S., Khataee A. R., Rasoulifard M. H., (2007Preparation and Investigation of Photocatalytic Properties of ZnO Nanocrystals: Effect of Operational Parameters and Kinetic Study, World Academy of Science, Engineering and Technology 29;

23. N Daneshvar, D Salari, A. R Khataee, 2003Photocatalytic degradation of azo dye acid red 14 in water: investigation of the effect of operational parameters, Journal of Photochemistry and Photobiology A: Chemistry 157111116

24. J De Laat, Truong Le G., Legube B., (2004A comparative study of the effects of chloride, sulfate and nitrate ions on the rates of decomposition of H2O2 and organic compounds by Fe(II)/H2O2 and Fe(III)/H2O2. Chemosphere, 55571523

25. Y Du, M Zhou, L Lei, 2006Role of the intermediates in the degradation of phenolic compounds by Fenton-like process, Journal of Hazardous Materials, 136859865

26. F Duarte, L. M Madeira, 2010Fenton and Photo-Fenton-Like Degradation

of a Textile Dye by Heterogeneous Process with Fe/ZSM-5 Zeolite, Separation Science and Technology, 4515121520

27. E Elmolla, M Chaudhuri, 2009Degradation of the antibiotics amoxicillin, ampicillin and cloxacillin in aqueous solution by photo-Fenton process. J. Hazard. Mater. 17214761481

28. S Esplugas, J Gimenez, S Contreras, E Pascual, M Rodriguez, 2002Comparison of different advanced oxidation processes for phenol degradation, Water Res., 36, 10341042

29. Fang HanVenkata Subba Rao Kambala, Madapusi Srinivasan, Dharmarajan Rajarathnam, Ravi Naidu, (2009Tailored titanium dioxide photocatalysts for the degradation of organic dyes in wastewater treatment: A review. Applied Catalysis A, General, 359 (1-2), 2009, 25-40.

30. P. C Fannin, C. N Marin, I Malaescu, N Stefu, P Vlazan, S Novaconi, S Popescu, 2011Effect of the concentration of precursors on the microwave absorbent properties of Zn/Fe oxide nanopowders; Journal of Nanoparticle Research; 13311319

31. J Feng, X Hu, P. L Yue, 2005Discoloration and mineralization of Orange II by using a bentonite clay-based Fe nanocomposite film as a heterogeneous photo-Fenton catalyst, Water Research, 398996

32. Y Gao, H Liu, 2005Preparation and catalytic property study of a novel kind of suspended photocatalyst of TiO2activated carbon immobilized on silicone rubber film, Mater. Chem. Phys. 92(2-3), 604-608.

33. U. I Gayaa, A. H Abdullaha, 2008Heterogeneous photocatalytic degradation of organic contaminants over titanium dioxide: A review of fundamentals, progress and problems, Journal of Photochemistry and Photobiology C: Photochemistry Reviews 9112

34. N Getoff, 2001Comparison of radiation and photoinduced degradation of pollutants in water: synergistic effect of O2, O3 and TiO2. A Short Review, Research on Chemical Intermediates, 27Numbers 4-5, 343-358.

35. Li Puma GBonob A., Krishnaiah D., Collin J.G., (2008Preparation of titanium dioxide photocatalyst loaded onto activated carbon support using chemical vapor deposition: A review paper, Journal of Hazardous Materials 157. 209219

36. P. R Gogate, A. B Pandit, 2004A review of imperative technologies for wastewater treatment II: hybrid methods, Advances in Environmental Research, 8553597

37. C Guillard, J Disdier, J Herrmann, M Lehaut, C Chopin, T Malato, S

Blanco, J., 1999Comparison of various titania samples of industrial origin in the solar photocatalytic detoxification of water containing 4chlorophenol, Catalysis Today, 54, (2,-3), 217-228.

38. A Hänel, P Moren, A Zaleska, J Hupka, 2010Photocatalytic activity of TiO2immobilized on glass beads, Physicochem. Probl. Miner. Process. 454956

39. H. K Hansen, P Nunez, D Raboz, I Schippacasse, R Grandon, 2007Electrocoagulation in wastewater containing arsenic: Comparing different process designs, ScienceDirect, Electrochimica Acta 5234643470

40. N. H Hai, N. D Phu, 2009Arsenic removal from water by magnetic Fe1xCoxFe2O4 and Fe1-yNiyFe2O4 nanoparticles; VNU Journal of Science, Mathematics- Physics 15-19.

41. J He, W Ma, W Song, J Zhao, X Qian, S Zhang, J. C Yu, 2005Photoreaction of aromatic compounds at a-FeOOH/H2O interface in the presence of H2O2: evidence for organic-goethite surface complex formation, Water Research, 39119128

42. M Hepel, S Hazelton, 2005Photoelectrocatalytic degradation of diazo dyes on nanostructured WO3 electrodes, Electrochimica Acta 50. 52785291

43. M. R Hoffmann, S. T Martin, W Choi, D. W Bahnemann, 1995Environmental Applications of Semiconductor Photocatalysis", Chemical Reviews, 95, 6996

44. K. H Hu, Yong Kui Cai Y.K., Li S., (2011Photocatalytic Degradation of Methylene Blue on MoS2/TiO2, Nanocomposite, Advanced Materials Research, Volumes 197198

45. HuaZi-Le, Shi, Jian-Lin, Zhang, Wen-Hua, Huang, Wei-Min, (2002Direct synthesis and characterization of Ti-containing mesoporous silica thin films, Materials Letters 53(4-5), 299-304.

46. S Kaniou, K Pitarakis, I Barlagianni, I Poulios, 2005Photocatalytic oxidation of sulfamethazine. Chemosphere 60372380

47. Jarnuzi GunlazuardiWinarti Andayani Lindu, (2005Photocatalytic degradation of pentachlorophenol in aqueous solution employing immobilized TiO2 supported on titanium metal, Journal of Photochemistry and Photobiology A: Chemistry, 173(1), 51-55.

48. Jiří ZitaJosef Krýsa, Urh Černigoj, Urška Lavrenčič-Štangar, Jaromír Jirkovský, Jiří Rathouský, (2011Photocatalytic properties of different TiO2 thin films of various porosity and titania loading, Catalysis Today,

161 (1), 29-34.

49. João RochaArtur Ferreira, Zhi Lin, Michael W. Anderson, (1998Synthesis of microporous titanosilicate ETS-10 from TiCl3 and TiO2: a comprehensive study, Micropor. Mesopor. Mater, 23(5-6), 253-263.

50. Jorge Medina-ValtierraManuel Sánchez-Cárdenas, Claudio Frausto-Reyes, Sergio Calixto, (2006Formation of smooth and rough TiO2 thin films on fiberglass by sol-gel method, J. Mex. Chem. Soc., 50(1), 8-13.

51. V Kavitha, K Palanivelu, 2004The role of ferrous ion in Fenton and photo-Fenton processes for the degradation of phenol, Chemosphere, 5512351243

52. S Kakuta, T Abe, 2009Photocatalysis for water oxidation by Fe2O3 nanoparticles embedded in clay compound: correlation between its polymorphs and their photocatalytic activities, Journal of Materials Science, 4411

53. A Khelifa, S Moulay, A. W Naceur, 2005Treatment of metal finishing effluents by the electroflotation technique, Desalination 1812733

54. Y Kondo, S Fujihara, 2011Solvothermal Synthesis of WO3 Photocatalysts and their Enhanced Activity, Key Engineering Materials (485

55. K. H Hu, Yong Kui Cai, Sai Li, (2011Photocatalytic Degradation of Methylene Blue on MoS2/TiO2 Nanocomposite, Advanced Materials Research, 197-198

56. K Kümmerer, 2009Antibiotics in the aquatic environment- A review-Part I. Chemosphere 75417434

57. B. G Kwon, D. S Lee, N Kang, J Yoon, 1999Characteristics of p-chlorophenol oxidation by Fenton's reagent.Water Res. 33 (9), 2110-2118.

58. C Lazau, C Ratiu, C Orha, R Pode, F Manea, 2011Photocatalytic activity of undoped and Ag-doped TiO2-supported zeolite for humic acid degradation and mineralization, Materials Research Bulletin 46, 11, 19161921

59. J. K Leland, A. J Bard, 1987Photochemistry of colloidal semiconducting iron oxide polymorphs, Journal of Physical Chemistry, 9150765083

60. Li Puma GBonob A, Krishnaiah D., Collin J.G., (2008Preparation of titanium dioxide photocatalyst loaded onto activated carbon support using chemical vapor deposition: A review paper, Journal of Hazardous Materials, 157209

61. J Li, H Xu, Y. -Y Zhu, L. -P Wang, J. -H Du, and C. -H Fabrication, and characterization of a novel TiO2 nanoparticle self-assembly membrane

with improved fouling resistance, Journal of Membrane Science, 326(2), 659666

62. X. Y Li, Y. H Cui, Y. J Feng, Z. M Xie, J. D Gu, 2005Reaction pathways and mechanisms of the electrochemical degradation of phenol on different electrodes. Water Res. 3919721981

63. L. L. P Lim, R. J Lynch, S. I In, 2009Comparison of simple and economical photocatalyst immobilisation procedures, Applied Catalysis A: General, 3652214221

64. Liu XinshengKerry Thomas. J., (1996Synthesis of microporous titanosilicates ETS-10 and ETS-4 using solid TiO2 as the source of titanium, Chem. Commun., 1214351436

65. T Lopez, M Alvarez, F Tzompantzi, M Picquart, Photocatalytic degradation of 2,4-dichlorophenoxiacetic acid and 246trichlorophenol with ZrO2 and Mn/ZrO2 sol-gel materials, J Sol-Gel Sci Techn (2006

66. M. C Lu, J. N Chen, C. P Chang, 1997Effect of inorganic ions on the oxidation of dichlorvos insecticide with Fenton's reagent. Chemosphere, 35(10), 2285-2293.

67. Y Mansourpanah, S. S Madaeni, A Rahimpour, A Farhadian, A. H Taheri, 2009Formation of appropiate sites on nanofiltration membrane surface for binding TiO2 photo-catalyst: Performance, characterization and fouling-resistant capability", Journal of Membrane Science, 330(1-2), 297306

68. R Molinari, M Mungari, E Drioli, Di Paola, A., Loddo, V., Palmisano L., Schiavello,M., (2000Study on a photocatalytic membrane reactor for water purification", Catalysis Today, 55(1-2), 7178

69. F. C. C Moura, M. H Araujo, R. C. C Costa, J. D Fabris, J. D Ardisson, W. A. A Macedo, R. M Lago, 2005Efficient use of Fe metal as an electron transfer agent in a heterogeneous Fenton system based on Fe0/Fe3O4 composites, Chemosphere, 6011181123

70. F. C. C Moura, G. C Oliveira, M. H Araujo, J. D Ardisson, W. A. A Macedo, R. M Lago, Highly reactive species formed by interface reaction between Fe-iron oxides particles: An efficient electron transfer system for environmental applications, Applied Catalysis A: General, 307, (2006

71. S Mozia, 2010Photocatalytic membrane reactors (PMRs) in water and wastewater treatment. A review", Separation and Purification Technology, 73(2), 7191

72. I Munoz, J Rieradevall, F Torrades, J Peral, X Domenech, 2005Environmental Assessment of Different Solar Driven Advanced

Oxidation Processes", Sol. Energy, 79, 369

73. R Munter, 2001Advanced Oxidation Processes- Current Status and Prospects, Proc. Estonian Acad. Sci. Chem. 50(2) 59

74. Y Nie, C Hu, L Zhou, J Qu, 2008An efficient electron transfer at the Fe0/ iron oxide interface for the photoassisted degradation of pollutants with H2O2, Applied Catalysis B; Environmental, 82151156

75. M. D Nikolaki, D Malamis, S. G Poulopoulos, C. J Philippopoulos, 2006Photocatalytical degradation of 13dichloro-2-propanol aqueous solutions by using an immobilized TiO2 photoreactor, Journal of Hazardous Materials 2006, 137(2), 1189-1196.

76. S. H Oh, J. S Kim, J. S Chung, E. J Kim, S. H Hahn, 2005Crystallization and Photoactivity of TiO2 Films Formed on Soda Lime Glass by a Sol-Gel Dip-Coating Process, Chem. Eng. Comm. 192 (3), 327-335.

77. S. T Ong, C. K Lee, Z Zainal, P. S Keng, S. T Ha, 2009Photocatalytic degradation of basic and reactive dyes in both single and binary systems using immobilized TiO2, Journal of Fundamental Sciences, 5(2), 88-93.

78. C Orbeci, I Untea, G Kopsiaftis, 2008The influence of inorganic species on oxidative degradation of 4chlorphenol by photo-Fenton type process, Revista de Chimie 59(9), 952-955.

79. C Orbeci, I Untea, M Dancila, D. S Stefan, 2010Kinetics considerations concerning the oxidative degradation by photo-Fenton process of some antibiotics. Environ. Eng. Manage. J., 915

80. Ortiz de la Plata GB., Alfano O. M., Cassano A. E., (2010Decomposition of 2-chlorophenol employing goethite as Fenton catalyst. I. Proposal of a feasible, combined reaction scheme of heterogeneous and homogeneous reactions, Applied Catalysis B: Environmental, 95113

81. J Papp, H. S Shen, R Kershaw, K Dwight, A Wold, 1993Titanium(IV) Oxide Photocatalysts with Palladium, Chem. Mater. 5, 284;

82. M Pera-titus, V Garcia-molina, M. A Banos, J Gimenez, S Esplugas, 2004Degradation of chlorophenols by means of advanced oxidation processes: a general review, Applied Catalysis B: Environmental, 47219256

83. J. J Pignatello, 1992Dark and photoassisted Fe3+ catalyzed degradation of chlorophenoxy herbicides by hydrogen peroxide, Environ.Sci.Technol. 26944951

84. J Qu, 2008Research progress of novel adsorption processes in water purification: A review; Journal of Environmental Sciences 20113

85. C Ratiu, ., F Manea, . C Lazau, ., I Grozescu, . C Radovan, ., J Schoonman,

., Electrochemical oxidation of p-aminophenol from water with boron-doped diamond anodes and assisted photocatalytically by TiO2-supported zeolite, Desalination 260 (2010) 51-56.

86. C Ratiu, F Manea, C Lazau, C Orha, G Burtica, I Grozescu, J Schoonman, Photocatalytically-assisted electrochemical degradation of p-aminophenol in aqueous solutions using zeolite-supported TiO2 catalyst, Chemical Papers 65 (3) 289-298 (2011

87. A Riga, K Soutsas, K Ntampegliotis, V Karayannis, G Papapolymerou, 2007Effect of System Parameters and of Inorganic Salts on the Degradation Kinetics of Procion Hexl Dyes. Comparison of H2O2/uv, Fenton, and photo-Fenton, TiO2/UV and TiO2/UV / H2O2 processes, Desalination, 2117286

88. R Pozzo, M Baltanás, A Cassano, Supported titanium oxide as photocatalyst in water decontamination: State of the art, Catalysis Today, 39 (3), 1997

89. O Rozas, D Contreras, M. A Mondaca, M Pérez-moya, H. D Mansilla, 2010Experimental design of Fenton and photo-Fenton reactions for the treatment of ampicillin solutions. J. Hazard. Mater. 17710251030

90. N Shimizu, C Ogino, Farshbaf Dadjour M., Murata T., (2007Sonocatalytic degradation of methylene blue with TiO2 pellets in water, Ultrasonics Sonochemistry, 14184190

91. E. M Siedlecka, A Wieckowska, P Stepnowski, 2007Influence of inorganic ions on MTBE degradation by Fenton's reagent, Journal of Hazardous Materials, 147497502

92. W Spacek, R Bauer, G Heisler, 1995Heterogeneous and homogeneous wastewater treatment Comparison between photodegradation with TiO2 and the photo-Fenton `reaction, Chemosphere 30477484

93. A. G Trovó, R. F. P Nogueira, A Agüera, A. R Fernandez-alba, S Malato, 2011Degradation of the antibiotic amoxicillin by photo-Fenton process-Chemical and toxicological assessment. Water Res. 4513941402

94. K Venkata, S Rao, M Subrahmanyam, P Boule, 2004Immobilized TiO2 photocatalyst during long-term use: decrease of its activity, Applied Catalysis B: Environmental 49239

95. M Vinita, R. P. J Dorathi, K Palanivelu, 2010Degradation of 246trichlorophenol by photo Fenton's like method using nano heterogeneous catalytic ferric ion, Solar Energy, 84, 1613-1618.

96. P Vlazan, M Vasile, 2010Synthesis and characterization CoFe2O4 nanoparticles prepared by the hydrothermal method; Optoelectronics and

Advanced Materials-Rapid Communications; 4; 13071309

97. XuHongwu, Zhang, Yiping, Navrotsky, Alexandra, (2001Enthalpies of formation of microporous titanosilicates ETS-4 and ETS-10, Microporous and Mesoporous Materials, 47, 2-3, 285-291.

98. H Zhou, D. W Smith, 2002Advanced technologies in water and wastewater treatment, J. Environ. Eng. Sci. 1247264

99. J. A Zazo, J. A Casas, A. F Mohedano, M. A Gilarranz, J. J Rodríguez, 2005Chemical Pathway and Kinetics of Phenol Oxidation by Fenton's Reagent, Environmental Science and Technology, 3992959302

100. H Wang, J Wang, 2007Electrochemical degradation of 4-chlorophenol using a novel Pd/C gas-diffusion electrode, Applied Catalysis B; Environmental, 775865

101. Y Yu, Y Ding, S Zuo, J Liu, 2011Photocatalytic Activity of Nanosized Cadmium Sulfides Synthesized by Complex Compound Thermolysis, International Journal of Photoenergy, 15

Chapter 6

TREATMENT TECHNOLOGIES FOR ORGANIC WASTEWATER

Chunli Zheng[1], Ling Zhao[2], Zhimin Fu[2], Xiaobai Zhou[3] and An Li[4]

[1]School of Energy and Power Engineering, Xi'an Jiaotong University, China

[2]College of Environment Resources of Inner Mongolia University, China

[3]The Environmental Monitoring Center of Jiangsu Province, Nanjing, China

[4]School of Petrochemical Engineering Lanzhou University of Technology, China

INTRODUCTION

Resources of Organic Wastewater

There are several contaminants in wastewater, with organic pollutants playing the major role. Many kinds of organic compounds, such as PCBs, pesticides, herbicides, phenols, polycylic aromatic hydrocarbons (PAHs), aliphatic and hetercyclic compounds are included in the wastewater, and industrial and agricultural production as well as the people living could be the source of organic wastewater endangering the safety of the water resource [1]. The wastewater of the farmland may contain high concentration of pesticides or herbicides; the wastewater of the coke plant may contain various PAHs; the wastewater of the chemical industry may contain various heterogeneity compounds, such as PCB, PBDE; the wastewater discharged by the food industry contains complex organic pollutants with high concentration of SS and BOD; and the municipal sewage contains different type of organic pollutants, such as oil, food, some dissolved organics and some surfactants. These organic pollutants in water can harm the environment and also pose health risks for humans.

Common Poisonous Substances in Organic Wastewater

The organic pollutants in the wastewater could be divided into two groups according to their biological degradation abilities. The organic pollutants with simple structures and good hydrophilicity are easy to be degraded in the environment. These organic pollutants, such as polysaccharide, methanol could be degraded by the bacteria, fungus and algae. However, some of them, such as acetone and methanol, could cause acute toxicity when existed in wastewater at a high concentration. On the other hand, the persistent organic pollutants, such as PAHs, PCBs, and DDT, are very slowly metabolized or otherwise degraded. Some of them, for example, the pesticides were widely used for several years. Although their concentration as well as the cute toxicity in the wastewater is lower than the soluble organic pollutant, they can be sequestered in sediment and exist for decades, and transport into the wastewater and then the food chain. The POPs are lipid soluble, and many of them mentioned above are carcinogenic, teratogenic, and neurotoxic. Since they are persistent, long way transported and toxic, these organic pollutants draw more attentions.

The classic poisonous substances in organic wastewater are as follow:

1.   Water organic matter

Water organic matter is the genetic name of the organic compounds in the sediment and wastewater. Generated from the residues of the animal, plants and microorganisms, the water organic matters could be divided into two categories: one is non - humic, which is composed of the various organic compounds of organisms, such as protein, carbohydrate, organic acids, etc., the other is a special organic compounds named humus. Water organic matter could affect the physical and chemical properties of the water, and could also influence the self-purification, degradation, migration and transformation process in the water.

1.   Formaldehyde

Formaldehyde is an organic compound with the formula CH_2O. The main sources of formaldehyde are organic synthesis, chemical industry, synthetic fiber, dyestuff, wood processing and the paint industry emissions of wastewater. With a strong reducibility, formaldehyde could easily combine with a variety of material, and is easy to be together. Formaldehy is a stimulus to skin and mucous membrane. It could enter the central nervous of human body and cause retinal damage

1.   Phenols

Phenols are a class of chemical compounds consisting of a hydroxyl group (-OH) bonded directly to an aromatic hydrocarbon group. The phenol in the wastewater mainly comes from the coking plant, refining, insulation material manufacturing, paper making and phenolic chemical plant. Phenol is known human carcinogen and is of considerable health concern, even at low concentration. Phenol also has potential to decrease the growth and the reproductive capacity of the aquatic organisms.

1.   Nitrobenzene

Nitrobenzene is an organic compound with the chemical formula $C_6H_5NO_2$. It is produced on a large scale as a precursor to aniline. In the laboratory, it is occasionally used as a solvent, especially for electrophilic reagents. Prolonged exposure may cause serious damage to the central nervous system, impair vision, cause liver or kidney damage, anemia and lung irritation[2]. Recent research also found nitrobenzene as a potential carcinogenic substance.

1.   PCBs

PCBs are biphenyl combined with 2 to 10 chlorine atoms. PCBs are widely used as dielectric and coolant fluids, for example in transformers, capacitors, and electric motors, and various kinds of PCBs could be found the wastewater of this factories[3]. PCBs are carcinogenic, and could accumulate in adipose tissue, causing brain, skin and the internal organs disease, and influence nerve, reproductive and immune system. PCBs also have shown toxic and mutagenic effects by interfering with hormones in the body. PCBs, depending on the specific congener, have been shown to both inhibit and imitate estradiol.

1.   PAHs

PAHs are recalcitrant organic pollutants consisting of two or more fused benzene rings in linear, angular, or cluster arrangements. PAHs occur in oil, coal, and tar deposits, and PAHs in the aquatic system could come from accidently leaking, atmosphere deposition and contaminated sediment release. The concentration of PAHs, especially the PAHs with high molecular weight, in the water is usually low in the water owing to their hydrophobia property, but they are still among the most problematic substances as they could accumulate in the environment and threaten the development of living organisms because of their acute toxicity, mutagenicity or carcinogenity [4].

1.   Organophosphorus pesticide

The wastewater of organophosphorus pesticide manufacturers often contains a high concentration of organophosphorus pesticide, intermedia productions and degradation productions, and the wastewater from the farmland could contain some of this pesticide since this substance could exist in the environment for a period of time. The discharge of water contained organophosphorus pesticide could cause serious environmental pollution. Some organophosphorus pesticides have acute poison on the people and livestock. In spite of the severe toxicity of the organophosphorus pesticide, it is easy to be degraded in the environment [5].

1.   Petroleum hydrocarbons

The petroleum hydrocarbons in the water system mainly come from the industrial wastewater and municipal sewage. The industry, such as oil exploration, oil manufacture, transportation and refining could produce the wastewater with a mixture of various petroleum hydrocarbons. The petroleum hydrocarbons are toxic towards aquatic living things, and they could also aggravate the water quality by forming a layer of oil film, which could decrease the oxygen exchange of the air and water body.

1.   Atrazine

Atrazine is the most widely used herbicide in conservation tillage systems, which are designed to prevent soil erosion. This chemical herbicide could stop pre- and post-emergence broadleaf and grassy weeds in dry farmland, and increase the production of the major crops [6]. The wastewater contained atrazine mainly comes from the chemical industry manufacturing this product and the farmlands which are over loaded. This substance could remain in the environment for a period of time, and it has been detected in the surface water and groundwater of many countries and regions. Atrazine could volatilize at high temperature and release poisonous gas such as carbon monoxide, nitrogen oxides, which could irritate people's skin, eyes and respiratory tract. Besides, atrazine also has a potential cause of birth defects, low birth weights and menstrual problems when consumed at concentrations below federal standards.

Environmental Hazards of Organic Wastewater

High mount of hydrophilic organic pollutants, such as organic matters, oil could consume a large amount of soluble oxygen. The acute toxicity and high quantity of oxygen demand could worsen the water quality and lead to great damage to the aquatic ecological system. However, their bad influence towards the environment will not last long, since they could easily be degraded by microorganisms.

The situation is different for the POPs, which have low water solubility, high accumulation capacity and potential carcinogenic, teratogenic, and neurotoxic properties. For example, many of the organochlorine pesticides cited above are carcinogenic, teratogenic, and neurotoxic. The dioxins and benzofurans are highly toxic and are extremely persistent in the human body as well as the environment. Several of the POPs, including DDT and its metabolites, PCBs, dioxins, and some chlorobenzene, can be detected in human body fat and serum years after any known exposures. Lindane (hexachlorocyclohexane), which was used for the treatment of body lice and as a broad-spectrum insecticide, could cause very high tissue levels, and could cause acute deaths when improperly used.

Many factors, such as the characters of the pollutants, the environmental factors (PH value, temperature etc.), aging process could affect the toxicity of the organic wastewater, and their long-term influence to the ecosystem deserve further investigation.

Monitoring Analysis Method of Poisonous Substances

1. Gross analysis

The amount of organic compounds in wastewater is generally evaluated by chemical oxygen demand (COD) test, biological oxygen demand (BOD) test, and (TOC) test.

The basis for the COD test is that nearly all organic compounds can be fully oxidized to carbon dioxide with a strong oxidizing agent under acidic conditions. The COD value is always measured by the acidic potassium permanganate method and potessium dichromate method, and could reflect the pollution degree of reducing matter in water, including ammonia and reducing sulfide, so in wastewater with high quantity of reducing matter, the COD value will overestimate the organic pollutants in the water.

BOD value is the amount of dissolved oxygen needed by aerobic biological organisms in a body of water to break down organic material present in a given water sample at certain temperature over a specific time period. The BOD value is most commonly expressed in milligrams of oxygen consumed per liter of sample during 5 days of incubation at 20 °C and is often used as a robust surrogate of the degree of organic pollution of water. This is not a precise quantitative test, although it is widely used as an indication of the organic quality of water.

TOC value is the mount of total carbon (water soluble and suspended in water) in the water. Using combustion during the assessment, this method could oxidize all the organic pollutants, and value reflects the amount of organic matter more directly than BOD_5 or COD.

The COD, BOD and TOC test could quickly reflect the organic pollution in the wastewater, however, they can't reflect the kinds of organic matter and composition of the water, and therefore cannot reflect the total amount of the same total organic carbon pollution caused by different consequences.

1. Chromatography-mass spectrometry method

Chromatography-mass spectrometry method is an advanced method to separate and define the organic pollutants in the waste water. Spectrometry is the collective term for a set of laboratory techniques for the separation of mixtures. The separation is based on differential partitioning between the mobile and stationary phases. The structure diversity of different components in the wastewater results in a different retention on the stationary phase and thus changing the separation. The mobile phase of the chromatography can be gas or liquid, so the chromatography can be divided into gas chromatography (GC) and liquid chromatography (LC).

The mass spectrometer could ionize the organism and shoot it through an electric field. Since the electric field could bend the path (trajectory) of lighter molecules more than that of heavy molecules, the organic matter of different mass would strike at different position (the position is fixed for each organic matter) in the detector. This method could identify and quantify organic pollutants. The combination of chromatography and mass spectrometry could offer complete information on the type of organic pollutants in a sample and the concentration of each pollutant in the sample.

BIOLOGICAL TREATMENT TECHNOLOGY OF ORGANIC WASTEWATER

Principle of the Biodegradation

Biodegradation is a process using microorganisms, fungi, green plants and their enzymes to remove the pollutants from natural environment or transform them harmless. Biodegradation could happen in nature world and is used in wastewater treatment in recent years since humanity strives to find sustainable ways to clean up contaminated water economically and safely.

Biodegradation of Organic Compounds

Chemical, physical and biological methods have been used to remove the organic compounds from the wastewater, and biological method has been paid much attention owing to its economic and ecologic superiority. The biodegradation rate and biodegradation degree of the organic substance partly depended on the characters of the substance. Some of the organic pollutants like organic matters, organophosphorus pesticide, which have relativity high water solubility and low acute toxicity, are bioavailable and easy to be degraded [7]. However, for some POPs and xenobiotic organic pollutants, such as polychlorinated biphenyls (PCBs), polyaromatic hydrocarbons (PAHs), heterocyclic compounds, pharmaceutical substances, which possess a higher bioaccumulation, biomagnification and biotoxicity properties, are reluctant to biodegradation in the nature condition. Organic material can be degraded aerobically with oxygen, or anaerobically, without oxygen [8].

Aerobic Biodegradation

The principle of the aerobic biodegradation is as follow: oxygen is needed by degradable organisms in their degradation at two metabolic sites, at the initial attack of the substrate and at the end of respiratory chain [9]. Bacteria and fungi could produce oxygenases and peroxidases they could help with the pollutant oxidization and get benefits from observing the energy, carbon and nutrient elements released during this process. A huge number of bacterial and fungal general possess the capability to release non-special oxidase and degrade organic pollutants. There are generally two types of relationships between the microorganism and organic pollutants: one is that the microorganisms use organic pollutant as sole source of carbon and energy; the other is that the microorganisms use a growth substrate as carbon and energy source, while another organic compound in the organic substrate which could not provide carbon and energy resource is also degraded, namely cometabolism.

The classic aerobic biodegradation reactors include activated sludge reactor and membrane bioreactor.

Activated Sludge Reactor

Activated sludge is a process for treating sewage and industrial wastewaters using air and a biological floc composed of bacteria and protozoans. This technique was invented by Ardern and Lockett at the beginning of last century and was considered as a wastewater treatment technique for larger cities as it required a more sophisticated mode of operation (Fig. 1) [10].

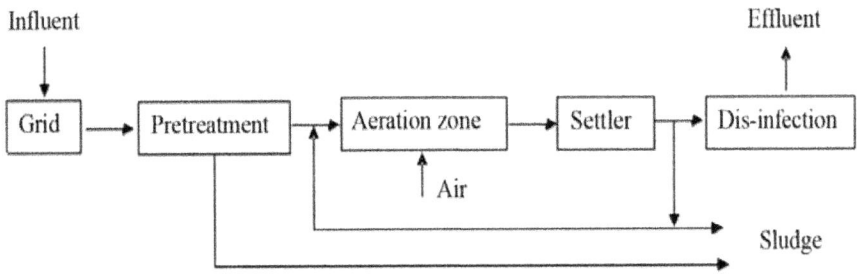

Figure 1: Fig 2-1 the scheme of the activated sludge reactor [1].

This process introduced air or oxygen into a mixture of primary treated or screened wastewater combined with organisms to develop a biological floc which reduces the organic content of the sewage, which is largely composed of microorganisms such as saprotrophic bacteria, nitrobacteria and denitrifying bacteria. With this biological floc, we could degrade the organic pollutant and bio-transform the ammonia in wastewater. Generally speaking, the process contained two steps: adsorption followed by biological oxidation.

The technique could effectively remove the organic matters, nitrogeneous matters, phosphate in the wastewater, when there is enough oxygen and hydraulic retention time. However, the contaminated water is always short of oxygen, which could cause sludge bulking, a great problem decrease the water quality of the effluent. The oxygen concentration could be increased by including aeration devices in the system, but research need to be done to find out the optimal value since aeration would cause an increase of the costs of the wastewater treatment plants. Researches are also required to deal with the excess activated sludge, the by-product of this process, with a relatively low cost.

Membrane Bioreactor

Membrane bioreactor (MBR) is the combination of a membrane process like microfiltration or ultrafiltration with a suspended growth bioreactor, and is now widely used for municipal and industrial wastewater treatment. The scheme of the reactor is showed in Fig. 2 [11].

The Principle of this technique is nearly the same as activated sludge process, except that instead of separation the water and sludge through settlement, the MBR method uses the membrane which is more efficient and less dependent on oxygen concentration of the water.

The MBR has a higher organic pollutant and ammonia removal efficiency in comparison with the activated sludge process. Besides, the MBR processes is capable to treat waste water with higher mixed liquor suspended solids (MLSS) concentrations compared to activated sludge process, thus reducing the reactor volume to achieve the same loading rate.

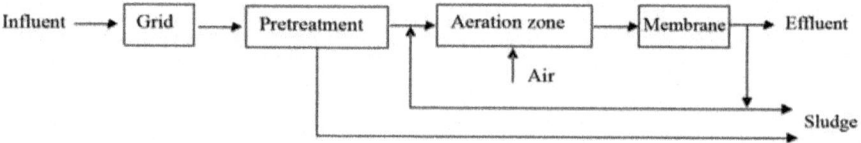

Figure 2: The scheme of the MBR reactor.

However, membrane fouling greatly affects the performance of this technique, since fouling leads to significantly increase trans-membrane pressure, which increased the hydraulic resistance time as well as the energy requirement of this reactor. Alternatively frequent membrane cleaning and replacement is therefore necessary, but it significantly increases the operating cost.

Anaerobic Biodegradation

Anaerobic degradation is a series of processes in which microorganisms break down biodegradable material in the absence of oxygen. The principle of the anaerobic degradation is as follow: first, the insoluble organic pollutant brakes down the into soluble substance, making them available for other bacteria; second, the acidogenic bacteria convert the sugars and amino acid into carbon dioxide, hydrogen, ammonia and organic acid; third, the organic acids convert into acetic acid, ammonia, hydrogen and carbon dioxide; finally, the methanogens convert the acetic acid into hydrogen, carbon dioxide and methane, a kind of gaseous fuel [9].

Anaerobic degradation processes have always been considered to be slow and inefficient, in comparison to aerobic degradation. However, the anaerobic degradation not only decreases the COD and BOD in the waste water, but also produces renewable energy. Moreover, the anaerobic bacteria could break down some persistent organic pollutants, such as lignin and high molecular weight PAH, which show little or no reaction to aerobic degradation. Besides, anaerobic processes could treat the wastewater with high loads of easy-to-degrade organic materials (wastewaters of the sugar industry, slaughter houses, food industry, paper industry, etc.) efficiently and costly. These advantages make investigation and application of anaerobic microbial mineralization in organic polluted water important.

Generally speaking, anaerobic reactor could be divided into anaerobic activated sludge process and anaerobic biological membrane process. The anaerobic activated sludge process includes conventional stirred anaerobic reactor, upflow anaerobic sludge blanket reactor, and anaerobic contact tank. The anaerobic biological membrane process includes fluidized bed reactor, anaerobic rotating biological contactor, anaerobic filter reactor. Upflow anaerobic sludge blanket reactor and anaerobic filter reactor are selected as the representative of the two kinds of reactors mentioned above.

Upflow Anaerobic Sludge Blanket Reactor (UASB)

The UASB system was developed in 1970s. No carrier is used to in the UASB system, and liquid waste moves upward through a thick blanket of anaerobic granular sludge suspended in the system. As shown in Fig. 3, mixing of sludge and wastewater is achieved by the generation of methane within the blanket as well as by hydraulic flow. And the triphase separator (gas, liquid, sludge biomass) could prevent the biomass loss of the sludge through the gas emission and water discharge. The advantage of this system are that 1) it contains a high concentration of naturally immobilized bacteria with excellent settling properties, and could remove the organic pollutants from wastewater efficiently; 2) a high concentrations of biomass can be achieved without support materials which reduces the cost of construction. These advantages would increase the efficient and stable performance of this system [12].

Anaerobic Biofilter

Anaerobic biofilter, so called anaerobic fixed film reactors, is a kind of high efficient anaerobic treatment equipment developed in 1960 s. These reactors use inert support materials to provide a surface for the growth of anaerobic bacteria and to reduce turbulence to allow unattached populations to be retained in the system (Fig 4). The organic matter of wastewater is degraded in the system, and produce methane gas, which will be released from the pool from the top [13].

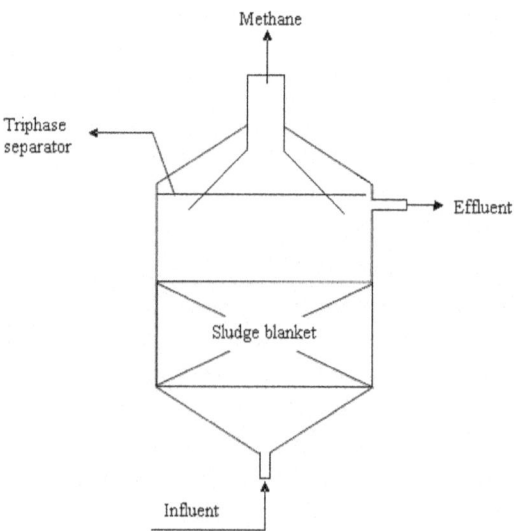

Figure 3: The scheme of upflow anaerobic sludge blanket reactor [1].

The advantages of this system are as follow: 1) the filler provides a large surface area for the growth of the microorganisms, and the filler also increases hydraulic retention time of the wastewater; 2) the system provides a large surface area for the interaction between the wastewater and film; 3) the fact that microorganisms grow on the filler reduces the run of the degraders. These advantages could increase the efficiency of this treatment, and guarantee the water quality of the effluent. The backward of this system is that the system could be blocked when dealing with high concentration organic water, especially in the water inlet parts. And no simple and effective way for filter washing has been developed yet.

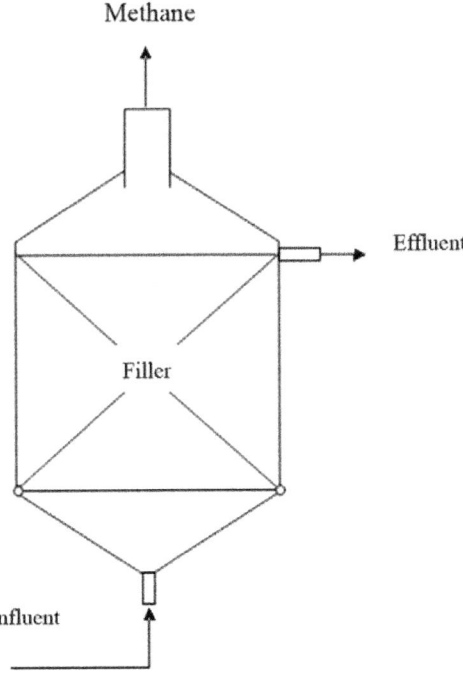

Figure 4: The scheme of the anaerobic biofilter [1].

Combination of the Aerobic and Anaerobic Biodegradation

Compared with the single anaerobic and aerobic reactors, the combination of the anaerobic and aerobic reactor is more efficient in organic pollutants degradation. The advantages of the combined system are as follow: 1) the anaerobic process could get rid of the organic matters and suspended solid from the wastewater, reduce the organic load of the aerobic degradation as well as the production of aerobic sludge, and finally reduce the volume of the reactors; 2) wastewater pretreated by anaerobic technology is more stable, indicating that anaerobic process could reduce the load fluctuation of the wastewater, and therefore decrease the oxygen requirement of the aerobic degradation; 3) the anaerobic process could modify the biochemical property of the wastewater, making the following aerobic process more efficient. Investigation showed that the wastewater from aerobic-anaerobic combined reactor are more stable and ready for degradation, indicating that this technical have a huge potential for application. The classic aerobic-anaerobic reactors include A/O reactor, A2/O reactor, oxidation ditch, constructed wetland.

Two classic aerobic biodegradation reactors, oxidation ditch and constructed wetland are introduced.

Oxidation Ditch

The oxidation ditch is a circular basin through which the wastewater flows. Activated sludge is added to the oxidation ditch so that the microorganisms will digest the organic pollutants in the water. This mixture of raw wastewater and returned sludge is known as mixed liquor. The rotating biological contactors could add oxygen into the flowing mixed liquor, and they could also increase surface area and create waves and movement within the ditches. Once the organic pollutant has been removed from the wastewater, the mixed liquor flows out of the oxidation ditch. Sludge is removed in the secondary settling tank, and part of the sludge is pumped to a sludge pumping room where the sludge is thickened with the help of aerator pumps [14]. Some of the sludge is returned to the oxidation ditch while the rest of the sludge is sent to waste.

The oxidation ditch is characterized by simple process, low maintain consumption, steady operation, and strong shock resistance. The effluent of the system has high water quality effluent with low concentration of organic pollutants, nitrogen and phosphorus. However, the problems of this reactor, such as sludge expansion, rising sludge and foam, are important factors which confines the development of this technique.

Constructed Wetland

A constructed wetland is an artificial wetland which could wetlands act as a biofilter, removing sediments and pollutants such as heavy metals and organic pollutants from the water. Constructed wetland is a combination of water, media, plants, microorganisms and other animals. Constructed wetlands are of two basic types: subsurface-flow and surface-flow wetlands [15].

Physical, chemical, and biological processes combine in wetlands to remove contaminants from wastewater. Besides absorbing heavy metals and organic pollutants (especially POPs) on the filler of the constructed wetland, plants can supply carbon and other nutrients such as nitrogen through their roots to for the growth and reproduction of the microorganisms. Plants could also pump oxygen to form an aerobic and anaerobic area in the deep level of constructed wetland to assist the breaking down of organic materials. The major reactor in constructed wetland was supposed to be microorganisms, and microorganisms and natural chemical processes are responsible for approximately 90 percent of pollutant removal, while, the plants remove about 7-10 percent of pollutants. In addition to organic pollutants, this device could remove the nitrogen and phosphorous in the wastewater and prevent eutrophication.

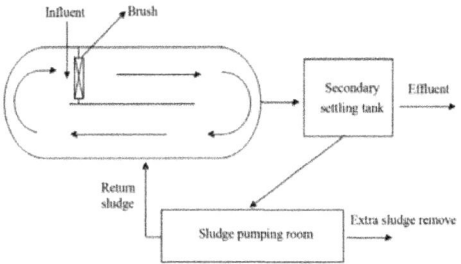

Figure 5: The scheme of the oxidation ditch [1].

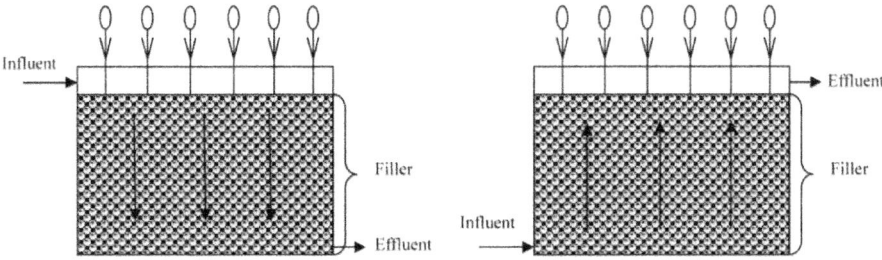

Figure 6: The scheme of the subsurface flow constructed wetland of different feeding pattern.

As an economical, easy management and ecological friendly reactor, constructed wetland is supposed to be a promising technique to treat the wastewater in developing country. However, this technique was not widely used up to now for 1) the plants couldn't adapt to heavy contaminated wastewater, which strikes its application scope; 2) the device of this technique demands large area of land; 3) the efficiency of this device relativity lower than other biological device such as activated sludge process and membrane bioreactor. Thus, efforts should be made in plants selection, device structure modification and multiple devices combination to enhance the adaption and efficiency of this technique.

CHEMICAL OXIDATION TECHNOLOGIES

Nowadays, due to the increasing presence of molecules, refractory to the microorganisms in the wastewater streams, the conventional biological methods cannot be used for complete treatment of the effluent and hence, introduction of newer technologies to convert it into less harmful or lower chain compounds which can be then treated biologically, has become imperative. Chemical oxidation technology is one of these newer technologies which use chemical oxidant (H_2O_2, O_3, ClO_2, K_2MnO_4, K_2FeO_4 and so on) oxide pollutant to slightly toxic, harmless substances or transform it into manageable form. However, Chemical oxidation technologies constitute the use of oxidizing agents such as ozone and hydrogen peroxide, exhibit lower rates of degradation. Therefore, advanced oxidation processes (AOPs) with the capability of exploiting the high reactivity of hydroxyl radicals in driving oxidation have emerged a promising technology for the treatment of wastewaters containing refractory organic compounds. Several technologies like Fenton, photo-Fenton, wet oxidation, ozonation, photocatalysis, etc. are included in the AOPs and their main difference is the source of radicals.

Chemical Oxidation Technologies under Normal Temperature and Pressure

This part aims at highlighting three different oxidation processes operating at ambient conditions viz. Fenton's chemistry (belonging to the class of AOPs) and ozonation, use of hydrogen peroxide (belonging to the class of chemical oxidation technologies).

Classification and Principle

Hydrogen Peroxide

Hydrogen peroxide (H_2O_2) is an environment friendly oxide which could oxidate organic pollutants efficiently and economically. The standard reduction potentials (1.77V, 0.88v) of hydrogen peroxide imply that it is a strong oxidant in both acidic and basic solutions [16]. It can oxidize many kinds of organic contaminants in wastewater directly. The very slow decomposition rate of hydrogen peroxide in the drinking water treatment, with mild operation conditions, can ensure a longer disinfection. Also, it can be utilized as a dechlorination agent (reductant) without organic halogen compounds production. Therefore, the hydrogen peroxide is the ideal drinking water pre-oxidant and disinfectant.

$$H_2O_2 + 2H^+ + 2e^- \rightarrow 2H_2O \quad E° = 1.77V$$

$HO_2^- + H_2O + 2e^- \rightarrow 3OH^-$ $E^\circ = 0.87V$

However, considering for the removal of organic compounds in wastewater, the reactivity of hydrogen peroxide is generally low and largely incomplete due to kinetics, in particular in acidic media. It can be enhanced by homogeneous and/or heterogeneous catalysts, the progress named wet hydrogen peroxide catalytic oxidation (WHPCO). WHPCO operates at temperatures in the 20-80°C range and atmospheric pressure.

Fenton

The Fenton's process has its origin in the discovery reported in 1894 that ferrous ion strongly promotes the hydrogen peroxide oxidation of tartaric acid. The mechanism of the Fenton's process is quite complex, and some papers can be found in the literature where tens of equations are used for its description. Nevertheless, it can be summarized by the following steps: first, a mixture of H_2O_2 and ferrous iron in acidic solution generates the hydroxyl radicals which will subsequently attack the organic compounds present in the solution [17].

$Fe^{2+} + H_2O_2 \rightarrow Fe^{3+} + HO^- + HO\bullet$

As iron (II) acts as a catalyst, it has to be regenerated, which seems to occur through the following scheme:

$Fe^{3+} + H_2O_2 \leftrightarrow Fe\text{-}OOH_2^+ + H^+$

$Fe\text{-}OOH_2^+ \rightarrow Fe^{2+} + HO_2\bullet$

The important mechanistic feature of the Fenton reaction is that in the outer-sphere single electron transfer from Fe^{2+} to H_2O_2 and generates hydroxyl radicals and hydroxide anions. Hydroxyl radicals are after fluorine atoms the most oxidizing chemical species. They are extremely powerful species to abstract one electron from an electron rich organic substrate or any other species present in the medium to form hydroxide anion. The oxidation potential of hydroxyl radicals has been estimated as +2.8 and +2.0V at pH 0 and 14, respectively. The high reactivity of $HO\bullet$ ensures that it will attack a wide range of organic compounds. Fenton reaction gives rise to CO_2 and the heteroatoms also form the corresponding oxygenated species such as NOx, SOx and POx, meaning that the carbons and heteroatoms of the organic substrate are converted to inorganic species. Equations illustrate the cyclic processes occurring in Fenton chemistry under aerobic conditions leading to the formation of CO_2.

$RH + HO\bullet \rightarrow R\bullet + H_2O$

$R\bullet + Fe^{3+} \rightarrow R^+ + Fe^{2+}$

$R^+ + H_2O \rightarrow ROH + H^+$

$R\bullet + Fe^{2+} \rightarrow$ products $+ Fe^{3+}$

$R\bullet + O_2 \rightarrow ROO\bullet$

$R\bullet + \bullet OOH \rightarrow RO\bullet + \bullet OH$

$ROO\bullet + RH \rightarrow ROOH + R\bullet$

$ROO\bullet + Fe^{2+} \rightarrow$ products $+ Fe^{3+}$

$ROO\bullet + Fe^{3+} \rightarrow$ products $+ Fe^{2+}$

The performance of Fenton oxidation application to wastewater treatment was based on the following parameters: operating pH, amount of ferrous ions, concentration of hydrogen peroxide, initial concentration of the pollutant, type of buffer used for pH adjustment, operating temperature and chemical coagulation. The optimum pH has been observed to be 3 in the majority of the cases. The pollutant removal efficiency increases with an increase in the dosage of ferrous ions and hydrogen peroxide. However, care should be taken while selecting the dosage, for high dosage leasing environmental question and high treatment cost. The optimum dosage is available in the open literature or required to establish in laboratory scale studies under similar conditions.

The conventional Fenton reaction that hydrogen peroxide in conjunction with an iron(II) salt to produce high fluxes of hydroxyl radicals is homogeneous catalytic reaction. Therefore, the application of conventional Fenton reaction is complicated by the problems typical of homogeneous catalysis, such as catalyst separation, regeneration, etc. It is necessary to control pH carefully to prevent precipitation of iron hydroxide. Thus, heterogeneous catalysts Fenton reaction, i.e., solids containing transition metal cations (mostly iron ions) have been developed and tested [18].

Ozonation

Ozone is one of the most powerful oxidants with an oxidation potential of 2.07 V. In acidic conditions, ozone undergoes selective electrophilic attack which occurs at particular parts of complexing agent with high electronic density. Under alkaline environment, ozone is catalyzed by OH^- in basic conditions to intermediate compounds such as superoxide, HO radicals and HO_2 radicals which are highly reactive. Apart from pH, the degradation of target compounds in the liquid phase corresponds to the amount and form (species) of oxidants present in a reactor [19].

$O_3 + OH^- \rightarrow HO_2^- + O_2$

$O_3 + OH^- \rightarrow HO\bullet + O_3^-$

$O_3 + HO_2^- \rightarrow HO_2\bullet + O_3\bullet$

The applications of ozonation for water treatment offer various advantages. Due to its short half-life of less than 10 min, the oxidant degrades most of pollutants rapidly. However, at pH 10, the half-life of ozone in solutions is less than 1 min. As a result, ozonation extensively consumes energy, thus reducing its treatment efficiency. Due to the improvement in ozone production from pure oxygen and the increase of its concentration in the feeding gas, an ozone generation with less cost may be economically attractive.

The performance of ozonation application to wastewater treatment was based on the following parameters: operating pH, ozone partial pressure, contact time and interfacial area, presence of radical scavengers, operating temperature, presence of catalyst, combination with other oxidation processes.

Very low reaction rates have been observed for the degradation of complex compounds or mixture of contaminants by ozonation alone. Catalyst such as BST catalyst TiO_2 fixed on alumina beads, Fe (II), Mn (II) can be used to increase the degradation efficiency. Heterogeneous catalytic ozonation has received increasing attention in recent years due to its potentially higher effectiveness in the degradation and the mineralization of refractory organic pollutants and a lower negative effect on water quality. The major advantage of a heterogeneous over a homogeneous catalytic system is the ease of catalytic retrieval from the reaction media. Results suggest that catalytic ozonation with MnOx/MZ, CoOx/MZ and $CuOx/Al_2O_3$ is a promising technique for the mineralization of refractory organic compounds in water [20].

Reactors

Typical reactor used for Hydrogen peroxide

Introduction of hydrogen peroxide into the waste stream is critical due to lower stability of hydrogen peroxide. An addition point should give large residence time of H_2O_2 in the pollutant stream, but due to the practical constraints and poor mixing conditions, it is not always possible to inject H_2O_2 in line and an additional holding tank is required. The simplest, faster and cheapest method for injection of hydrogen peroxide is gravity feed system. Pump feed systems can also be used, but it requires regular attention.

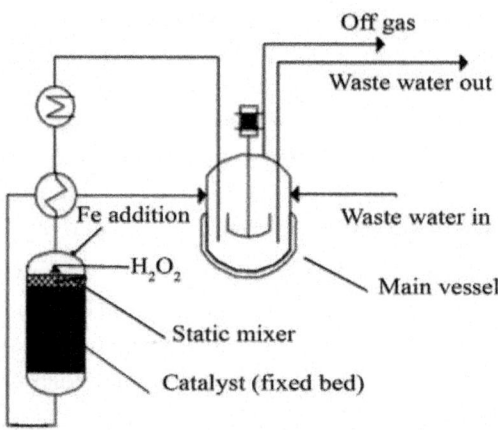

Figure 7: Typical reactor used for WHPCO technology.

Figure 7 reports a simplified flow diagram of the WHPCO technology for the treatment of olive oil milling waste water using Fe-ZSM-5 solid catalysts. H_2O_2 is added progressively at the top of a fixed bed catalytic reactor (before a static mixer), in order to maximize its local concentration. An iron solution is added on the top of the reactor to maintain catalyst activity constant. The feed solution is recirculated to and from a tank in order to have good turbulence in the catalyst bed, but also to guarantee the necessary total residence time to obtain the required level of removal of phytotoxic chemicals.

Typical Reactor used for Fenton Oxidation

A batch Fenton reactor essentially consists of a nonpressurized stirred reactor with metering pumps for the addition of acid, base, a ferrous sulfate catalyst solution and industrial strength (35-50%) hydrogen peroxide. The reactor vessel should be coated with an acid-resistant material, because the Fenton reagent is very aggressive and corrosion can be a serious problem. pH of the solution must be adjusted at 6, usually iron hydroxide is formed. For many organic pollutants, the ideal pH for the Fenton reaction is between 3 and 4, and the optimum catalyst to peroxide ratio is usually 1:5 wt/wt. Addition of reactants are done in the following sequence: dilute sulfuric acid catalyst in acidic solutions, pH adjusting agent (adjustment of pH at 3-4) and lastly added hydrogen peroxide slowly. Effluent of the Fenton reactor (Oxidation tank) is fed into a neutralizing tank for adjusting the pH (adjustment of pH at 9), then the stream followed by a flocculation tank and a solid-liquid separation tank for removing the precipitate. A schematic representation of the Fenton oxidation treatment has been shown in Figure 8 [21].

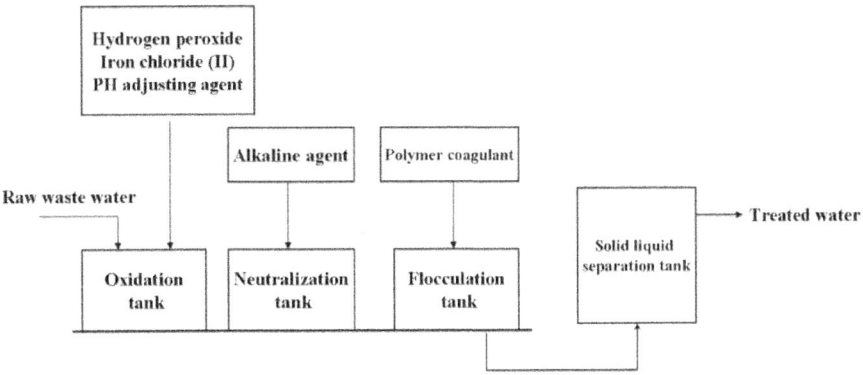

Figure 8: Typical reactor used for fenton oxidation.

Typical Reactor used for Ozonation

Ozone transfer efficiency should be maximized by increasing the interfacial area of contact (reducing the bubble size by using small size ozone diffusers such as porous disks, porous glass diffusers, ceramic membranes) and increasing the contact time between the gas and the water (increase the depths in the contactor, optimum being 3.7 to 5.5 m) [21].

Application

Hydrogen Peroxide

Hydrogen peroxide has been used in the industrial effluent treatment for detoxification of cyanide, nitrite and hypochlorite, for the destruction of phenol aromatics, formaldehyde, removal of sulfite, thiosulfate and sulfide compounds.

However, the application of hydrogen peroxide alone for wastewater treatment applications present major problems such as very low rates for applications involving complex materials, stability of H_2O_2 and mass transfer limitations. Hence, use of hydrogen peroxide alone does not seem to be a recommendable option for industrial wastewater treatment.

WHPCO process has been proposed for a variety of agro-food and industrial effluents: removal of dyestuffs from textile, treat sewage sludge, purify wastewater from pharmaceutical and chemical production, dumping site, or from cellulose production and pre-treat water streams from food-processing industries (olive oil mills, distilleries, sugar refineries, coffee production, tanneries, etc.) [22].

Fenton

Fenton process can significantly remove recalcitrant and toxic organic compounds, and increase the biodegradability of organic compounds. Leachate quality in terms of organic content, odor, and color can be greatly improved following Fenton treatment. Fenton's reagent has been used quite effectively for the treatment and pre-treatment of leachate from composting of different wastes. Reported COD removal efficiencies range from 45% to 85%, and reported final BOD_5/COD ratio can be increased from less than 0.10 initially to values ranging from 0.14 to more than 0.60, depending on leachate characteristic and dosages of Fenton reagents. Color and odor in leachate can also be reduced considerably. The decolorization efficiency is as high as 92% in Fenton treatment of a mature leachate[23]. The optimal conditions for Fenton reaction were found at a ratio $[Fe^{2+}]/[COD]$ equal to 0.1. Both leachates were significantly oxidized under these conditions in terms of COD removal 77-75% and BOD_5 removal 90-98%. Fenton's reagent was found to oxidize preferably biodegradable organic matter of leachate [24]. Pirkanniemi et al. (2007) [25] tested the Fenton's oxidation to degrade complexing agents such as N-bis[2-(1,2-dicarboxyethoxy) ethyl)] glysine (BCA5), N-bis[2-(1,2-dicarboxyethoxy)ethyl]aspartic acid (BCA6) and EDTA from bleaching wastewater. It was reported that an almost complete removal of EDTA was attained at its concentration of 76 mM.

Ozonation

Ozone can be used for treatment of effluents from various industries relating to pulp and paper production (bleaching and secondary effluents), Shale oil processing, production and usage of pesticides, dye manufacture, textile dyeing, production of antioxidants for rubber, pharmaceutical production etc.

Beltrán et al. (2006) [26] reported that ozonation alone improved the removal of succinic acid up to 65% at pH 7 with an initial concentration of 339 mM. Decolorization of dye Methylene Blue can be achieved by ozonation. The COD of basic dyestuff wastewater was reduced to 64.96% and decolorization was observed under basic conditions (pH 12), complete Methylene Blue degradation occurring in 12 min. The decolorization time decrease linearly with the increase in ozone concentration. For example, increasing ozone concentration from 4.21 g/m^3 to 24.03 g/m^3 in the gas phase reduces the decolorization time of 400 mg/L dye concentration by about 88.43% [27].

Chemical Oxidation Technologies under High Temperature and Pressure

Classification and Principle

Wet Air Oxidation (WAO)

WAO is based on the oxidizing properties of air's oxygen. Typical conditions for wet oxidation range from 180 °C and 2 MPa to 315 °C and 15 MPa. Residence times may range from 15 to 120 min, and the chemical oxygen demand (COD) and total organic carbon (TOC) removal may typically be about 75-90%. Insoluble organic matter is converted to simpler soluble organic compounds without emissions of NOx, SO_2, HCl, dioxins, furans, fly ash, etc.

$$O_2 + 4H^+ + 4e^- \rightarrow 2H_2O \quad E° = +1.23V$$
$$O_2 + 2H_2O + 4e^- \rightarrow 4OH^- \quad E° = +0.40V$$

Catalytic Wet Air Oxidation (CWAO)

Organic pollutant is impossible to obtain a complete mineralization of the waste stream by WAO, since some low molecular weight oxygenated compounds (especially acetic and propionic acids, methanol, ethanol, and acetaldehyde) are resistant to oxidation. Organic nitrogen compounds are easily transformed into ammonia, which is also very stable in WAO conditions. Therefore, WAO is a pre-treatment of liquid wastes which requires additional treatment. The use of catalysts (WACO) allows to use milder reaction conditions but especially to promote conversion of the reaction intermediates (for example, acetic acid and ammonia) which are very difficult to convert in the absence of catalysts, as mentioned above.

Though it varies with type of wastewater, the operating cost of CWAO is about half that of non-catalytic WAO due to milder operating conditions and shorter residence time. Although the homogenous catalysts, e.g. dissolved copper salts, are effective, an additional separation step is required to remove or recover the metal ions from the treated effluent due to their toxicity, and accordingly increases operational costs. Thus, the development of active heterogeneous catalysts has received a great attention because a separation step is not necessary. Various solid catalysts including noble metals, metal oxides, and mixed oxides have been widely studied for the CWAO of aqueous pollutants. To further decrease the reaction temperature and pressure, intensive oxidants are added and form Wet Peroxide Oxidation (WPO), WHPCO is belonging to WPO.

Reactors

WAO Reactors

The experimental set up consisted mainly of a reactor and a condenser. It was equipped with suitable measuring devices, such as thermocouple, rotameter and pressure gauge. The material of construction for reactor is titanium. The top of the reactor is connected to a reflux condenser with a stainless steel flange. The reactor was equipped with a heating jacket and a gas sparger. The gas (air or oxygen) entered the reactor through the titanium sparger. Air or oxygen bubbled out through the sparger at high speed and thus ensured proper agitation [28].

CWAO Reactors

Homogeneous catalysts for CWAO are usually transition metal cations, such as Cu and Fe ions. Industrial homogenous CWAO processes have been developed such as the Ciba-Geigy/Garnit process working at high temperature (300 °C), and the LOPROX Bayer process working with oxygen below 200 °C in the presence of iron ions. Common two-phase reactor types used in homogeneous CWO include bubble columns, jet-agitated reactors, and mechanically stirred reactor vessels.

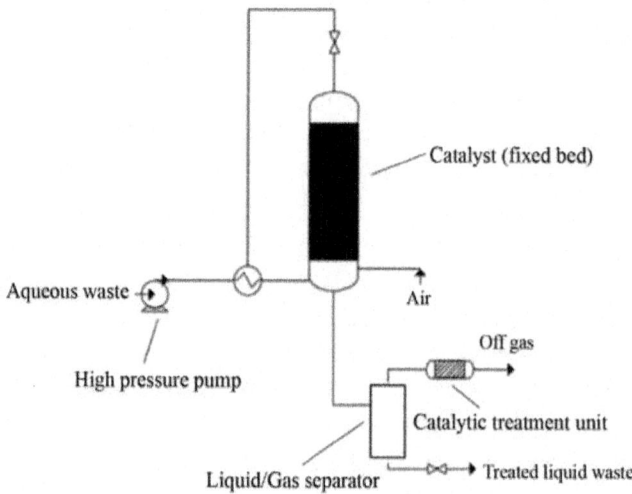

Figure 9: Typical reactor used for WACO process.

Figure 9 reports a simplified flow diagram of a WACO process which consists mainly of a high-pressure pump, an air or oxygen compressor, a heat-exchanger, a high-pressure (fixed bed) reactor and a downstream separator. The simplest reactor design is usually a cocurrent vertical bubble column with a height-todiameter ratio in the range of 5-20. A catalytic unit for the treatment of the off-gas is also typically necessary [25].

Application

WAO is not used as a complete treatment method, but only as a pretreatment step where the wastewater is rendered to nontoxic materials and the COD is reduced for the final treatment. For integrated WAO-biological treatment process, more detailed studies concerning the WAO pretreatment step are necessary for the design of a rational and efficient integrated process. The WAO process has been subjected to numerous investigations by researchers in the past decades as a pretreatment step before the biological treatment [29-32].

Pretreatment of Afyon (Turkey) alcaloide factory wastewater, a typical high strength industrial wastewater (COD= 26.65 kg/m^3; BOD$_5$= 3.95 kg/m^3) was carried out by WAO process. Experimental results indicated that over 26% COD removal of the wastewater could be achieved in 2.0 h of reaction time at 150°C; 0.65 MPa and with an airflow rate of 1.57×10^5 m^3/s. BOD$_5$/COD ratio is increased from 0.15 to 0.4. The experimental data also revealed that the pressure and temperature effects on the COD removal were important. The COD removal was observed to increase with an increase in both pressure and temperature. Maximum COD removal was obtained at around pH 7.0 [28].

As considering for using the CWAO process to treat Afyon (Turkey) alcaloide factory wastewater, results indicate that the presence of catalyst increases the COD removal. The COD removal for 2.0 h reaction time increased from 25.7% without catalyst to 33.2% with 0.25 kg/m^3 catalyst. While the BOD$_5$/COD ratio is increased to over 0.4 [28].

CWAO makes a promising technology for the treatment of refractory organic pollutants (phenolic compounds, carboxylic acids, N-containing compounds) in industrial wastewaters, such as Olive oil mill wastewater, Kraft bleaching plant effluents, Coke plant wastewater, Textile wastewater, Alcohol-distillery wastewater, Landfill leachate, Pulp and paper bleaching liquor, Heavily organic halogen polluted industrial wastewater and so on [33].

ADSORPTION TECHNOLOGY

Principle of Adsorption Technology

Adsorption offers a cleaner technology, free from sludge handling problems and produces a high quality effluent. Over the last few decades, adsorption has gained importance as an effective purification and separation technique used in water and wastewater treatment. Adsorption is the process by which a solid adsorbent can attach a component dissolved in water to its surface and form an attachment via physical or chemical bonds, thus removing the component from the fluid phase. Adsorption is used extensively in industrial processes for many purposes of separation and purification. The removal of metals, coloured and colourless organic pollutants from industrial wastewater are considered an important application of adsorption processes using suitable adsorbents.

Adsorption is nearly always an exothermic process. We can distinguish between 2 types of adsorption process depending on which of these two force types plays the bigger role in the process. Adsorption processes can be classified as either physical adsorption (van der Waals adsorption) or chemisorption (activated adsorption) depending on the type of forces between the adsorbate and the adsorbent.

Physical adsorption occurs quickly and may be mono-molecular (unimolecular) layer or monolayer, or 2, 3 or more layers thick (multi-molecular). As physical adsorption takes place, it begins as a monolayer. It can then become multi-layer, and then, if the pores are close to the size of the molecules, more adsorption occurs until the pores are filled with adsorbate. Accordingly, the maximum capacity of a porous adsorbent can be more related to the pore volume than to the surface area.

Chemisorption involves the formation of chemical bonds between the adsorbate and adsorbent is a monolayer, often with a release of heat much larger than the heat of condensation. Chemisorption from a gas generally takes place only at temperatures greater than 300 °C, and may be slow and irreversible. Most commercial adsorbents rely on physical adsorption; while catalysis relies on chemisorption.

Development of Adsorption Materials

Activated Carbon

Activated carbon is by far the most common adsorbent used in wastewater treatment. Since, during adsorption, the pollutant is removed by accumulation at the interface between the activated carbon (absorbent) and the wastewater (liquid phase) the adsorbing capacity of activated carbon is always associated with very high surface area per unit volume. Activated carbon can be manufactured from carbonaceous material, including coal (bituminous, subbituminous, and lignite), peat, wood, or nutshells (i.e., coconut). The manufacturing process consists of two phases, carbonization and activation. The carbonization process includes drying and then heating to separate by-products, including tars and other hydrocarbons, from the raw material, as well as to drive off any gases generated. The carbonization process is completed by heating the material at 400–600°C in an oxygen-deficient atmosphere that cannot support combustion. Powdered activated carbon is made up of crushed or ground carbon particles, 95–100% of which will pass through a designated mesh sieve or sieves. Granular activated carbon can be either in the granular form or extruded. It is designated by sizes such as 8×20, 20×40, or 8×30 for liquid phase applications and 4×6, 4×8 or 4×10 for vapor phase applications.

Activated Alumina

Activated alumina had been used in the treatment of wastewater and its adsorption capability for the removal of both organic and inorganic compounds was found to be favoured by a specific surface area, pore structure, ionic strength and chemical inertness. It can be produced from the mixtures of amorphous and gamma alumina prepared by the dehydration of $Al(OH)_3$ under low-temperatures of 300-600°C, with surface areas in the range of 250-350 m^2/g.

Research conducted on the use of microporous alumina pillared montmorillonite (clay) and mesoporous alumina aluminium phosphate as adsorbents had shown successful removal of fluoride, arsenic, selenium, beryllium, 2,4-chlorophenol, 2,4,6- trichlorophenol, pentachlorophenol, and also pesticides such as: molinate, propazine and atrazine from waster water. The removal efficiency of the pillared clay material for the herbicide was found to be higher than that of the mesoporous aluminium phosphate due to the substitution of the alkyl lateral chains of the aluminium phosphate during the sorption of s-triazines and the increase of P/Al ratio during the adsorption of propachlor.

Zeolites

The drawback suffered by activated carbon due to its high regeneration cost and production cost has lead to the application of zeolites as an alternative adsorbent. Zeolites are a group of natural or synthetic hydrated aluminosilicate minerals which contain both alkaline and alkali-earth metals. It has been used as an adsorbent, molecular sieve, ion-exchangers and catalysts in the past decades, because their chemical properties and large effective surface area gives them superior adsorptive qualities. There are several types of zeolites such as MCM-22, ZSM-5, ZSM-22, BETA, and Y. Their adsorption equilibrium had been studied and showed that the synthetic zeolites have higher adsorption capacity than the natural zeolites for the removal of ink, dyes and polluted wastewater.

Peat

Peat and other biomass materials have been used previously in the treatment of wastewater containing heavy metals and organic compounds. Peat is a yellow to dark brown residue, which occurs during the first stage of coal formation. It is composed of partly carbonizing materials such as decayed trees and peats bogs that have accumulated in water–saturated environments and swamps. The main constituents of peat moss are lignin, humic acid and cellulose. In addition, the surface functional groups of peat include aldehydes, carboxylic acids, ketones, alcohols, ethers and phenolic hydroxides, which are all involved in the adsorption of pollutants. In addition, its polar nature is responsible for its specific adsorption potential for dissolved metals and polar organic compounds.

Natural Materials

Natural materials that are available in large quantities, or certain waste products from industrial or agricultural operations, may have potential as inexpensive adsorbents. The abundance, availability, and low cost of agricultural byproducts make them good adsorbents for the removal of various pollutants from wastewaters. Agricultural waste biomass currently is gaining importance. In this perspective rice husk, which is an agro-based waste, has emerged as an invaluable source for the utilization in the wastewater treatment. Rice husk contains ~20% silica, and it has been reported as a good adsorbent for the removal of heavy metals, phenols, pesticides, and dyes. The adsorptive capacity of rice husk silica had been evaluated by Grisdanurak et al. [34] and its adsorption capacity for chlorinated volatile organic compounds was found to be higher than that of commercial mordenite and activated carbons. It has been utilized for solving disposal problems and also as an adsorbent in treating organic wastewaters. The adsorption potential of this biomass for adsorbing phenol from aqueous solution was found to be depended on the pH, contact time and the initial phenol concentration. This result shows that phenol was adsorbed to a lesser extent at higher pH values. Phenol forms salts, which readily ionize leaving negative charge on the phenolic group while, its present on the adsorbent prevents the removal of phenolateions. In addition, the percentage adsorption of phenol for this test also decreases as the initial phenol concentration increases. The adsorption capacity determined for this test was 0.886 mg/g for phenol and the equilibrium data was fitted successfully by the Freundlich model [35].

Polymeric

Polymeric adsorbents are non-functionalized organic polymers which are capable of removing organics from water. The principle is quite simple. Wastewater is passed through a column containing the polymeric adsorbent. The organic materials are retained on the resin while water and some simple salts pass through. When the resin is fully loaded, the organics are stripped from the resin with solvents or caustic. The organic material may be concentrated by orders of magnitude in some cases. The following recommendations are those being used at the present time. The regenerants used are not the only ones possible. The choice of regenerant (solvent) usually depends on the availability at the particular location.

Adsorption Equipment and Their Applications

Granular activated carbon systems are generally composed of carbon contactors, virgin and spent carbon storage, carbon transport systems, and carbon regeneration systems. The carbon contactor consists of a lined steel column or a steel or concrete rectangular tank in which the carbon is placed to form a "filter" bed. A fixed bed downflow column contactor is often used to contact wastewater with granular activated carbon. Wastewater is applied at the top of the column, flows downward through the carbon bed, and is withdrawn at the bottom of the column. The carbon is held in place with an underdrain system at the bottom of the contactor. Provisions for backwash and surface wash of the carbon bed are required to prevent buildup of excessive headloss due to accumulation of solids and to prevent the bed surface from clogging.

There are two basic types of water filters: particulate filters and adsorptive/reactive filters. Particulate filters exclude particles by size, and adsorptive/reactive filters contain a material (medium) that either adsorbs or reacts with a contaminant in water. The principles of adsorptive activated carbon filtration are the same as those of any other adsorption material. The contaminant is attracted to and held (adsorbed) on the surface of the carbon particles. The characteristics of the carbon material (particle and pore size, surface area, surface chemistry, etc.) influence the efficiency of adsorption [36].

The characteristics of the chemical contaminant are also important. Compounds that are less water-soluble are more likely to be adsorbed to a solid. A second characteristic is the affinity that a given contaminant has with the carbon surface. This affinity depends on the charge and is higher for molecules possessing less charge. If several compounds are present in the water, strong adsorbers will attach to the carbon in greater quantity than those with weak adsorbing ability.

OTHER TECHNOLOGIES

Solvent Extraction

Solvent extraction is a common form of chemical extraction using organic solvent as the extractant. It is commonly used in combination with other technologies, such as solidification/stabilization, incineration, or soil washing, depending upon site-specific conditions. Solvent extraction also can be used as a stand alone technology in some instances. Organically bound metals can be extracted along with the target organic contaminants, thereby creating residuals with special handling requirements. Traces of solvent may remain within the treated soil matrix, so the toxicity of the solvent is an important consideration.

Solvent extraction method has many advantages, such as less investment in equipment, easy to operate and lower consumption. Moreover, the major pollutants can be effectively recycled by solvent extraction method. The extraction method is widely used in a variety of organic waste, such as phenol, organic carboxylation acids, organic phosphorus nitrogen, organic sulfonic acid, organic amine, etc. Solvent extraction has been shown to be effective in treating sediments, sludges, and soils containing primarily organic contaminants such as PCBs, VOCs, halogenated solvents, and petroleum wastes. The process has been shown to be applicable for the separation of the organic contaminants in paint wastes, synthetic rubber process wastes, coal tar wastes, drilling muds, wood-treating wastes, separation sludges, pesticide/insecticide wastes, and petroleum refinery oily wastes.

Adopting solvent extraction treatment technology for organic wastewater, the most important thing is to choose the right process flow specifically for specially appointed pollution. For the general flow, most difficult degradation of the pollutants are removed after the process of solvent extraction. Crafts residue mainly contain some pollutants which are not extractive and dissolved, and they would meet the emissions standards through the secondary treatment regeneration (such as biochemistry, chemical oxidation, etc.).

Incineration

Incineration involves the combustion of the organic (carbon-containing) solids present in wastewater solids and biosolids to form carbon dioxide and water. The temperature in the combustion zone of furnaces is typically 1023K to 1143K. The solids that remain at the end of the process are an inert material commonly known as ash. Either undigested wastewater solids or biosolids may be incinerated. The terms thermal oxidation and combustion may be used interchangeably with incineration.

Incineration takes advantage of the fuel value of wastewater treatment residual solids (referred to as sludge) and biosolids. In some cases, the energy recovered from this process has been used in heat exchangers and waste heat boilers to save on energy use at

the wastewater treatment plant. For example, in Montreal, a portion of the biosolids generated at the facility are incinerated, while the remaining portion is pelletized. Waste heat from the biosolids that are incinerated is used in the thermal dryers that produce fertilizer pellets. In Europe, there is a trend to use biosolids as a fuel source in dedicated power generation facilities. In addition, incineration results in a large reduction in volume and mass in comparison to other alternatives and options. The mass of solids in the ash that results from the inceration process is approximately 10% of that of the biosolids fed into the incinerator. This reduces the mass and volume requiring disposal.

There are two common incineration technologies for wastewater solids and biosolids: fluidized bed incinerators and multiple hearth incinerators. Fluidized bed incinerators are steel cylinders lined with refractory bricks to withstand the high operating temperatures of the unit. Multiple hearth incinerators consist of a series of refractory brick hearths, stacked vertically. A rotating shaft through the centre of the hearths supports rake arms for each hearth, thereby facilitating drying and incineration. Solids are usually fed through at the top hearth and are directed to successive inner or outer dropholes as they move down through the hearths. Most of the ash is discharged from the bottom hearth.

Over the years, incineration technologies have evolved considerably and regulations and procedures have continually been enhanced to protect human and animal health and the environment. A considerable amount of scientific study has been undertaken to support the development of the regulations, and ongoing research contributes to the continuous improvement of this practice. However, some segments of the public still have concerns that incineration may be unsafe because of perceptions related to outdated technology and to experiences with incineration of other materials such as hazardous waste, municipal solid waste and medical waste.

Photocatalysis

To date, the most widely applied photocatalyst in the research of water treatment is the Degussa P-25 TiO_2 catalyst. This catalyst is used as a standard reference for comparisons of photoactivity under different treatment conditions. The fine particles of the Degussa P-25 TiO_2 have always been applied in a slurry form. This is usually associated with a high volumetric generation rate of reactive oxygen species as proportional to the amount of surface active sites when the TiO_2 catalyst in suspension. On the contrary, the fixation of catalysts into a large inert substrate reduces the amount of catalyst active sites and also enlarges the mass transfer limitations. Immobilization of the catalysts results in increasing the operation difficulty as the photon penetration might not reach every single surface site for photonic activation. Thus, the slurry type of TiO_2 catalyst application is usually preferred. With the slurry TiO_2 system, an additional process step would need to be entailed for post-separation of the catalysts. This separation process is crucial to avoid the loss of catalyst particles and introduction of the new pollutant of contami-nation of TiO_2 in the treated water [37]. The catalyst recovery can be achieved through process hybridization with conventional sedimentation [38], cross-flow filtration [39] or various membrane filtrations [40].

Natural clays have been used intensively as the support for TiO_2 owing to their high adsorption capacity and cost-effectiveness. The use of the photocatalytic membranes has been targeted owing to the photocatalytic reaction can take place on the membrane surface and the treated water could be continuously discharged without the loss of photocatalyst particles. To broaden the photoresponse of TiO_2 catalyst for solar spectrum, various material engineering solutions have been devised, including composite photocatalysts with carbon nanotubes [41], dyed sensitizers [42], noble metals or metal ions incorporation [43], transition metals and non-metals doping [44].

Ultrasonic

High-frequency ultrasound is a mechanical wave, with a shorter wavelength, the energy concentration characteristics, its application mainly on the basis of energy major, along a straight line features of these two started. The 20th century, the early 90s, some scholars have begun to study abroad, such as the ultrasonic degradation of organic pollutants in water. Ultrasound technology is simple, efficient, non-polluting or less polluting characteristics, in recent years the development of a new type of water treatment technology. It combines advanced oxidation, pyrolysis, supercritical oxidation technology in one, and the degradation of speed, be able to water of harmful organic compounds into CO_2, H_2O, inorganic ions or organic toxic than the original readily biodegradable organic matter, and therefore in dealing with difficult Bio-degradation of organic contaminants has significant advantages.

TREATMENT PROCESSES OF VARIOUS INDUSTRIAL OR-GANIC WASTEWATERS

Coking Plant

Coke, produced by the pyrolysis of natural coals, is an indispensable material for most of the metallurgical facilities. During coking, coal decomposes into gases, liquid and solid organic compounds. Coke wastewater contains high concentration of ammonia, phenols, thiocyanate, cyanide and lower amounts of other toxic compounds, such as polyaromatic hydrocarbons (PAHs), e.g. naphthalene, and heterocyclic nitrogenous compounds, e.g. quinoline. The individual concentration of the contaminants depends on the quality of coal and the properties of the coking process.

Coke wastewater handling usually consists of a series of physico-chemical treatments reducing the concentration of ammonia, cyanide, solids and other substances, followed by different biological treatments, mainly activated sludge process. The application of two or three consecutive activated sludge systems is particularly favored as readily biodegradable substrates like phenol can be removed in the first step. Phenols, which contribute to the greatest extent to the total COD in coke wastewater, are not only highly toxic and carcinogenic compounds, but also inhibit advantageous biological processes like nitrification. Under optimal circumstances, thiocyanate degradation can also be achieved in the first activated sludge step.

The influent concentrations of NH_4^+-N, phenols, COD and thiocyanate (SCN⁻) in the wastewater ranged between 504 and 2340, 110 and 350, 807 and 3275 and 185 and 370 mg/L, respectively. A laboratory-scale activated sludge plant composed of a 20 L volume aerobic reactor followed by a 12 L volume settling tank and operating at 35 °C was used to study the biodegradation of coke wastewater. Maximum removal efficiencies of 75%, 98% and 90% were obtained for COD, phenols and thyocianates, respectively, without the addition of bicarbonate. The concentration of ammonia increased in the effluent due to both the formation of NH_4^+ as a result of SCN⁻ biodegradation and to organic nitrogen oxidation. A maximum nitrification efficiency of 71% was achieved when bicarbonate was added, the removals of COD and phenols being almost similar to those obtained in the absence of nitrification [45]. An anaerobic-anoxic-aerobic (A(1)-A(2)-O) and an anoxic-aerobic (A/O) biofilm system were used to treat coke-plant wastewater. At same or similar levels of HRT, the two systems had almost identical COD and NH_3 removals, but a different organic-N removal. Set-up of an acidogenic stage benefited for the removal of organic-N and the A(1)-A(2)-O system was more useful for total nitrogen removal than the A-O system [46]. Newly studies for treatment of coking wastewaters are listed. Chu et al. investigated coking wastewater treatment by an advanced Fenton oxidation process using iron powder and hydrogen peroxide. The results showed that higher COD and total phenol removal rates were achieved with a decrease in initial pH and an increase in H_2O_2 dosage. At an initial pH of less than 6.5 and H_2O_2 concentration of 0.3 M, COD removal reached 44-50% and approximately 95% of total phenol removal was achieved at a reaction time of 1 h. The oxygen uptake rate of the effluent measured at a reaction time of 1 h increased by approximately 65% compared to that of the raw coking wastewater. This indicated that biodegradation of the coking wastewater was significantly improved. Several organic compounds, including bifuran, quinoline, resorcinol and benzofuranol were removed completely as determined by GC-MS analysis. The advanced Fenton oxidation process is an effective pretreatment method for the removal of organic pollutants from coking wastewater. This process increases biodegradation, and may be combined with a classical biological process to achieve effluent of high quality [47].

Bioaugmented zeolite-biological aerated filters (Z-BAFs) were designed to treat coking wastewater containing high concentrations of pyridine and quinoline and to explore the bacterial community of biofilm on the zeolite surface. The investigation was carried out for 91 days of column operation and the treatment of pyridine, quinoline, total organic carbon (TOC), and ammonium was shown to be highly efficient by bioaugmentation and adsorption. This bioaugmented Z-BAF method was shown to be an alternative technology for the treatment of wastewater containing pyridine and quinoline or other N-heterocyclic aromatic compounds [48].

Textile Wastewater

Dyes and pigments have been utilized for coloring in the textile industry for many years. Several types of textile dyes are available for use with various types of textile materials. Textile wastewater contains dyes damages the esthetic nature of water and reduces light penetration through the water's surface, and also the photosynthetic activity of aquatic organisms. It also contains toxic and potential carcinogenic substances. Therefore it must be adequately treated before they can discharge into receiving water bodies. There are several applied treatment methods for textile effluents, involving biological, physical or chemical methods and combinations of these. Among the different technologies that can be applied for the treatment of textile wastewaters, Coagulation-flocculation (CF) and Activated Sludge Process (ASP) are widely used as they are efficient and simple to operate. Generally, these processes can be applied alone to remove suspended colloidal particles or as pre-treatment prior to Ultrafiltration (UF), Nanofiltration (NF) or Reverse Osmosis (RO) respectively for dissolved organic substances removal, decolorization and desalination.

Biological treatment resulted in a high percent reduction in chemical oxygen demand (COD), total Kjeldahl nitrogen (TKN), and total phosphorus (TP), and in a moderate decrease in color. The process was found to be independent of the variations in the anoxic time period studied; however, an increase in solids retention time (SRT) improved COD and color removal, although it reduced the nutrient (TKN and TP) removal efficiency. Furthermore, combined treatment (biological treatment and Fenton oxidation) resulted in enhanced color reduction [49].

The treatability of textile wastewaters in a bench-scale experimental system, comprising an anaerobic biofilter, an anoxic reactor and an aerobic membrane bioreactor (MBR) was evaluated by S. Grilli et al. The MBR effluent was thereafter treated by a nanofiltration (NF) membrane. The proposed system was demonstrated to be effective in the treatment of the textile wastewater. The MBR system achieved a good COD (90-95%) removal; due to the presence of the anaerobic biofilter, also effective color removal was obtained (70%). The addition of the NF membrane allowed the further improvement in COD (50-80%), color (70-90%) and salt removal (60-70% as conductivity). In particular the NF treatment allowed the almost complete removal of the residual color and a reduction of the conductivity such as to achieve water quality suitable for reuse [50].

Typical contaminants of wool textile effluents are heavy metal complexes with azo-dyes. One of the most representative heavy metals is chromium. In aquatic environments chromium can be present as Cr(III) and/or Cr(VI), mainly depending on pH and redox conditions; the two forms behave quite differently, since Cr(III) is much less soluble and therefore less mobile than Cr(VI). The heavy metal can not be removed by activated sludge effectively. The constructed wetlands (CWs) in full-scale systems and in pilot plants evidenced good performances for several elements, including chromium. Donatella et al investigated the fate of Cr(III) and Cr(VI) in a full-scale subsurface horizontal flow constructed wetland planted. The reed bed operated as post-treatment of the effluent wastewater from an activated sludge plant serving the textile industrial district. Removals of Cr(III) and Cr(VI) was 72% and 26%, respectively. The mean Cr(VI) outlet concentration was 1.6 ± 0.9 g/l and complied with the Italian legal limits for water reuse [51].

Food and Fermentation Wastewater

Food processing and fermentation industries have being experiencing a significant growth in China. Wastewater streams discharged from these industries are generally characterized with high strength organic and nutrient contents, e.g., COD 10000 mg/L, TN 600 mg/L, and tend to bring serious water environment contamination if discharged without proper treatment. The conventional treatment of this kind of high strength wastewater is anaerobic/ aerobic activated sludge processes.

Recent years, considerable concern has been focused on the development of the anaerobic membrane bioreactor (AMBR), which is an anaerobic reactor coupled with a membrane filtration unit. The viability of the AMBR treating high-concentration food wastewater depended upon feedwater organic concentration, loading rate, HRT, SRT, hydraulic shearing effect and membrane properties. The HRT kept at 60 h, SRT was designed for 50 days. The effluent COD removal achieved above 90% at loading rate of 2.0 kg/m^3/d and above 80% at a loading of 2.0-4.5 kg/m^3/d. The membranes all exhibited high efficiency in removal of SS, color, COD and bacteria, reaching 499.9%, 98%, 90%, and 5 logs, respectively [52].

Wang et al. [53] applied an anoxic/aerobic membrane bioreactor (MBR) to simultaneous removals of nitrogen and carbon from food processing wastewater. The system is proposed to be applied jointly with anaerobic pre-treatment. In order to simulate the quality from anaerobic pre-treatment, raw wastewater taken from a food processing factory was fed to the system after dilution. By continuous runs under appropriate operational conditions, COD, NH_4^+-N and TN removal was over 94, 91 and 74%, respectively. The anoxic reactor and aerobic MBR contributed 40-63 and 29-46% to COD removal, and 31-43 and 47-64% to NH_4^+-N removal, respectively. The maximum volumetric COD and TN loadings as high as 3.4 kg COD/m^3/day and 1.26 kg N/m^3/day were achieved.

Food processing and fermentation wastewaters can be characterized as nontoxic because they contain few hazardous compounds, have high BOD_5 and much of the organic matter in them consists of simple sugars and starch. Hence, this high-carbohydrate wastewater is the most useful for industrial production of hydrogen. Food Wastewaters obtained from four different food-processing industries had COD of 9 g/L (apple processing), 21 g/L (potato processing), and 0.6 and 20 g/L (confectioners A and B). Biogas produced from all four food processing wastewaters consistently contained 60% hydrogen, with the balance as carbon dioxide. COD removals as a result of hydrogen gas production were generally in the range of 5-11%. Overall hydrogen gas conversions were 0.7-0.9 L H_2/L-wastewater for the apple wastewater, 0.1 L/L for Confectioner-A, 0.4-2.0 L/L for Confectioner B, and 2.1-2.8 L/L for the potato wastewater [54].

Hydrogen yields were 0.61-0.79 mol/mol for the food processing wastewater (Cereal), ranged from 1 to 2.52 mol/mol for the other samples. A maximum power density of $8177mW/m^2$ (normalized to the anode surface area) was produced using the two-chambered MFC and the Cereal wastewater (diluted 10 times to 595 mg COD/L), while at the same time the final COD was reduced to lower 30 mg/L (95% removal). Although more studies are needed to improve hydrogen yields, these results suggest that it is possible to link a MFC to biohydrogen to recover energy from food processing wastewaters, providing a new method to offset wastewater treatment plant operating costs [55].

Pharmaceutical Wastewater

The pharmaceutical manufacturing industry produces a wide range of products to be used as human and animal medications. Treatment of pharmaceutical wastewater is troublesome to reach the desired effluent standards due to the wide variety of the products produced in a drug manufacturing plant, thus, variable wastewater composition and fluctuations in pollutant concentrations. The substances synthesized in a pharmaceutical industry are structurally complex organic chemical that are resistant to biological degradation. Soluble COD removal efficiency is about 62% at 30°C. Therefore, there is a need for advanced oxidation methods. As the process costs may be considered the main obstacle to their commercial application. Cost-cutting approaches have been proposed, such as combining AOP and biological treatment.

Fenton's oxidation is very effective method in the removal of many hazardous organic pollutants from wastewaters. Fenton's oxidation can also be an effective pretreatment step by transforming constituents to by-products that are more readily biodegradable and reducing overall toxicity to microorganisms in the downstream biological treatment processes.

Optimum pH was determined as 3.5 and 7.0 for the first (oxidation 30 min) and second stage (coagulation 30 min) of the Fenton process, respectively. For all chemicals, COD removal efficiency was highest when the molar ratio of H_2O_2/Fe^{2+} was 150-250. At H_2O_2/Fe^{2+} ratio of 155, 0.3M H_2O_2 and 0.002M Fe^{2+}, Fenton process provided 45-65% COD removal (influent COD 35000-40000 mg/L) [56].

Real pharmaceutical wastewater containing 775 mg dissolved organic carbon (3324 mg COD) per liter was treated by a solar photo-Fenton/biotreatment. The photo-Fenton treatment time (190 min) and H_2O_2 dose (66 mM) necessary for adequate biodegradability of the wastewater. And biological treatment was able to reduce the remaining dissolved organic carbon to less than 35 mg/L. Overall dissolved organic carbon degradation efficiency of the combined photo-Fenton and biological treatment was over 95%, of which 33% correspond to the solar photochemical process and 62% to the biological treatment [57]. Due to the high COD concentration in pharmaceutical wastewaters, anaerobic processes have been made to utilize, such as upflow anaerobic sludge blanket (UASB) reactor, anaerobic filter (AF), anaerobic continuous stirred tank reactor (CSTR) and a hybrid reactor combining UASB and AF. The COD reduction of anaerobic process treating pharmaceutical wastewater containing macrolide antibiotics was 70-75%, at a total HRT of 4 d and OLR of 1.86 kg $COD/m^3/d$ [58].

The two-phase anaerobic digestion (TPAD) system comprised a CSTR and a UASBAF reactor, working as the acidogenic and methanogenic phases, respectively. The wastewater was high in COD, varying daily between 5789 and 58,792mg/L, with a wide range of pH from 4.3 to 7.2. Almost all the COD was removed by the TPAD-MBR system, leaving a COD of around 40mg/L in the MBR effluent, at respective HRTs of 12, 55 and 5 h. The pH of the MBR effluent was found in a narrow range of 6.8-7.6, indicating that the MBR effluent can be directly discharged into natural waters. As demonstrated by an overall COD removal efficiency of more than 99% [59].

Sugar Refinery Wastewater

Sugar refineries generate a highly coloured effluent resulting from the regeneration of anion-exchange resins (used to decolourize sugar liquor). This effluent represents an environmental problem due to its high organic load, intense colouration and presence of phenolic compounds. The colored nature of the effluent is mainly due to (1) the presence of melanoidins, that are brown polymers formed by the Maillard amino-carbonyl reaction and (2) the presence of thermal and alkaline degradation products of sugars (e.g. caramels). Most of the organic matter present in the effluent can be reduced by conventional biological treatments but the colour is hardly removed by these treatments.

The remaining colour can lead to a reduction of sunlight penetration in rivers and streams which in turn decreases both photosynthetic activity and dissolved oxygen concentrations causing harm to aquatic life. *P. chrysosporium* can remove color and total phenols from the sugar refinery effluent. A rotating biological contactor (RBC) containing *P. chrysosporium* immobilized on polyurethane foam (PUF) disks was operated with optimized decolourization medium, in continuous mode with a retention time of 3 days. During the course of operation the color, total phenols and chemical oxygen demand were reduced by 55, 63 and 48%, respectively. Addition of glucose was obligatory and the minimum glucose concentration was found to be 5 g/L [60].

Wastewater obtained from Guangxi Nanning sugar refinery (COD 86.02

g/L) is first diluted by 100 times, then treated by adding amphiphilic flocculants (CMTMC) mg/L at pH 6.6, COD removal to reached to 95%. The wastewater color changed from fuscous brown to buff yellow. After flocculation and purification, the treated water could reach the national first level discharge standards. (GB8978-88, China) [61].

Sugar refinery wastewater containing high organic load can be used as carbon sources for hydrogen production by microorganisms. As reported pretreated sugar refinery wastewater was used for the production of hydrogen by *Rhodobacter sphaeroides* O.U.001. Hydrogen was produced at a rate of 0.001 L hydrogen/h/L culture in 20% dilution of the wastewater. To adjust the carbon concentration to 70 mM and nitrogen concentration to 2 mM, sucrose or L-malic acid was added as carbon source and sodium glutamate was added as nitrogen source to the 20% dilution of SRWW. By these adjustments, hydrogen production rate was increased to 0.005 L hydrogen/h/L culture [62].

THE COST ACCOUNTING OF DIFFERENT ORGANIC WASTEWATER TREATMENT

The cost of organic wastewater treatment includes two parts: the capital expenditure and the operation expenditure. The total cost relates to the characters of the influent, the technique we selected, the characters of the effluent, the time cost during the treatment etc. In this section, the pollutants are divided into degradable and reluctant ones. Some typical wastewater was selected in each group, and the feasible methods to treat it and their cost were discussed.

The Degradable Organic Pollutants

Wastewater with degradable organic pollutants usually comes from domestic sewage, food processing, breeding industry etc. This wastewater has high BOD, and could break down in the nature condition, given enough time. Most of the techniques could be used to treat the degradable organic pollutants, and biological methods are favorite because of their efficiency and economic properties.

Sewage is one of the most important sources of degradable organic pollutants, which contributes to 37.5% of total COD in China in 2011. Therefore, sewages are treated before discharge in order to reduce the impact of the pollutants to the environment. Several biological methods, including aerobic biodegradation, activated sludge reactor, membrane bioreactor, constructed wetland etc., have been used in the sewages treatment, and their efficiency and cost have been compared. Taking the research of Song as an example[63], in response to the characteristics of decentralized domestic sewage, several treatment technologies including biogas purification tank, constructed wetland, viewing earthworm ecological, high rate algal pond, membrane bioreactor and integrated treatment equipment were applied to the domestic sewage, and their efficiency and cost were calculated and showed in table 1.

Table 1: The load, cost accounting and effluent of different treatments

Treatment	Load (m³/d)	Capital expenditure (10⁴Yuan/ m³)	Operation expenditure (10⁴Yuan/ m³)	Quality of the effluent (GB18919- 2002)
Biogas purification tank	20-200	0.06-0.08	0.02-0.05	2nd grade
Constructed wetland	30-3100	0.06-0.2	0.05-0.2	1st grade B
Viewing earthworm ecological	2-12	0.7-2.0	0.5-1.2	1st grade B
High rate algal pond	-	-	-	1st grade B
Membrane bioreactor	5-100000	0.19-1.0	0.25-1.05	1st grade B
Integrated treatment equipment	20-	1.0-1.5	0.27-0.8	1st grade A

According to the technologies, the biogas purification tank, constructed wetland, viewing earthworm ecological and high rate algal pond were characterized by low investment, operating cost, and convenient management. The membrane bioreactor and integrated treatment equipment had the higher operating cost, and the need for professional management, which could be used in the area with higher economic development and stricter effluent qualities.

The industrial waste water from agricultural and sideline food processing industry contain high concentration of organics and suspended substance. Food wastewater is composited of natural organic matters (such as protein, fat, sugar, starch), so they are of low toxicity and high BOD/COD value (up to 0.84). Physical (such as adsorption, air flotation), chemical (flocculation) and biological methods (aerobic biodegradation, activated sludge reactor, sequencing batch reactor, oxidation pond) could be used to remove the pollutants. Most of the physical and chemical techniques are costly and need secondary treatment, therefore, food wastewater was mainly treated by biological methods. The cost varied greatly with the characters of the influent. Longda food industry compared the load and cost of oxidation pond and sequencing batch reactor, results were shown in the table 2 [64].

Table 2: The load, cost accounting of oxidation pond and sequencing batch reactor

Treatment	Design capacity (m³/d)	Wastewater quantity (m³/a)	Total cost (Yuan/m³)	Electricity consumption (kwh/m³)
Oxidation pond	6500	1985300	0.56	0.335
Sequencing batch reactor	4500	1114200	0.455	0.25

The Reluctant Organic Pollutants

The reluctant organic pollutants, including benzene series, pharmaceutical intermediates, pesticide etc., mainly come from paper making industry, chemical industry, printing and dyeing wastewater, mechanical manufacturing industry, and agriculture [65]. This kind of wastewater is reluctant to biodegradation either owed to its toxicity or stable structure, therefore, their disposal usually costs higher than degradable ones.

The paper making wastewater reaches 10% of total industrial water. This kind of wastewater contained high concentration and complex structure pollutants, such as lignin, cellulose, hemicellulose, monosaccharide, and could cause serious pollution. The traditional two-stage biochemical treatment has relativity low cost, but the effluent could hardly meet the discharge standard of China owing to its high COD and chroma. The advanced oxidation technique could remove the pollutants from paper making wastewater efficiently, without any secondary pollution. However, the H_2O_2 used in this method is very expensive, which affects the application and extension of this technology [66]. Flocculation is another efficient method for paper making wastewater treatment, and its COD remove rate could reach 95% at the optimal condition, and the flocculants could be reused after treatment. The cost of this technique is in the middle of the two methods mentioned above (around 1.5-2 Yuan/m³).

The printing and dyeing wastewater contains of much refractory biodegradable organism with extremely high chrome, therefore, it is hard to be efficiently treated with biological technique [67]. Advanced oxidation could degrade the organisms and reduce the toxicity of this wastewater, but it is too expensive to be used to deal with a great amount of dyeing wastewater. The membrane separation technique could also obtain high pollutants remove rate, but the high cost of the membrane and the energy also hinder the technique from widely application. The flocculation is the most common used technique owing to its moderate price and basically satisfactory results. Partial related with the character of the wastewater, the cost of the flocculation treatment ranges from 3 yuan/m³ to 5 yuan/m³. Some researchers suggested that the combination of the flocculation technique with other techniques, such as Fenton, biological technique could reduce the cost without affecting the effluent quality.

Generally speaking, among all the techniques, biological technique costs the lowest if the pollutants are degradable. The flocculation and adsorption techniques could dispose of the wastewater at a moderate price, but the flocculant and adsorbent need secondary treatment for reuse. Membrane separation and the advanced oxidation could remove pollutants efficiently, but they are costly.

CONCLUSION

The treatment technologies for organic wastewater at present were reviewed. That a variety of technologies such as biological treatment, chemical oxidation technologies, adsorption technology and the others were introduced. At last, the cost accounting of different organic wastewater treatments was discussed.

REFERENCES

1.   W. E Wu, H. G Ge, K. F Zhang, Wastewater biological treatment technology. Chemical Industry Press (CIP) Publishing: BeiJing, 2003In Chinese].

2.   K. S Ju, R. E Parales, 2010Nitroaromatic Compounds, from Synthesis to BiodegradationMicrobiol. Mol. Biol. R. 74250272

3.   Van den Berg MBirnbaum L, Bosveld ATC, et al. (1998Toxic Equivalency Factors (TEFs) for PCBs, PCDDs, PCDFs for Humans and Wildlife. Environ. Health Perspect. 106775792

4.   R. C Sims, M. R Overcash, 1983Fate of Polynuclear Aromatic Compounds (PNAs) in Soil- Plant Systems. Residue Reviews 88168

5.   C. N Pope, 1999Organophosphorus pesticides: Do they all have the same Mechanism of Toxicity? J. Toxicol. Env. Heal. B. 2161181

6.   J Aislabie, G Lloydjones, 1995A Review of Bacterial-Degradation of Pesticides.Aus. J. Soil Res. 33925942

7.   J. G Leahy, R. R Colwell, 1990Microbial-Degradation of Hydrocarbons in the Environment.Microbiol. R. 54305315

8.   J. P Scott, D. F Ollis, 1995Integration of Chemical and Biological Oxidation Processes For Water Treatment: Review and RecommendationsEnviron. Prog. 1488103

9. Pedro JJAWalter AI. Bioremediation and Natural Attenuation: Process Fundamentals and Mathematical Models.Copyright © 2006John Wiley & Sons, Inc.

10. E. U Low, H. A Chase, M. G Milner, 2000Uncoupling of Metabolism to Reduce Biomass Production in the Activated Sludge ProcessWat. Res. 3432043212

11. F. N Ahmed, C. Q Lan, 2012Treatment of Landfill Leachate Using Membrane Bioreactors: A ReviewDesalination2874154

12. G Leitinga, Hulshoff Pol L W (1991UASB-process design for various types of wastewaters, Water Sci. Techol. 2487107

13. G Kassab, M Halalsheh, A Klapwijk, M Fayyad, J. B Van Lier, 2010Sequential Anaerobic-Aerobic Treatment for Domestic Wastewater-A Review. Bioresour. Technol. 10132993310

14. Y Peng, H Hou, S Wang, Y Cui, Y Zhiguo, 2008Nitrogen and Phosphorus Removal in Pilot-Scale Anaerobic-Anoxic Oxidation Ditch SystemJ Environ Sci 204398403

15. W. T Mook, M. H Chakrabarti, M. K Aroua, et al2012Removal of total ammonia nitrogen (TAN), nitrate and total organic carbon (TOC) from aquaculture wastewater using electrochemical technology: A reviewDesalination285113

16. G Busca, S Berardinelli, C Resini, 2008Technologies for the Removal of Phenol from Fluid Streams: A short review of recent developmentsJ. Hazard. Mater. 160265288

17. J Herney-ramirez, M. A Vicente, L. M Madeira, 2010Heterogeneous Photo-Fenton Oxidation with Pillared Clay-based Catalysts for Wastewater Treatment: A reviewAppl. Catal., B. 981026

18. S Navalon, M Alvaro, H Garcia, 2010Heterogeneous Fenton Catalysts Based on Clays, Silicas and ZeolitesAppl. Catal., B. 99126

19. Sillanpää METKurniawan TA, Lo W (2011Degradation of Chelating Agents in Aqueous Solution Using Advanced Oxidation Process (AOP) Chemosphere8314431460

20. D Li, J Qu, 2009The Progress of Catalytic Technologies in Water Purification: A review,J. Environ. Sci. 21713719

21. P. R Gogate, A. B Pandit, 2004A Review of Imperative Technologies for Wastewater Treatment I: Oxidation Technologies at Ambient ConditionsAdv. in Environ. Res. 8501551

22. S Perathoner, G Centi, 2005Wet Hydrogen Peroxide Catalytic Oxidation (WHPCO) of Organic Waste in Agro-food and Industrial StreamsTop.

Catal. 3314

23. Y Deng, J. D Englehardt, 2006Treatment of Landfill Leachate by the Fenton ProcessWater Res. 402036833694

24. D Trujillo, X Font, A Sanchez, 2006Use of Fenton Reaction for the Treatment of Leachate from Composting of Different WastesJ. Hazard. Mater. B. 138201204

25. K Pirkanniemi, S Metsärinne, M Sillanpää, 2007Degradation of EDTA and Novel Complexing Agents in Pulp and Paper Mill Process and Wastewaters by Fenton's Reagent. J. Hazard. Mater. 147556561

26. F. J Beltrán, Araya JFG, Giráldez I, Masa FJ (2006Kinetics of Activated Carbon Promoted Ozonation of Succinic Acid in WaterInd. Eng. Chem. Res. 4530153021

27. K Turhan, I Durukan, S. A Ozturkcan, Z Turgut, 2012Decolorization of Textile Basic Dye in Aqueous Solution By OzoneDyes Pigment. 92897901

28. Y Ka-car, E Alpay, V. K Ceylan, 2003Pretreatment of Afyon Alcaloide Factory's Wastewater by Wet Air Oxidation (WAO), Water Res. 3711701176

29. N Kawabata, H Urano, 1985Improvement of Biodegradability of Organic Compounds by Wet Oxidation. Mem. Fac. Eng. Des, Kyoto Inst. Technol. Ser. Sci. Technol. 346471

30. S. H Lin, T. S Chuang, 1994Wet Air Oxidation and Activated Sludge Treatment of Phenolic WastewaterJ. Environ. Sci. Health A. 29354764

31. S. H Lin, S. J Ho, 1996Treatment of Desizing Wastewater by Wet Air OxidationJ. Environ. Sci. Health A. 31235566

32. D Mantzavinos, R Hellenbrand, I. S Metcalfe, A. G Livingston, 1996Partial Wet Oxidation of P-coumaric Acid: Oxidation Intermediates, Reaction Pathways and Implications for Wastewater TreatmentWater Res. 301229692976

33. K-H Kim, S-K Ihm, 2011Heterogeneous Catalytic Wet Air Oxidation of Refractory Organic Pollutants in Industrial Wastewaters: A reviewJ. Hazard. Mater. 1861634

34. N Grisdanurak, S Chiarakorn, J Wittayakun, 2003Utilization of Mesoporous Molecular Sieves Synthesized from Natural Source Rice Husk Silica for Chlorinated Volatile Organic Compounds (CVOCs) Adsorption. Korean J. Chem. Eng. 20950955

35. A H Mahvi, A Maleki, A Eslami, 2004Potential of Rice Husk and Rice Husk Ash for Phenol Removal in Aqueous SystemsAm. J. Appl. Sci.

1321326

36. Focus technology go ltd (2011Water Treatment System (Active Carbon Filter)Zhangjiagang Beyond Machinery Co. Ltd.

37. G Yang, C C Li, C J (2007Electrofi Ltration of Silica Nanoparticle-containing Wastewater Using Tubular Ceramic Membranes. Sep. Purif. Technol. 58159165

38. P Fernandez-ibanez, J Blanco, S Malato, 2003Application of the Colloidal Stability of TiO2 Particles for Recovery and Reuse in Solar Photocatalysis.Water Res. 3731803188

39. T E Doll, F H Frimmel, 2005Cross-flow Microfiltration with Periodical Back-was Hing for Photocatalytic Degradation of Pharmaceutical and Diagnostic Residues-evaluation of the Long-term Stability of the Photocatalytic Activity of TiO2. Water Res. 39847854

40. X Zhang, , A J Du, , P Lee, , D D Sun, , J O Leckie, (2008) Ti, 2 Nanowire Membrane for Concurrent Filtration and Photocatalytic Oxidation of Humicacid in Water. J. Memb. Sci. 313: 44-51.

41. Y Yu, J C Yu, J G Yu, 2005Enhancement of Photocatalytic Activity of Mesoporous TiO2 by Using Carbon NanotubesAppl. Catal. A: Gen. 289186196

42. K Vinodgopal, D E Wynkoop, P V Kamat, 1996Environmental Photochemistry on Semiconductor Surfaces: Photosensitized Degradation of a Textile Azo Dye, Acid Orange 7, on TiO2 Particles Using Visible Light. Environ. Sci. Technol. 3016601666

43. M Ni, M Leung, K H Leung, D Y C Sumathy, K (2007A Review, and Recent Developments in Photocatalytic Water-splitting Using TiO2 for Hydrogen Production. Renew. Sust. Energy Rev. 11401425

44. A Fujishima, , X Zhang, , D A Tryk, (2008) Ti, 2 Photocatalysis and Related Surface Phenomena. Surf. Sci. Rep. 63: 515-582.

45. I Vázquez, J Rodríguez, E Maranón, L Castrillón, Y Fernández, 2006Simultaneous Removal of Phenol, Ammonium and Thiocyanate from Coke Wastewater by Aerobic Biodegradation. J. Hazard. Mater. 137317731780

46. Y. M Li, G. W Gu, I Zhao, H. Q Yu, Y. L Qiu, Y. Z Peng, 2003Treatment of Coke-plant Wastewater by Biofilm Systems for Removal of Organic Compounds and NitrogenChemosphere5269971005

47. L Chu, J Wang, J Dong, H Liu, X Sun, 2012Treatment of Coking Wastewater by an Advanced Fenton Oxidation Process Using IronPowder and Hydrogen Peroxide. Chemosphere. 86409414

48. Y Bai, Q Sun, R Sun, D Wen, X Tang, 2011Bioaugmentation and Adsorption Treatment of Coking Wastewater Containing Pyridine and Quinoline Using Zeolite-Biological Aerated FiltersEnviron. Sci. Technol. 4519401948

49. P Fongsatitkul, P Elefsiniotis, A Yamasmit, N Yamasmit, 2004Use of Sequencing Batch Reactors and Fenton's Reagent to Treat a Wastewater from a Textile IndustryBiochem. Eng. J. 213213220

50. S Grilli, D Piscitelli, D Mattioli, S Casu, A Spagni, 2011Textile Wastewater Treatment in a Bench-scale Anaerobic-biofilm Anoxic-aerobic Membrane Bioreactor Combined with NanofiltrationJ. Environ. Sci. Heal A-Tox. Hazard. Subst. Environ. Eng. 461315121518

51. D Fibbi, S Doumett, L Lepri, L Checchini, C Gonnelli, E Coppini, M. D Bubba, 2012Distribution and Mass Balance of Hexavalent and Trivalent Chromium in a Subsurface, Horizontal Flow (SF-h) Constructed Wetland Operating as Post-treatment of Textile Wastewater for Water reuseJ. Hazard. Mater. 199-200: 209 EOF216 EOF

52. Y He, P Xu, C Li, B Zhang, 2005High-concentration Food Wastewater Treatment by an Anaerobic Membrane BioreactorWater Res. 3941104118

53. Y Wang, X Huang, Q Yuan, 2005Nitrogen and Carbon Removals from Food Processing Wastewater by an Anoxic/aerobic Membrane BioreactorProcess Biochem. 4017331739

54. S. W Van Ginkel, S. E Oh, B. E Logan, 2005Biohydrogen Gas Production from Food Processing and Domestic WastewatersInt. J. Hydrogen Energ. 30 (15), 1535 EOF1542 EOF

55. S. E Oh, B. E Logan, 2005Hydrogen and Electricity Production from a Food Processing Wastewater Using Fermentation and Microbial Fuel cell TechnologiesWater Res. 3946734682

56. H Tekin, O Bilkay, S. S Ataberk, T. H Balta, I. H Ceribasi, F. D Sanin, F. B Dilek, U Yetis, 2006Use of Fenton Oxidation to Improve the Biodegradability of a Pharmaceutical WastewaterJ. Hazard. Mater. B. 136258265

57. C Sirtori, A Zapata, I Oller, 2009Decontamination Industrial Pharmaceutical Wastewater by Combining Solar Photo-Fenton and Biological TreatmentWater Res. 43661668

58. S Chelliapan, T Wilby, P. J Sallis, 2006Performance of an Up-flow Anaerobic Stage Reactor (UASR) in the Treatment of Pharmaceutical Wastewater Containing Macrolide AntibioticsWater Res. 40507516

59. Z Chen, N Ren, A Wang, Z-P Zhang, Y Shi, 2008A Novel Application of

TPAD-MBR System to the Pilot Treatment of Chemical Synthesis-based Pharmaceutical WastewaterWater Res. 4233853392

60. C Guimaraes, P Porto, R Oliveira, M Mota, 2005Continuous Decolourization of a Sugar Refinery Wastewater in a Modified Rotating Biological Contactor with Phanerochaete Chrysosporium Immobilized on Polyurethane Foam DisksProcess Biochem. 402535540

61. S Li, P Zhou, P Yao, 2010Preparation of O-Carboxymethyl-N-Trimethyl Chitosan Chloride and Flocculation of the Wastewater in Sugar RefineryJ. Appl. Polym. Sci. 11627422748

62. M Yetis, GuÈ nduÈz U, Eroglu I (2000Photoproduction of Hydrogen from Sugar Refinery Wastewater by Rhodobacter sphaeroides O.U. 001, Int. J. Hydrogen Energ. 2510351041

63. X. K Song, Y. L Shen, N Jiao, 2012Analysis on Decentralized Domestic Sewage Treatment Technologies. Environmental Science and Technology. 2536871In Chinese)

64. Q. L Yan, 2008Treatment of agricultural and sideline products processing wastewater with the SBR technique. Ocean university of china: 312

65. J. W Patterson, 2008Industrial wastewater treatment technologySecond Edition. Butterworth Publishers, Stoneham, MA. USA.

66. D. M Yang, B Wang, 2010Application of advanced oxidation processes in papermaking wastewater treatmentChina pulp and paper. 2976973In Chinese)

67. Q. Y Yu, 2011Advances in the treatment of printing and dyeing wastewater. Industrial Safety and Environmental Protection. 3784143In Chinese)

Chapter 7

RELATIONSHIP OF ALGAE TO WATER POLLUTION AND WASTE WATER TREATMENT

Bulent Sen[1], Feray Sonmez[1], Ozgur Canpolat[1], Mehmet Tahir Alp[2] and Mehmet Ali Turan Kocer[3]

[1]University of Firat, Faculty of Fisheries Department of Aquatic Basic Sciences Elazığ, Turkey

[2]University of Mersin, Faculty of Fisheries Department of Aquatic Basic Sciences Mersin, Turkey

[3]Mediterranean Fisheries Research Production and Training Institue, Antalya, Turkey

INTRODUCTION

Pollution of surface water has become one of the most important environmental problems. Two types of large and long-lasting pollution threats can be recognized at the global level: on the one hand, organic pollution leading to high organic content in aquatic ecosystems and, in the long term, to eutrophication. It is a well-known fact that polluted water can reduce water quality thus restricting use of water bodies for many purposes.

Organic pollution occurs when large quantities of organic compounds from many sources are released into the receiving running waters, lakes and also seas. Organic pollutants originate from domestic sewage (raw or treated), or urban run off, industrial effluents and farm water. Organic pollution could negatively affect the water quality in many ways. During the decomposition process of organic water dissolved oxygen in the water may be used up greater rate than it can be replenished thus, giving rise to oxygen depletion which causes severe consequences on the aquatic biota. Organic effluents also frequenlty contain large quantities of suspended solid which reduce the light available to photosynthetic organisms mainly algae. In addition organic wastes from people and animals may also rich in disease causing (pathogenic) organisms [1,2,3].

ALGAE AND WATER POLLUTION

Algae are the main the primary producers in all kinds of water bodies and they are involved in water pollution in a number of significant ways. Firstly, enrichments of the algal nutrients in water through organic effluents may selectively stimulate the growth of algal species producing massive surface growths or 'blooms' that in turn reduce the water quality and affect its use. However, certain algae flourished in water polluted with organic wastes play an important part in "self-purification of water bodies". Some pollution algae may frequently are toxic to fish and also mankind and animals using polluted water. In fact, algae can play significant part of food chain of aquatic life, thus whatever alters the number and kinds of algae strongly affects all organisms in the chain including fish.

Algae are also known to be causes of tastes and odors in water [4]. In fact, a large number of algae are associated with tastes and odors that vary in type. Certain diatoms, blue-green algae and coloured flagellates (particularly Chrysophyta and Euglenophyta) are the best known algae to pose such problems in water supplies, but green algae may also be involved. Some algae produce an aromatic odor resembling to that of particular flowers or vegetables. In addition, a spicy, a fishy odor and a grassy odor can also be produced by odor algae [5,6].

ALGAE AS BIOINDICATORS

Bioindicator organisms can be used to identify and qualify the effects of pollutants on the environment. Bioindicators can tell us about the cumulative effects of different pollutants in the ecosystem and about how long a problem may persist. Although indicator organisms can be any biological species that defines a trait or characteristics of the environment, algae are known to be good indicators of pollution of many types for the following reasons.

- algae have wide temporal and spatial distribution.
- many algal species are avaliable all the year.
- response quickly to the charges in the environment due to pollution.
- Algae are diverse group of organisms found in large quantities.
- easier to detect and sample.
- The presence of some algae are well correlated with particular type of pollution particularly to organic pollution

Algae of many kinds are really good indicators of water quality and many lakes are characterized based on their dominant phytoplankton groups. Many desmids are known to be present in oligotrophic waters whilst a few species frequently occurs in eutrophic bodies of water [7]. Similarly, many blue-green algae occurs in nutrient-poor waters, while some grow well in organically polluted waters [8]. The ecosystem approach to water quality assessment also include diatom species and accociations used as indicators of organic pollution. Five algal species were selected as indicators of the degree of pollution in rivers in England. *Stigeoclonium tenue*is presentat the down streammargin of the heavily polluted part of a river, *Nitzschia palea and Gomphonema parvulum* always appear to be dominant in the mild pollution zone whilst *Cocconeis* and *Chamaesiphon* are reported to occur in unpolluted parts of the stream or in repurified zone[9].

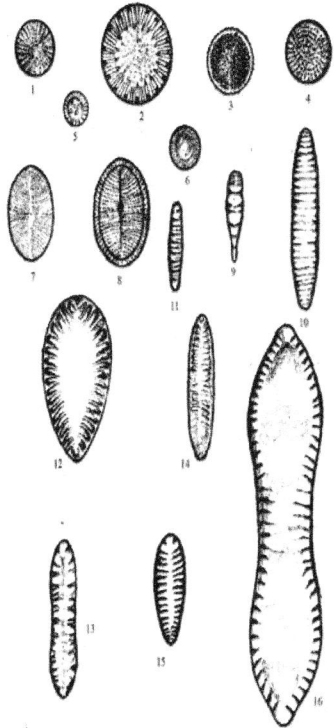

Plate 1: 1.Stephanodiscus hantzschii Grunow 2.Cyclotella comta (Ehrenberg) Kützing 3.Thalassiosira weissflogii (Grunow) G.Fryxel & Hasle 4.Aulacoseira distans (Ehrenberg) Simonsen 5.Cyclotella ocellata Pantocsek 6. C. kützingiana Thwaites 7.Cocconeis pediculus Ehrenberg 8.C.placentula Ehrenberg 9. Meridion circulare (Greville) C. Agardh 10. Diatoma vulgaris var. lineare Grunow 11. D. tenuis C.Agardh 12. Surirella ovalis Brébisson 13. S. ovata var. apiculata W. Smith 14. S. linearis W. Smith 15. S. minuta Brébisson 16. Cymatopleura solea (Brébisson) W. Smith [14].

Navicula accomoda is stressed to be a good indicator of sewage/organic pollution as the species comfortably occur in the most heavily polluted zones in which other species can not occur. The same hold true for species and varietes of *Gomphonema*[10] which is commonly found in highly organically polluted water.*Amphora ovalis* and *Gyrosigma attenuatum* are also introduced as good examples of diatoms to be affected by high organic content of water [11]. A list of more than 850 algal taxa was published based on the reports of considerable number of authors. According to this list, many algal genera have species that grow well in water containing a high concentration of organic wastes. Green algae *Chlamydomonas*, *Euglena*, diatoms, *Navicula*,*Synedra* and blue- green algae *Oscillatoria* and *Phormidium* are emphasized to tolerate organic pollution [12]. At species level, *Euglena viridis* (Euglenophyta), *Nitzschia palea* (Bacillariophyta),*Oscillatoria limosa*, *O.tenuis*, *O.princeps* and *Phormidium uncinatum* (Cyanophyta) are reported to be present than any other species in organically polluted waters [13]. Some diatom taxa in a stream polluted with the waste water of a slaugher house are given in Plate 1-4 [14].

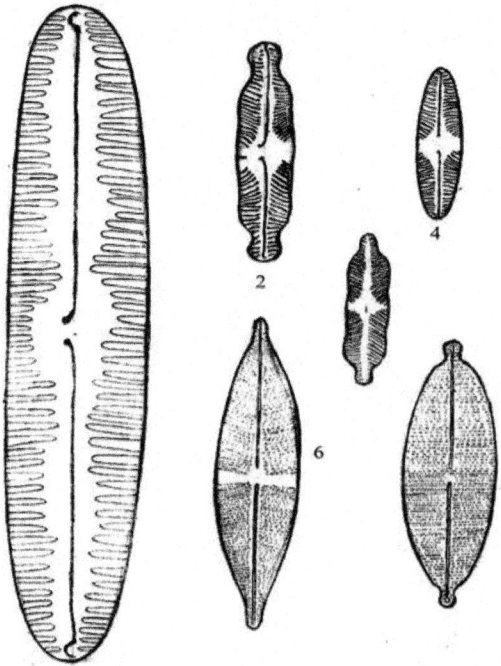

Plate 2: 1. *Pinnularia viridis* (Nitzsch) Ehrenberg 2. *P. biceps* W. Gregory 3. *Stauroneis phoenicenteron*(Nitzsch) Ehrenberg 4. *Pinnularia brebissonii* (Kützing) Rabenhorst 5. *Craticula ambigua*(Ehrenberg) D. G. Mann 6. *Pinnularia mesolepta* (Ehrenberg) W. Smith [14].

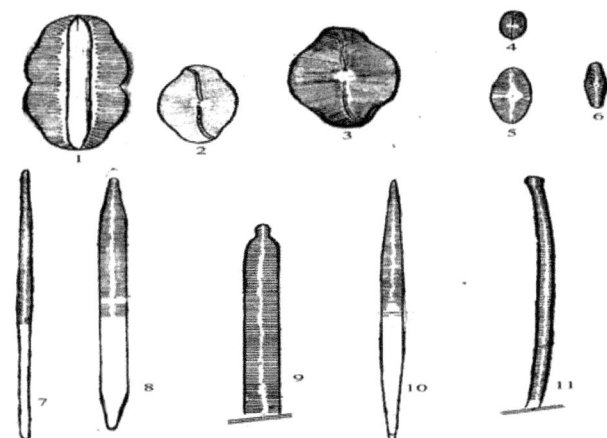

Plate 3: 1. *Rhopalodia gibba* (Ehrenberg) Otto Müller 2-3. *Eucocconeis flexella* (Kützing) Meister 4.*Achnanthidium minutisimum* (Kützing) Czarnecki 5. *Eucocconeisquadratarea* (Østrup) Lange-Bertalot 6. *Achnanthes marginulata* Grunow 7. *Ulnaria delicatissima* var. *angustissima* (Grunow) M. Aboal & P. C. Silva 8. *U. acus* (Kützing) M. Aboal 9. *U. amphirhyncus* (Ehrenberg) Compère & Bukhtiyarova 10. *U.danica* (Kützing) Compère & Bukhtiyarova 11. *U. biceps* (Kützing) P. Compère [14].

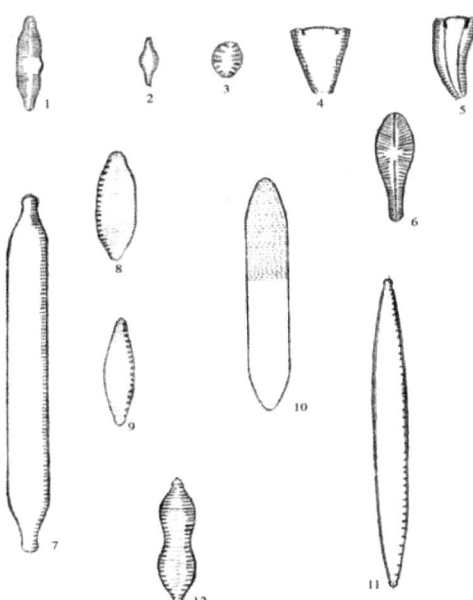

Plate 4: 1.*Fragilaria capucina* var. *vaucheriae* (Kützing) Lange-Bertalot 2.*Pseudostaurosira brevistrita*(Grunow) D.M.Williams & Round 3.*Staurosirella pinnata* (Ehrenberg) D.M.Williams & Round 4.*Gomphonema truncatum* Ehrenberg

5.*Rhoicosphenia abbreviata* (C. Agardh) Lange-Bertalot 6.*Gomphonema olivaceum* (Hornemann) Brébisson 7.*Nitzschia sublinearis* Hustedt 8.*N. umbonata*(Ehrenberg) Lange-Bertalot 9. *N.hantzschiana* Rabenhorst 10. *Tryblionella angustata* W. Smith 11.*Nitzschia linearis* (C. Agardh) W. Smith12. *N. constricta* (Kützing) Ralfs. [14].

Algae are also good indicators of clean water since many species occur insistently and predominately in the clean water zone of the streams. However it is more satisfactory to emphasize the presence or absence of several species of clean water algae rather than of any one species to define the clean water zone. Approximately 46 taxa has been announced as representatives of the clean water algae including many diatoms, several flagellates and certain green and blue-green algae [12]. However it is emphasized that minute flagellates are better indicators of clean water than many larger algae. A few of the clean water algae are planktonic whilst many are benthic, attached to substrata at the bottom or sides of the running waters.

There are many studies by various authors emphasizing the relationships of algae to clean water. A community composed of the diatom *Cocconeis* and the blue-green alga *Chamaesiphon* is claimed to be present in the portion of the stream which has returned to normal following purification of a polluted condition [9]. Kolkwitz listed 61 diatoms, 42 green algae, 41 pigmented flagellates, 23 blue-green algae, and 5 red algae as organisms of oligosaprobic and /or unpolluted zones and Lackey found 77 species of planktonic algae in the clean water portion of a small stream, 40 of which were absent in the polluted area[15,16]. The flagellates *Chromulina rosanoffi, Mallomonas caudata,* the green algae*Ulothrix zonata* and *Microspora amoena* are also reported as oligosaprobic zone organisms [17]. Two groups of algae, Cryptophyta and Chrysophyta, are reported to be indicators of clean and/or unpolluted water as the members of these algal groups tend to occur in abundance, oppositely reacting adversely to pollution [18]. The absence of blue-green algae was also accepted an indication of clean water [19].

USE OF ALGAE IN SAPROBIEN SYSTEM

The classic scheme for the interpretation of streams ecological conditions based on the biota was first introduced by Kolkwitz and Marsson [20]. They defined five zones based on the degree of pollution and proposed the use of aquatic organisms as indicators of different pollution and/or recovery zones of rivers which were polluted with organic matter such as sewage. However more recently Werner proposed nine different zones in the saprobic system in a stream organically polluted [21]. Survey of the saprobic zones and the corresponding communities are given Table 1. Pollution zones proposed in that saprobient system were basically termed "Coprozoic" "Polysaprobic", "Mesosaprobic", "Oligosaprobic" and "Katharobic". Each zone was different in chemical and physical characteristics and containing characteristic species. He listed indicator species of these zones except the last one which is infact clean water.

Polysaprobic zone was characterized by almost complete absence of algae except for blue-green alga*Arthrospira* (*Spirulina*) *jenneri* and green alga *Euglena viridis*. Bacteria and Protozoa were the most common groups in this zone. The preponderance of blue-green algae (Cyanophyta) was characteristic of alfa-mesosaprobic zone while diatoms (Bacillariophyta) and green algae (Chlorophyta) were dominant organisms in beta-mesosaprobic zone. Peridiniales (Dinophyta) and Charales (Charophyta) occurred in any quantity only in the oligosaprobic zone. In the same zone, the bacterial count was low but there was a great variety of plants and animals (including fish) in considerable numbers.

USE OF ALGAE IN WASTEWATER TREATMENT

Recently, algae have become significant organisms for biological purification of wastewater since they are able to accumulate plant nutrients, heavy metals, pesticides, organic and inorganic toxic substances and radioactive matters in their cells/bodies [22-25]. Biological wastewater treatment systems with micro algae have particularly gained importance in last 50 years and it is now widely accepted that algal wastewater treatment systems are as effective as conventional treatment systems. These spesific features have made algal wastewaters treatment systems an significant low-cost alternatives to complex expensive treatment systems particularly for purification of municipal wastewaters.

In addition, algae harvested from treatment ponds are widely used as nitrogen and phophorus suplement for agricultural purpose and can be subjected to fermentation in order to obtain energy from metane. Algae are also able to accumulate highly toxic substances such as selenium, zinc and arsenic in their cells and/or bodies thus eliminating such substances from aquatic enviroments. Radiation is also an important type of pollution as some water contain naturally radioactive materials, and others become radioactive through contamination. Many algae can take up and accumulate many radioactive minerals in their cells even from greater concentrations in the water [12]. MacKenthunemphasized that*Spirogyra* can accumulate radio-phosphorus by a factor 850.000 times that of water [26]. Considering all these abilities of algae to purify the polluted waters of many types, it is worth to emphasize that algal technology in wastewater treatment systems are expected to get even more common in future years. Wastewater treatment which is applied to improve or upgrade the quality of a wastewater involves physical, chemical and biological processes in primary, secondary or tertiary stages. Primary treatment removes materials that will either float or readily settle out by gravity. It includes the physical processes of screening, commination, grit removal, and sedimentation. While the secondary treatment is usually accomplished by biological processes and removes the soluble organic matter and suspended solids left from primary treatment. Tertiary or advanced treatment is process for purification in which nitrates and phosphates, as well as fine particles are removed [27]. However initial cost as well as operating cost of wastewater treatment plant including primary, secondary or advanced stages is highly expensive [28].

It is well known that algae have an important role in self purification of organic pollution in natural waters [29]. Moreover, many studies revealed that algae remove nutrients especially nitrogen and phosphorus, heavy metals, pesticides, organic and inorganic toxins, pathogens from surrounding water by accumulating and/or using them in their cells[30,31,32,33,34,35,36,37,38]. Also, studies showed that algae may be used successfully for wastewater treatment as a result of their bioaccumulation abilities [39].

Table 1: Aquatic communities representing various zones of pollution. Survey of the saprobic zones and the corresponding communities[21]

Zone I.	Coprozoic zone
	the bacterium community
	the *Bodo* community
	both communities
Zone II.	α-Polysaprobic zone
	Euglena community
	Rhodo-Thio bacterium community
	Pure Chlorobacterium community
Zone III.	β-Polysaprobic zone
	Beggiatoa community
	Thiothrix nivea mommunity
	Euglena community

Zone IV.	γ-Polysaprobic zone
	Oscillatoria chlorina community
	Sphaerotilus natans community
Zone V.	δ-Mesosaprobic zone
	Ulothrix zonata community
	Oscillatoria benthonicum community (O.brevis, O.limnosa, O.splendidawith O.subtilissima, O.princeps and O.tenuis present as associate species)
	Stigeoclonium tenue community
Zone VI.	β-Mesosaprobic zone
	Cladophora fracta community
	Phormidium community
Zone VII.	γ-Mesosaprobic zone
	Rhodophyceae community (Batrachospermum moniliforme or Lemanea fluviatilis)
	Chlorophyceae community (*Cladophora glomerata or Ulothrix zonata* (clean-water type))
Zone VIII.	Oligosaprobic zone
	Chlorophyceae community (*Draparnaldia glomerat*a)
	Pure Meridion circulare community
	Rhodophyceae community (Lemanea annulata, Batrahcospermum vagum or Hildenbrandia rivularis)
	Vaucheria sessiis community
	Phormidium inundatum community
Zone IX.	Katharobic zone
	Chlorophyceae community (Chlorotylium cataractum and Draparnaldia plumosa)
	Rhodophyceae community (*Hildenbrandia rivularis*)
	Lime-incrusting algal communities (*Chamaesiphon polonius* and various *Calothrix* species)
a, b, c, d, e	= as alternatives
1, 2, 3	= as differences in degree

ADVANTAGES OF USE OF ALGAE IN WASTEWATER TREATMENT

There are a symbiotic relation among bacteria and algae in aquatic ecosystems. Algae support to aerobic bacterial oxidation of organic matter producing oxygen via photosynthesis whilst released carbon dioxide and nutrients in aerobic oxidation use for growth of algal biomass. Considering ammonium, carbon dioxide and orthophosphate as main nutrient sources, Oswald determined that oxygen release ratio is 1.5 g O_2/1 g algal biomass [40]. Grobbelaar et al. reported to oxygen release ratio of 1.9 g O_2/1 g algal biomass [41]. Arceivala, accounting latitude, climate and atmospheric conditions, calculated that 4-6% of mean daily solar radiation reaching on treatment pond in 40°N latitude use for new biomass production and production rate of algal biomass may reach 80 kg O_2/1 ha-day [42]. Most of nitrogen in algal cell bound to proteins which compose to 45-60% of dry weight and phosphorus is essential for synthesis of nucleic acids, phospholipids and phosphate esters. Algae using nitrogen and phosphorus in growth may remove to nutrients load of wastewater from a few hours to a few days [43].

In comparison to common treatment systems, oxidation ponds supporting growth of some species may be effective of nutrient removal (Fig. 1). Increasing dissolved oxygen concentration and pH cause for phosphorus sedimentation, ammonia and hydrogen sulphur removal. High pH in algal ponds also leads to pathogen disinfection [44]. Removal efficiency of heavy metals by algae shows changes among species. In fact, studies showed that chrome by *Oscillatoria*, cadmium, copper and zinc by*Chlorellavulgaris*, lead by *Chlamydomonas* and molybdenum by *Scenedesmus chlorelloides* may remove successfully [45,46,47,48,49]. Although algae have adaptation ability to sub-lethal concentrations, accumulation of heavy metals in cells may be potentially toxic effects to the other circles of food web [50].

ALGAL-BACTERIAL PONDS

Algal-bacterial pond is water body which is designed to keep and improve of wastewater in a certain time. Although wastewater is treated in pond via physical, chemical and biological processes and/or mechanical processes like aeration, there are also ponds completely based on processes of natural conditions. Ponds, where stabilization of dissolved compounds and suspended solids is in completely aerobic conditions, are named "oxidation ponds". When stabilisation in anaerobic or facultative conditions, ponds are named "waste stabilisation ponds". Stabilisation pond systems are assessed in different types: facultative, anaerobic, aeration and maturation ponds. Common pond type which utilizes from algae is facultative stabilisation ponds. Facultative ponds are designed for purposes such as decrease of waste retention time, achieve of effective treatment or algal culture (Fig. 2). Algal photosynthesis and bacterial decomposition is principal mechanism of algal-bacterial ponds. The processes including oxidation, settling, sedimentation, adsorption, disinfection in the ponds are results of symbiotic relation between algae and bacteria populations [51].

Facultative ponds (usually 2.5m in depth) are systems where effluent quality improves between 5 and 30 days depending on factors such as climate, temperature, wind and surface area [52]. There are three main zones in such ponds; two upper zones with oxygen whilst anaerobic conditions prevail in bottom. Algal photosynthesis and atmospheric diffusion are main oxygen source. Wastes are stabilized by aerobic bacteria in upper zone and by facultative bacteria in intermediate depths while degraded by anaerobic bacteria in bottom zone [53]. Zooplankton controls to excessive bacterial growth and algal blooms through grazing as well as contributing to carbon dioxide production for algal photosynthesis. Food web in a facultative pound is given in Figure 2.

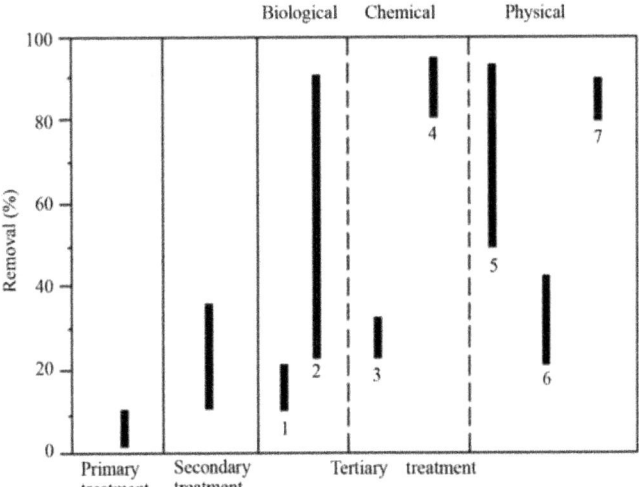

Figure 1: Removal efficiency of organic nitrogen in treatment methods [44]. 1-nitrification, 2-oxidation pond, 3-chemical coagulation, 4-chlorination, 5-ammonia removal, 6-filtration, 7-reverse osmosis.

Acceptable effluent quality is the most important advantage of facultative ponds though low operation and maintenance costs. However there are some disadvantages such as high land costs, odour problem in high waste loading, loss of nitrogen to atmosphere, limiting the nutrient reuse by phosphate sedimentation also limiting of irrigation potential by salinity increase during high evaporation period [54]. Although temperature largely affects retention time of wastewater, facultative ponds are widely used in different climate regimes. For example there are more than 3000 facultative ponds in Germany and France and 7000 in United States [53].

HIGH RATE ALGAL PONDS

Municipal wastewater treatments with high rate algal ponds were first proposed by Oswald and Golouke and thereinafter were used in many parts of the world [55,56]. High rate algal pond is usually shallow (20-50 cm) and is equipped with mechanical aeration and mixing by means of paddle wheels. High oxygen level resulting from photosynthesis and aeration allows to low retention times in these ponds. Removal rates of high rate algal ponds are almost similar to conventional treatment methods but may also be more efficient with lower retention time. In fact biochemical oxygen demand (BOD) up to 90% and more than 80% of nitrogen and phosphorus are treated in high rate algal ponds in a few days. However required time for treatment of biochemical oxygen demand up to 90% using by conventional activated sludge and bio filtration techniques, which are highly expensive secondary treatment methods, is between five and eight hours during which lower ratio of nitrogen and phosphorus may be removed. Further, construction and energy costs are highly lower and land requirement is half the required for facultative ponds [57]. It is a well-known fact that only a small amount of nitrogen and phosphorus are removed in active sludge and bio filtration techniques, In addition active sludge and bio filtration techniques require expensive chemicals and complex systems.

Cost of harvest in high rate algal ponds may be most important problem. Thus sedimentation of algae with flocculating is aimed when the wheels are stopped for harvest. In addition growth of resistant algal species to sinking such as *Chlorella, Euglena, Chlamydomonas* and *Oscillatoria* is undesired algae in the ponds. *Scenedesmus* or *Micractenium*, non-preferred species due to their cell morphology for grazing, are dominant in well mixed ponds [40]. Harvested algae may use for industrial and agricultural use as well as effluent in aquaculture (Fig. 3).

ADVANCED INTEGRATED WASTEWATER PONDS

Advanced integrated wastewater pond systems are an adaptation of waste stabilisation ponds systems based on a series of four advanced ponds: A facultative pond; a high rate algal pond; an algal settling pond and finally a maturation pond for solar disinfection and pathogen abatement. The first pond in series is a facultative pond with depth of 4 to 5 m containing a digester pit, which functions much like an anaerobic pond while surface zone remains aerobic. Effluent of the facultative pond flows to the high rate algal pond for remove to dissolved organic matter and nutrients, then to settling pond with residence time of one or two days for sedimentation of algae and suspended solids. The last unit is maturation ponds where treated water is exposed to the sun and wind leading to natural oxygenation and solar disinfection, and thus an inactivation of pathogens [58].

Wastewater Treatment and Reclamation Plant in St. Helena, California, by US Department of Energy built of formed earth rather than of reinforced concrete in the early 1960s (Fig. 4). The total pond area needed is much larger than that needed for a conventional plant, but ponds should still cost only one-third to one-half as much to build. Another important advantage of the plant is the small amount of sludge they produced. For example, during nearly 3 decades of operation, St. Helena's wastewater treatment plant has never had to remove residue and measured less than 1 meter of residue had accumulated at the bottom of the deep digester pit [59].

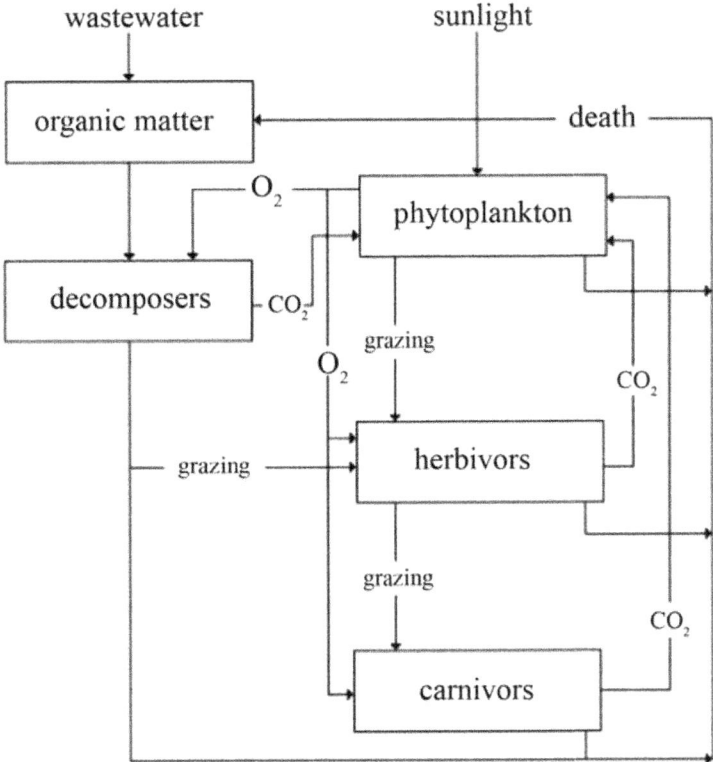

Figure 2: Food web in facultative wastewater treatment pond [51].

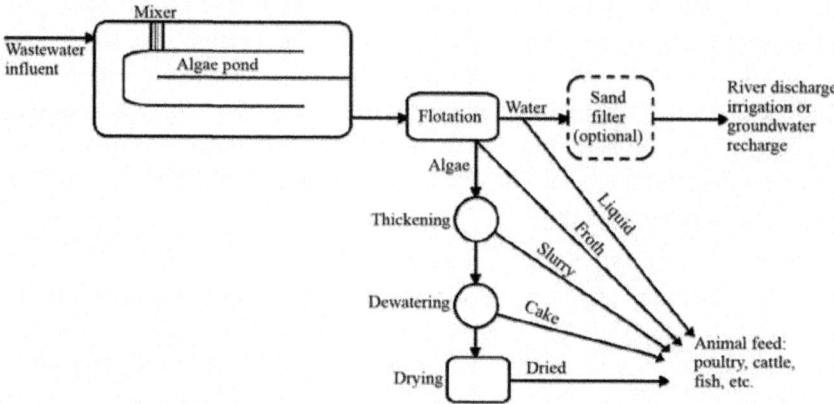

Figure 3: Flow Scheme of the Accelerated Photosynthetic Process for Waste-Water Treatment and Algal Protein Production [56].

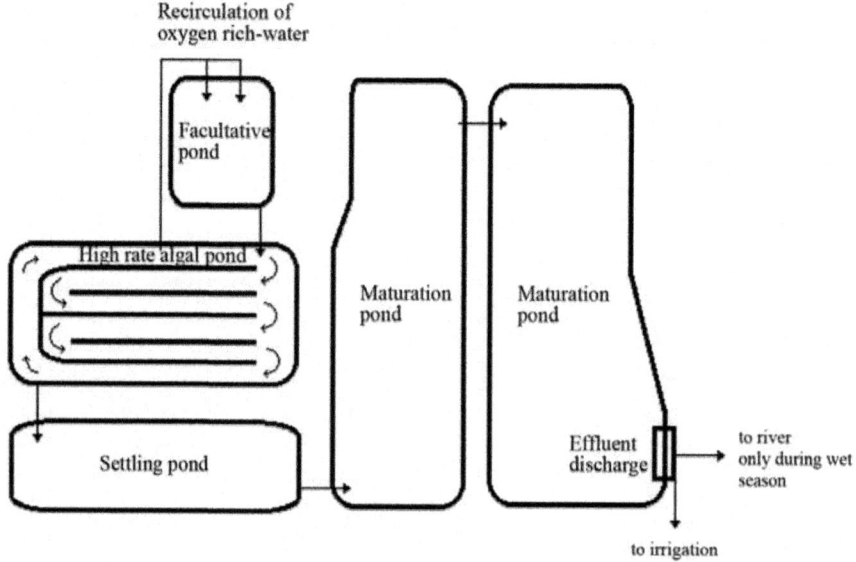

Figure 4: Diagram of St. Helena advanced integrated pond system [59].

Comparison of algae involved wastewater treatment systems is given at table 2. Many useful criteria have been used in the table for more constructive and trustable comparison of wastewater treatment systems connected to algae.

Table 2: Comparison of wastetreatment ponds in terms of various criteria

Criteria	Bacteria-Algae Pond	High rate Algae Pond	Integrated Pond
Depth	2.5 m	0.2-0.5 m (20-50 cm)	4-5 m
Salinity	increase	-	-
Retention time	low	lower (because O2 level is high)	high
Land required	high	low	low
Odor problem	occur	not occur	not occur
Loss of N to atmosphere	occur	not occur	not occur
Operation/maintenance cost	low	lower (5 folds lower)	high
PO4 sedimantation	occur	not occur	not occur
Time required for treatment	5-30 days	a few days	5-6 days
Energy requirement	low	low	high
Efficiency quality	low	high	high
BOD removal	fair	good	good
Suspended solid removal	fair	good	good
Harvesting cost		high	low

CONCLUSION

The water flows from lands into aquatic environments contribute enormous amounts of organic matters and plants nutrients to the aquatic systems which give rise to eutrophication and pollution. With increased urbanization the need for sewage treatment plants (STP) became more important. Wastewater treatment which is applied to improve or upgrade the quality of a wastewater involves physical, chemical and biological processes in primary, secondary or tertiary stages.More sewage plants are designed to remove solids (primary process), followed by a secondary process which involves either activated sludge or trickling filters to reduce the Biological Oxygen Demand (BOD). Removal of the nutrients left after secondary treatment is possible by a variety of processes, one of which involves growth and harvesting of algae from the effluents: others involve ion exchange electro chemical, electrodialysis, reserve osmosis, distillation, chemical precipitation as tertiaryprocesses. However initial cost as well as operating cost of wastewater treatment plant including primary, secondary or advanced stages is highly expensive.

Recently, algae have become significant organisms for biological purification of wastewater since they are able to accumulate plant nutrients, heavy metals, pesticides, organic and inorganic toxic substances and radioactive matters in their cells/bodies with their bioaccumulation abilities. Particularly, biological wastewater treatment systems with micro algae have gained great importance in last 50 years and it is now widely accepted that algal wastewater treatment systems are as effective as conventional treatment systems.Removal rates of particularly high rate algal ponds are almost similar to conventional treatment methods but it is more efficient with lower retention time.With these spesific features algal wastewater treatment systems can be accepted as an significant low-cost alternatives to complex expensive treatment systems particularly for purification of municipal wastewaters.

REFERENCES

1.   European Inland Fisheries Advisory Commission, Working Party on Water Quality Criteria for European freshwater fish (EIFAC),1980Relationship of Algae to Water Pollution and Waste Water TreatmentEIFAC Technical Paper 37, FAO, Rome.

2.   Xu, S., Nirmalakhandan, N.,. Use of QSAR models in predicting joint effects in multi-component mixtures of organic chemicals.Water Res.19982391 EOF

3.   R. Altenburger, T. Backhaus, W. Boedeker, M. Faust, M. Scholze, L. H. Grimme, Relationship of Algae to Water Pollution and Waste Water TreatmentEnviron. Toxicol. Chem.20002341 EOF

4.   E. A. Sigworth, Control of odor and taste in water supplies. J.Amer. Water Wks. Assn.19574915071521

5.   B. A. Adams, The role of actinomycetes in producing earthy tastes and smells in potable water. Dept. Of Public Wks., Roads and Transport Congress 1933Paper 4London, England.

6.   J. K. Silvey, A. W. Roach, Actinomycetes may cause tastes and odors in water supplies. Public Wks. Mag.1956

7.   A. J. Brook, Relationship of Algae to Water Pollution and Waste Water TreatmentLimnol and Oceannog 1965403411

8.   T.. Braarud, A. phytoplankton, of. survey, polluted. the, of. waters, Oslo. Inner, Fjord, Hvalraadets Skrifter, Scientific Results of Marine Biological Research194528281142

9.   R. W. Butcher, Pollution and re-purification as indicated by the algae. Fourth Internat. Congress for Microbiology, 1949Rept. of Proc. 149150

10.   R. E. M. Archibald, Diversity in some South Africation diatom associations and its relation to water quality1972Water Research 612291238

11. R. Patrick, Relationship of Algae to Water Pollution and Waste Water TreatmentBot. Rev.194814473524

12. C. M. A. Palmer, rating. composite, algae. of, organic. tolerating, pollution, J. Phycology.196957882

13. C. M. Palmer, Relationship of Algae to Water Pollution and Waste Water TreatmentCastle House Publications Ltd.1980p.

14. B. Şen, B. Topkaya, M. T. Alp, F. Özrenk, Madde. Organik, Kirlenen. ile, Çay. . Bir, Çayı. Selli, İçindeki. Elazığ, ve. Kirlilik, Üzerine. Algler, Araştırma. Bir, I. I. , Ulusal Ekoloji ve Çevre Kongresi Bildiriler Kitabı,1995s.599610Ankara,

15. R. Kolkwitz, Saprobien. Oekologie der, Uber die Bezelhungen der Wasser-organismen zur Umwelt. Schriftenreihe des Vereins für Wasser-, Border-, und Lufthygiene19504p.

16. J. B. Lackey, Stream and richment microbiota. Public Health Report 19561708718

17. H. Liebmann, Frischwasser. Handbuch der, Abwasserbiologie. R. und, München. Oldenburg, Germany, 1951p.

18. J. B. Lackey, Two groups of flagellated algae serving as indicators of clean water. J. Am. Water Wks. Assn. 19413310991110

19. G. W. Rafter, Relationship of Algae to Water Pollution and Waste Water TreatmentVan Nostrand Co., N.Y. 1900

20. R. Kolkwitz, M.. Marsson, Okologie der Pflanzichen Saprobien. Ber. Deutsch. Bot. Ges 190826505519

21. D.. Werner, Relationship of Algae to Water Pollution and Waste Water TreatmentBotanical monographs. California pres. 197713pp.

22. Kalesh NS, Nair SM The Accumulation Levels of Heavy Metals (Ni, Cr, Sr, & Ag) in Marine Algae from Southwest Coast of İndia.Toxicological & Environmental Chemistry 2005872135146

23. N. Jothinayagi, C. Anbazhagan, Heavy Metal Monitoring of Rameswaram Coast by Some Sargassum species. American-Eurasian Journal of Scientific Research 2009427380

24. M. T. Alp, B. Sen, O. Ozbay, Heavy Metal Levels in Cladophora glomerata which Seasonally Occur in the Lake Hazar. Ekoloji, 20781317doi:ekoloji.2011

25. M. T. Alp, O. Ozbay, M. A. Sungur, Determination of Heavy Metal Levels in Sediment and Macroalgae (Ulva sp. and Enteromorpha sp.) on the Mersin Coast 2011. Ekoloji 21, 82, 47 EOF55 EOF (2012).

26. Kenthun. K. M. Mac, wastes. Radioactive, . Chapt, n The Practice of Water Pollution Biology. U.S. Dept. Interior, Fed. Water Pol. Contr. Admin., Div. of Tech. Support. U.S. Printing Office1969

27. R. L. Droste, Relationship of Algae to Water Pollution and Waste Water TreatmentJohn Wiley and Sons, New York1997

28. Oswald,W.J.Relationship of Algae to Water Pollution and Waste Water Treatment1995311218

29. B. Şen, Nacar. V. ve, Su Kirliliği ve Algler. Fırat Havzası I. Çevre Sempozyumu Bildiriler Kitabı.198840521

30. K. R. Reddy, Fate of Nitrogen and Phosphorus in a Wastewater Retention Reservoir Containing Aquatic Macrophytes. Journal of Environmental Quality, 198312113741

31. R. J. Craggs, W. H. Adey, K. R. Jenson, John. M. S. St, F. B. Green, W. J. Oswald, Relationship of Algae to Water Pollution and Waste Water TreatmentWater Science and Technology199633719198

32. P. D. Rose, G. A. Boshoff, R. P. van Hille, L. C. Wallace, K. M. Dunn, J. R. Duncan, An integrated algal sulphate reducing high rate ponding process for the treatment of acid mine drainage wastewater, Biodeg. 1998924757

33. H. Guha, K. Jayachandran, F. Mauresse, Kinetics of chromium (VI) reduction by atype strain Shewanella alga under different growth conditions, Environmental Pollution2001115220918

34. P. Kaewsarn, Q. Yu, . I. I. Cadmium, from. removal, solutions. aqueous, pretreated. by, of. biomass, alga. marine, sp. Padina, Environmental, Pollution2001112220913

35. N. F. Y. Tam, J. P. K. Wong, Y. S. Wong, Relationship of Algae to Water Pollution and Waste Water TreatmentEnviron. Pol.200111418592

36. K. Weber, B. Probes, K. Lyvansky, F. Kredl, I. Beryl, Removal of biogenic elements, polychlorinated diphenyls and heavy metals during the biogical final treatment of wastewaters. Acta Microbiol. Pol. 19813025558

37. S. Shashirekha, L. Uma, G. Subramanian, Phenol degradation by marine cyanobacterium Phormodium valderianum, J. Indust.Microbiol. Biotechnol.199719213033

38. B. J. Lloyd, G. L. Frederick, Parasite removal by waste stabilisation pond systems and the relationship between concentrations in sewage and prevalence in the community, Water Science and Technology2000; 42 10 37586 .

39. W. J. Oswald, The role of microalgae in liquid waste treatment and

reclamation. In: C.A. Lembi and J.R. Waalnd (eds). Relationship of Algae to Water Pollution and Waste Water TreatmentCambridge University Press1988a40331

40. W. J. Oswald, Microalgae and Relationship of Algae to Water Pollution and Waste Water TreatmentBiotechnology, M.A. Borowitzka and L.J. Borowitzka (eds).Cambridge University Press, New York1988b357394

41. J. U. Grobbelaar, D. J. Soeder, E.. Stengel, Modelling algal production in large outdoor cultures and waste treatment systems, Biomass199021297314

42. S. J. Arceivala, Simple waste treatment methods. Metu Eng. Fac. Pub.1973 44 Ankara.

43. Lovaie, and De La Noüe, J. Hyperconcentrated cultures of Scenedesmus obliquus: A new approach for wastewater biological tertiary treatment, Water Res 198519143742

44. G. Laliberte, D. Proulx, De Pauw, and De La Noüe, J.,. Algal Technology in Wastewater Treatment. In: H. Kausch and W. Lampert (eds.), Advances in Limnology. E. Schweizerbart'sche Verlagsbuchhandlung, Stuttgart1994283382

45. D. S. Filip, T. Peters, V. D. Adams, E. J. Middlebrooks, Residual heavy metal removal by an algae-intermittent sand filtration system1979Water Res. 13305313

46. A. Nakajima, T. Horikoshi, T. Sakaguchi, Studies on the accumulation heavy metal elements in biological system XVII. Selective accumilation of heavy metal ions by Chlorella vulgaris. Eur. J. App. Microbiol. Biotechnol.1981127683

47. Y. P. Ting, E. Lawson, I. G. Prince, Uptake of cadmium and zinc by alga Chlorella vulgaris: Part I. İndividual ion species. Biotechnol. Bioeng. 19893499099

48. J. M. Hassett, J. C. Jennett, J. Smith, E., Relationship of Algae to Water Pollution and Waste Water TreatmentAppli. Environ. Microbiol1981411097106

49. T. Sakaguchi, A. Nakajima, T. Horikoshi, Studies on the accumulation heavy metal elements in biological system XVIII. Accumilation of molybdenum by green microalgae. Eur. J. App. Microbiol. Biotechnol.1981128489

50. G. H. Wikfors, R. Ukeles, Growth and adaptation of estaurine unicellular algae in media with excess copper, cadmium and zink and effect of metal contaminated algal food on Crassostrea virginica larvae. Mar. Ecol. Prog.

Ser.19827191206

51. L. G. Rich, Low Maintenance Mechanically Simple Wastewater Treatment Systems. McGraw-Hill, New York,1980

52. T. H. Y. Tebbutt, Relationship of Algae to Water Pollution and Waste Water Treatmentth ed, Butterworth-Heinemann, Oxford1998

53. D. D. Mara, S. W. Mills, H. W. Person, G. P. Alabaster, stabilization. Waste, A. ponds, alternatives. viable, small. for, treatment. community, Journal. systems, Water. of, Environmental, Management1992617278

54. NRC,. Microbial Processes: Promising Technologies for Developing Countries.National Academy Press, Washington D.C.1979p.

55. W. J. Oswald, C. G. Golueke, The high rate pond in waste disposal. Devel. Indust. Microb.1963411219

56. G. Shelef, R. Moraine, G. Oron, Photosynthetic Biomass Production from Sewage. Arch. Hydrobiol. Beih. 11314

57. I. Esen, K.. Puskas, Relationship of Algae to Water Pollution and Waste Water TreatmentWaste Management1991115965

58. W. J. Oswald, Introduction to Advanced Integrated Ponding Systems. Water Science and Technology199124517

59. DOE,. Alternative Wastewater Treatment: Advanced Integrated Pond Systems, US Departmant of Energy, Office of Energy Efficiency and Renewable Energy1993Technical Information Program Document No: DOE/CH100093246Washington. 8p.

Chapter 8

DETERMINATION OF TRACE METALS IN WASTE WATER AND THEIR REMOVAL PROCESSES

Asli Baysal[1, 2], Nil Ozbek[1] and Suleyman Akman[1]

[1]Istanbul Technical University, Science and Letters Faculty, Chemistry Dept., Istanbul, Turkey

[2]Yeditepe University, Faculty of Engineering and Architecture, Department of Genetics and Bioengineering, Kayisdagi- Istanbul, Turkey

INTRODUCTION

Since the second part of 20th century, there has been growing concern over the diverse effects of heavy metals on humans and aquatic ecosystems. Environmental impact of heavy metals was earlier mostly attributed to industrial sources. In recent years, metal production emissions have decreased in many countries due to strict legislation, improved cleaning/purification technology and altered industrial activities. Today and in the future, dissipate losses from consumption of various metal containing goods are of most concern. Therefore, regulations for heavy metal containing waste disposal have been tightened [1].

A significant part of the anthropogenic emissions of heavy metals ends up in wastewater. Major industrial sources include surface treatment processes with elements such as Cd, Pb, Mn, Cu, Zn, Cr, Hg, As, Fe and Ni, as well as industrial products that, at the end of their life, are discharged in wastes. Major urban inputs to sewage water include household effluents, drainage water, business effluents (e.g. car washes, dental uses, other enterprises, etc.), atmospheric deposition, and traffic related emissions (vehicle exhaust, brake linings, tires, asphalt wear, gasoline/oil leakage, etc.) transported with storm water into the sewerage system. For most applications of heavy metals, the applications are estimated to be the same in nearly all countries, but the consumption pattern may be different. For some applications which during the last decade has been phased out in some countries, there may, however, today be significant differences in uses [2-4].

Most common sources of heavy metals to waste and/or waste water are [1]; (i) Mining and extraction; by mining and extraction a part of the heavy metals will end up in tailings and other waste products. A significant part of the turn over of the four heavy metals with mining waste actually concerns the presence of the heavy metals in waste from extraction of other metals like zinc, copper and nickel. It should, however, be kept in mind that mining waste is generated independent of the subsequent application of the heavy metal. (ii) Primary smelting and processing; a minor part of the heavy metals will end up in waste from the further processing of the metals. (iii) Use phase; a small part of the heavy metals may be lost from the products during use by corrosion and wear. The lost material may be discharged to the environment or end up in solid waste either as dust or indirectly via sewage sludge. (iv) Waste disposal; the main part of the heavy metals will still be present when the discarded products are disposed off. The heavy metals will either be collected for recycling or disposed of to municipal solid waste incinerators (MSWI) or landfills or liquid waste. A minor part will be disposed of as chemical waste and recycled or landfilled via chemical waste treatment. (v) vulconic erruptions. (vi) fossil fuel combustion. (vii) agriculture (viii) erosions (ix) metallurgical industries. Actually metal pollutants are neither generated nor compleletely eliminated; they are only transferred from one source to another. Their chemical forms may be changed or they are collected and immobilized not to reach the human, animals or plants.

The term *heavy metal* has never been defined by any authoritative body such as The International Union of Pure and Applied Chemistry (IUPAC). It has been given such a wide range of meanings by different authors that it is effectively meaningless. No relationship can be found between density (specific gravity) and any of the various physicochemical concepts that have been used to define "heavy metals" and the toxicity or ecotoxicity attributed to *heavy metals*. The term bioavailability is more appropriate to define the potential toxicity of metallic elements and their compounds. Bioavailability depends on biological parameters as well as the physicochemical properties of metallic elements, their ions, and compounds. These in turn depend upon the atomic structure of the metallic elements. Thus, any classification of the metallic elements to be used in scientifically based legislation must itself be based on the periodic table or some subdivision of it. In conclusion, heavy metals commonly used in industry and generically toxic to animals and to aerobic and anaerobic processes, but all of them are not dense nor entirely metallic. Includes As, Cd, Cr, Cu, Pb, Hg, Ni, Se, Zn. All of them pose a number of undesired properties that affect humans and the environment [5].

Effluents from textile, leather, tannery, electroplating, galvanizing, pigment and dyes, metallurgical and paint industries and other metal processing and re-fining operations at small and large-scale sector contain considerable amounts of toxic metal ions [4]. The toxic metals and their ions are not only potential human health hazards but also to another life forms. Toxic metal ions cause physical discomfort and sometimes life-threatening illness including irrevers-ible damage to vital body system [6]. From the eco-toxicological point of view, the most dangerous metals are mercury, lead, cadmium and chromium(VI). In many instances, the effect of heavy metals on human is not well understood. Metal ions in the environment bioaccumulate and are biomagnified along the food chain. Therefore, their toxic effect is more pronounced in animals at high-er trophic levels. Mine tailing and effluents from non-ferrous metals industry are the major sources of heavy metals in the environment. Among commonly used heavy metals, Cr(III), Cu, Zn, Ni and V are comparatively less toxic then Fe and Al. Cu is mainly employed in electric goods industry and brass produc-tion. Major applications for Zn are galvanization and production of alloys. Cadmium has a half-life of 10–30 years and its accumulation in human body affects kidney, bone and also causes cancer and its use is increasing in indus-trial applications such as electroplating and making pigments and batteries. Chromium compounds are nephrotoxic and carcinogenic in nature. As a result of increasing awareness about the toxicity of Hg and Pb, their large-scale use by various industries has been either curtailed or eliminated. An effluent treat-ment facility within the industry discharging heavy metals contaminated efflu-ent will be more efficient than treating large volumes of mixed wastewater in a general sewage treatment plant. Thus it is beneficial to devise separate treat-ment procedures for scavenging heavy metals from the industrial wastewater [4,6,7].

The analysis of wastewater for trace and heavy metal contamination is an important step in ensuring human and environmental health. Wastewater is regulated differently in different countries, but the goal is to minimize the pollution introduced into natural waterways. In recent years, metal production emissions have decreased in many countries due to heavy legislation, improved production and cleaning technology. A variety of inorganic techniques can be used to measure trace elements in waste water including flame atomic absorption spectrometry (FAAS) and graphite furnace (or electrothermal) atomic absorption spectrometry (GFAAS or ETAAS), inductively coupled plasma optical emission spectrometry (ICP-OES) and inductively coupled plasma mass spectrometry (ICP-MS). Depending upon the number of elements to be determined, expected concentration range of analytes and the number of samples to be run, the most suitable technique for business requirements can be chosen.

Several industrial wastewater streams may contain heavy metals such as Sb, Cr, Cu, Pb, Zn, Co, Ni, etc. The toxic metals, existing in high or even in low concentrations, must be effectively treated/removed from the wastewaters. Among the various treatment methods applied to remove heavy or trace metals, chemical precipitation process has been the most common technology. The conventional heavy metal removal process has some inherent shortcomings such as requiring a large area of land, a sludge dewatering facility, skillful operators and multiple basin configuration. In recent years, some new processes such as biosorption, neutralization, precipitation, ion exchange, adsorption etc. have been developed and extensively used for the heavy metal removal from wastewater.

In this chapter the novel and common methods for the determination of trace heavy metals in waste water and their removal processes are explained.

TOXICITY EFFECTS OF SOME HEAVY METALS IN THE WASTEWATER [1,3,5]

All heavy metals are effected to human and environment by different ways. For example; lead in the environment is mainly particulate bound with relatively low mobility and bioavailability. Lead does, in general, not bioaccumulate and there is no increase in concentration of the metal in food chains. Lead is not essential for plant or animal life. Of particular concern for the general population is the effect of lead on the central nervous system. Lead has been shown to have effects on haemoglobin synthesis and anaemia has been observed in children at lead blood levels above 40 µg/dl. Lead is known to cause kidney damage. Some of the effects are reversible, whereas chronic exposure to high lead levels may result in continued decreased kidney function and possible renal failure. The evidence for carcinogenicity of lead and several inorganic lead compounds in humans is inadequate. Classification of The International Agency for Research on Cancer (IARC) is class 2B which is the agent (mixture) is possibly carcinogenic to humans. The exposure circumstance entails exposures that are possibly carcinogenic to humans. In the environment, lead binds strongly to particles, such as soil, sediment and sewage sludge. Because of the low solubility of most of its salts, lead tends to precipitate out of complex solutions. It does not bioaccumulate in most organisms, but can accumulate in biota feeding primarily on particles, e.g. mussels and worms. These organisms often possess special metal binding proteins that removes the metals from general distribution in their organism. Like in humans, lead may accumulate in the bones. One of the most important factors influencing the aquatic toxicity of lead is the free ionic concentration and the availability of lead to organisms. Lead is unlikely to affect aquatic plants at levels that might be found in the general environment. Mercury is a peculiar metal. Most conspicuous is its fluidity at room temperature, but more important for the possible exposure of man and the environment to mercury are two other properties:

1. Under reducing conditions in the environment, ionic mercury changes to the uncharged elemental mercury which is volatile and may be transported over long distances by air.

2. Mercury may be chemically or biologically transformed to methylmercury and dimethylmercury, of which the former is bioaccumulative and the latter is also volatile and may be transported over long distances. Mercury is not essential for plant or animal life. The organic forms of mercury are generally more toxic to aquatic organisms than the inorganic forms. Aquatic plants are affected by mercury in the water at concentrations approaching 1 mg/litre for inorganic mercury, but at much lower concentrations of organic mercury.

Cadmium and cadmium compounds are, compared to other heavy metals, relatively water soluble. They are therefore also more mobile in e.g. soil, generally more bioavailable and tends to bioaccumulate. Cadmium is not essential for plant or animal life. Cadmium is readily accumulated by many organisms, particularly by microorganisms and molluscs where the bioconcentration factors are in the order of thousands. In aquatic systems, cadmium is most readily absorbed by organisms directly from the water in its free ionic form Cd (II). The acute toxicity of cadmium to aquatic organisms is variable, even between closely related species, and is related to the free ionic concentration of the metal. Cadmium interacts with the calcium metabolism of animals. In fish it causes lack of calcium (hypocalcaemia), probably by inhibiting calcium uptake from the water. Effects of long-term exposure can include larval mortality and temporary reduction in growth.

Chromium occurs in a number of oxidation states, but Cr(III) (trivalent chromium) and Cr(IV) (hexavalent chromium) are of main biological relevance. There is a great difference between Cr(III) and Cr(VI) with respect to toxicological and environmental properties, and they must always be considered separately. Chromium is similar to lead typically found bound to particles. Chromium is in general not bioaccumulated and there is no increase in concentration of the metal in food chains. Contrary to the three other mentioned heavy metals, Cr(III) is an essential nutrient for man in amounts of 50 - 200 µg/day. Chromium is necessary for the metabolism of insulin. It is also essential for animals, whereas it is not known whether it is an essential nutrient for plants, but all plants contain the element. In general, Cr(III) is considerably less toxic than Cr(VI). Cr(VI) has been demonstrated to have a number of adverse effects ranging from causing irritation to cancer. Hexavalent chromium is in general more toxic to organisms in the environment that the trivalent chromium. Almost all the hexavalent chromium in the environment is a result of human activities. Chromium can make fish more susceptible to infection; high concentrations can damage and/or accumulate in various fish tissues and in invertebrates such as snails and worms. Reproduction of the water flea Daphnia was affected by exposure to 0.01 mg hexavalent chromium/litre. Hexavalent chromium is accumulated by aquatic species by passive diffusion. In general, invertebrate species, such as polychaete worms, insects, and crustaceans are more sensitive to the toxic effects of chromium than vertebrates such as some fish. The lethal chromium level for several aquatic and terrestrial invertebrates has been reported to be 0.05 mg/litre.

Copper can be found in many wastewater sources including, printed circuit board manufacturing, electronics plating, plating, wire drawing, copper polishing, paint manufacturing, wood preservatives and printing operations. Typical concentrations vary from several thousand mg/l from plating bath waste to less than 1 ppm from copper cleaning operations. Copper can be found in many kinds of food, in drinking water and in air. Because of that we absorb eminent quantities of copper each day by eating, drinking and breathing. The absorption of copper is necessary, because copper is a trace element that is essential for human health. Although humans can handle proportionally large concentrations of copper, too much copper can still cause eminent health problems. When copper ends up in soil it strongly attaches to organic matter and minerals. As a result it does not travel very far after release and it hardly ever enters groundwater. In surface water copper can travel great distances, either suspended on sludge particles or as free ions [8].

Nickel is a naturally occurring element widely used in many industrial applications for the shipbuilding, automobile, electrical, oil, food and chemical industries. Although it is not harmful in low quantities, nickel is toxic to humans and animals when in high concentrations. Nickel can be present in wastewater as a result of human activities. Sources of nickel in wastewater include ship cruise effluents, industrial applications and the chemical industry [9].

Arsenic is found in wastewater from electronic manufactures making gallium arsenide wafers and electronic devices. It also can be found in silicon semiconductor operations that use high dose arsenic implants. Other sources of arsenic are ground water in agricultural areas where arsenic was once used as an insecticide. Most environmental arsenic problems are the result of mobilization under natural conditions. However, mining activities, combustion of fossil fuels, use of arsenic pesticides, herbicides, and crop desiccants and use of arsenic additives to livestock feed create additional impacts. Arsenic exists in the −3, 0, +3 and +5 oxidation states. Each of them have diffrenet toxic effect both human and environment.

Actually all chemicals, including even essential elements, drugs and in fact water, are toxic above (and below) their limiting values. However, some elements such as arsenic lead, cadmium, mercury, described as toxic, are known to be toxic for living beings at any concentration and they are not asked to be taken in to the body even in ultratrace levels.

DETERMINATION TECHNIQUES OF HEAVY METALS IN WASTE WATER SAMPLES

In order to determine the heavy trace metals, there are many inorganic techniques such as FAAS, ETAAS, ICP-OES, ICP-MS as well as anodic stripping and recently laser induced breakdown spectroscopy (LIBS). Each technique has its own advantages and disadvantages which will be discussed in this chapter.

Actually all the steps of an analysis, namely (i) representative sampling, (ii) to prevent analyte loss e.g. its sorption on vessel wall, (iii) contamination from the environment, wares, chemicals added to the sample, (iv) transfer the sample to the lab, (v) treatment of sample prior to analysis (leaching, extraction, preconcentration/separation of the analytes, (vi) choose of the method considering its limitations, (vii) calibration of the vessels, instrument etc, (viii) preparation of sample, all solutions, standards correctly and appropriately, (ix) to test the accuracy of the method using Certified Reference Materials (CRM), (x) evaluation of results statistically and reporting are all the rings of a chain. Each step is important and potential source of error if not applied conveniently. The weakest ring of the chain limits the accuracy and quality of the results. If it is broken, the analysis collapses. Therefore, all the steps of an analysis should be performed with caution. A problem or error even in one of those steps causes the result to be wrong. As in all analyses, sample preparation step is the most important one which should be completed quickly, easy and safely. Waste water samples may contain particulates or organic materials which may require pretreatment before spectrometric analysis. In order to analyze total metal content of a sample, concentration of metals inorganically and organically bound, dissolved or particulated materials should be found.

As stated in Standard Methods, samples which are colorless and transparent, having a turbidity of <1 NTU (Nephelometric Turbidity Unit), no odor and single phase may be analyzed directly or, if necessary, after enrichment by atomic absorption spectrometry (flame or electro thermal vaporization) or inductively coupled plasma spectrometry (atomic emission or mass spectrometry) for metals without digestion. For further verification or if changes in existing matrices are encountered, comparison of digested and undigested samples should be done [10].

If samples have particulates and only the dissolved metals will be analyzed, filtration of sample and analyzing of filtrate will be enough. To be on the safe side, if particulates involved, convenient digestion procedures are suggested. Since different filtration procedures produce different blank values, it is always suggested to study with a blank solution. If only the metal contents of particulates are asked to be determined, then the sample is filtered and the filter is digested and analyzed.

In order to reduce interferences, organic matrix of the sample should be destroyed by digestion as well as metal containing compounds are decomposed to obtain free metal ions which can be determined by atomic absorption spectrometry (AAS) or inductively coupled plasma (ICP) more conveniently. The procedures for destroying organic material and dissolving heavy metals fall into three groups; wet digestion by acid mixtures prior to elemental analysis, dry ashing, followed by acid dissolution of the ash and microwave assisted digestion [11]. In Standard Methods if metal concentration is around 10-100, it is advised to digest 10 mL of sample. For less metal concentrations, sample volume could be around 100 mL for subsequent enrichment [10]. For most digestion procedures, nitric acid is used which is an acceptable matrix for both flame and electrothermal atomic absorption and ICP-MS [12]. For nitric acid digestion, 100 mL of sample is heated in a beaker with 5 mL concentrated nitric acid. Boiling should be prevented and addition of acid should be repeated til a light colored, clear solution is obtained [10]. Sometimes, if the samples involve readily oxidasable organic matters, mixtures of HNO_3- H_2SO_4 or HNO_3- HCl may be used. Samples with high organic contents, mixtures of HNO_3- $HClO_4$ or HNO_3-H_2O_2 or HNO_3-$HClO_4$-HF can be used. The latter is especially important for the dissolution of particulate matter. For samples which have high organic content, dry ashing may be favored. Wet digestion systems are performed either with a reflux or in a beaker on a laboratory hot plate. These methods are temperature limited because of the risk of contaminants from the air, laboratory equipment etc. Also there may be lost of volatile elements (As, Cd, Pb, Se, Zn and Hg etc.). Temperature limitation can be overcome by closed pressure vessels, i.e. microwave digestion. Closed systems allow high pressures above atmosphere to be used. This allows boiling at higher temperatures and often leads to complete dissolution of most samples [13]. In the American Society for Testing and Materials (ASTM) Standards (D1971-11) 'Standard practices for digestion of water samples for determination of metals by FAAS, ETAAS, ICP-OES or ICP-MS' for waste water samples it is advised to use, 100 volume of sample: 5 volume HCl: 1 volume HNO_3 is put to microwave digestion vessels for 30 minutes at 121°C and 15 psig [14]. A comparision of digestion techniques is given in Table 1.

Table 1: Comparision of Digestion Techniques

	Wet Ashing	**Microwave Digestion**
Time Consumption	Slow	Rapid
Temperature	Low	High
Pressure	Atmospheric	Above Atmospheric
Operator Skills	High	Moderate
Safety	Corrosive-explosive reagents	Corrosive-explosive reagents
Operating Cost	Low	High
Environmental Effect	High	Low
Analyte Loss&Contamination	High	Low

After choosing the effective sample preparation step, most useful techniques were explained below, like atomic absorption spectrometry (AAS), inductively coupled plasma optical emission spectrometry (ICP-OES), inductively coupled plasma mass spectrometry (ICP-MS), laser induced breakdown spectroscopy (LIBS) and anodic stripping.

Atomic Absorption Spectrometry

Atomic Absorption Spectrometry (AAS) is an analytical method for quantification of over 70 different elements in solution or directly in solid samples. Procedure depends on atomization of elements by different atomization techniques like flame (FAAS), electrothermal (ETAAS), hydride or cold vapor. Each atomization technique has its advantages and limitations or drawbacks. A comparison of several AAS techniques is given in Table 2.

Two types of flame are used in FAAS: (i) air/acetylene flame, (ii) nitrous oxide/acetylene flame. Flame type depends on thermal stability of the analyte and its possible compounds formed with flame concomitants. Temperature formed in air-acetylene flame is around 2300°C whereas acetylene-nitrous oxide (dinitrogen oxide) flame is around 3000°C [15]. Generally with air/ acetylene flame antimony, bismuth, cadmium, calcium, cesium, chromium, cobalt, copper, gold, itidium, iron, lead, lithium, magnesium, manganese, nickel, palladium, platinium, potassium, rhodium, ruthenium, silver, sodium, strontium, thallium, tin and zinc can be determined. On the other hand for refractory elements such as aluminium, barium, molybdenum, osmium, rhenium, silicon, thorium, titanium and vanadium, nitrous oxide/acetylene flame should be used [10]. But some elements like vanadium, zirconium, molibdenium and boron has lower sensitivity in the determination by FAAS because the temperature is insufficient to break down compounds of these elements. Samples should be in solution form, or digested to be detected by FAAS. Typical detection limits are around ppm range and sample analysis took 10-15 seconds per element [16]. The block diagram of FAAS and GFAAS is depicted in Figure 1. Generally, hollow cathode lamps as source, flame or graphite furnace as an atomizer, grating as a wavelength selector and photomultiplier as a detector are used.

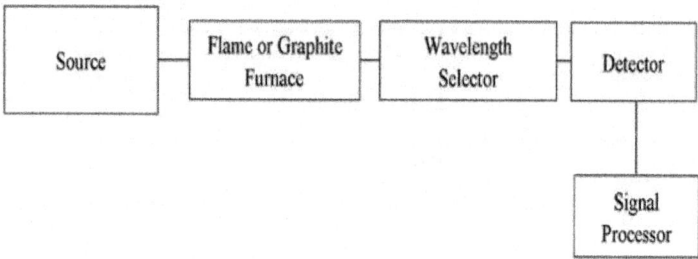

Figure 1: Block Diagram for FAAS and GFAAS.

Mahmoud et al. determined Cr, Mn, Fe, Co, Ni, Cu, Zn, Cd and Pb by FAAS after enrichment with chemically modified silica gel N-(1-carboxy-6-hydroxy) benzylidenepropylamine (SiG-CHBPA) [16]. Afkhami et al. determined Cd in water samples after cloud point extraction in Triton X-114 without adding chelating agents [18]. Mohamed et al. determined chromium species based on the catalytic effect of Cr(III) and/or Cr(VI) on the oxidation of 2-amino-5-methylphenol (AMP) with H_2O_2 by FAAS [19]. Mahmoud et al. pre-concentrated Pb(II) by newly modified three alumina–physically loaded-dithizone adsorbents then determined by FAAS [20]. Cassella et al. prepared a minicolumn packed with a styrene-divinylbenzene resin functionalized with (S)-2-[hydroxy-bis-(4-vinyl-phenyl)-methyl]- pyrrolidine-1-carboxylic acid ethyl ester to determine Cu in water samples [21]. Carletto et al. used 8-hydroxyquinoline-chitosan chelating resin in an automated on-line preconcentration system for determination of Zn(II) by FAAS [22]. Gunduz et al. preconcentrated Cu and Cd using TiO_2 core-Au shell nanoparticles modified with 11-mercaptoundecanoic acid and analysed their slurry [23].

ETAAS is basically same as FAAS; the only difference is flame is replaced by graphite tube which can be heated up to 3000 °C for atomization. Since sample is atomized in a much smaller volume the atoms density will be higher, its detection limit is much more than FAAS, around ppb range. Graphite furnace program typically consists of four stages; drying for evaporation of solvent; pyrolysis for removal of matrix constituents; atomization for generation of free gaseous atoms of the analyte; cleaning for removal of residuals in high temperature. Generally samples are liquids, but there are some commercial solid sampling instruments also. Analyze took 3-4 minutes per element. 50 and more elements can be analyzed by GFAAS [15].

Burguera et al. determined of beryllium in natural and waste waters using on-line flow-injection preconcentration by precipitation dissolution for electrothermal atomic absorption spectrometry. They used a precipitation method quantitatively with NH_4OH-NH_4Cl and collected in a knotted tube of Tygon without using a filter then the precipitate was dissolved with nitric acid injected to graphite furnace [24]. Baysal et al. accomplished to preconcentrate Pb by cobalt/pyrrolidine dithiocarbamate complex $(Co(PDC)_2)$. For this purpose, at first, lead was coprecipitated with cobalt/pyrrolidine dithiocarbamate complex formed using ammonium pyrrolidine dithiocarbamate (APDC) as a chelating agent and cobalt as a carrier element. The supernatant was then separated and the slurry of the precipitate prepared in Triton X-100 was directly analyzed [25].

Hydride generation atomic absorption spectrometry is a technique for some metalloid elements such as arsenic, antimony, selenium as well as tin, bismuth and lead which are introduced to instrument in gas phase. Hydride is generated mostly by adding sodium borohydride to the sample in acidic media in a generator chamber. The volatile hydride of the analyte generated is transferred to the atomizer by inert gas where it is atomized. The oxidation state of the metaloid is very improtant so before introducing to the hydride system, specific metalloid oxidation state should be produced. This method lowers limit of detection (LOD) 10-100 times [15,27]

Coelho et al, presented a simple procedure was developed for the direct determination of As(III) and As(V) in water samples by flow injection hydride generation atomic absorption spectrometry (FI–HG–AAS), without pre-reduction of As(V) [24]. Cabon and Madec determined antimony in sea water samples by continuous flow injection hydride generation atomic absorption spectrometry. After continuous flow injection hydride generation and collection onto a graphite tube coated with iridium, antimony was determined by graphite furnace atomic absorption spectrometry [28]. Yersel et al. developed a seperation method with a synthetic zeolite (mordenite) was developed in order to eliminate the gas phase interference of Sb(III) on As(III) during quartz furnace hydride generation atomic absorption spectrometric determination [29]. Anthemidis et al. determined arsenic (III) and total arsenic in water by using an on-line sequential insertion system and hydride generation atomic absorption spectrometry [31]. Erdogan et al. determined inorganic arsenic species by hydride generation atomic absorption spectrometry in water samples after preconcentration/separation on nano ZrO_2/B_2O_3 by solid phase extraction [31]. Korkmaz et al. developed a novel silica trap for lead determination by hydride generation atomic absorption spectrometry. The device consists of a 7.0cm silica tubing which is externally heated to a desired temperature. The lead hydride vapor is generated by a conventional hydride-generation flow system. The trap is placed between the gas–liquid separator and silica T-tube; the device traps analyte species at 500 °C and releases them when heated further to 750 °C. The presence of hydrogen gas is required for revolatilization; O_2 gas must also be present [32].

Cold vapour atomization technique is used for the determination of mercury which is the only element to have enough vapour pressure at room temperature. Method is based on converting mercury into Hg^{+2}, followed by reduction of Hg^{+2} with tin(II)chloride or borohydride. Then produced elementel mercury swept into a long-pass absorption tube along with an inert gas. Absorbance of this gas at 253.7 nm determines the concentration. Detection limit is around ppb range. Beside to inorganic mercury compounds, organic mercury compounds are problematic as they cannot be reduced to the element by sodium tetrahydroborate, and particularly not by stannous chloride. So it is advised to apply an appropriate digestion methode prior to the actual determination [15].

Kagaya et al. managed to determine organic mercury, including methylmercury and phenylmercury, as well as inorganic mercury by cold vapor atomic absorption spectrometry (CV-AAS) by adding sodium hypochloride solution [33]. Pourreza and Ghanemi developed a novel solid phase extraction for the determination of mercury. The Hg(II) ions were retained on a mini-column packed with agar powder modified with 2-mercaptobenzimidazole. The retained Hg(II) ions were eluted and analysed by CV-AAS [34]. Sahan and Sahin developed for on-line solid phase preconcentration and cold vapour atomic absorption spectrometric determination of Cd(II) in aqueous samples. Lewatit Monoplus TP207 iminodiacetate chelating resin was used for the separation and preconcentration of Cd(II) ions at pH 4.0 [35].

However, qualitative analysis cannot be made by AAS because a specific hollow cathode lamp (HCL) is used for each element. Therefore, elements should be determined one by one which make a qualitative analysis almost impossible. In addition, non-metals cannot be determined because their atomic absorption wavelengths are in far UV range which is not suitable for analysis due to absorption of air components.

Since 2004, new generation high resolution continuum source atomic absorption spectrometer (HR-CS-AAS) which is equipped with high intensity xenon short-arc lamp, high resolution double monochromator, CCD detector are produced. The continuous source lamp emits radiation of intensity at least an order of magnitude above that of a typical hollow cathode lamp (HCL) over the entire wavelength range from 190 nm to 900 nm. With these instruments, aside from the analysis line, the spectral environment is also recorded simultaneously, which shows noises and interferences effecting analysis. Improved simultaneous background correction and capabilities to correct spectral interferences, increase the accuracy of analytical results. With high resolution detector, interferences are minimized through optimum line separation. With these instruments, not only metals and non-metals e.g. F, Cl, Br, I, S, P can be determined by their hyperfine structured diatomic molecular absorbances. There are various papers for fluoride determination by GaF [36], SrF [37], AlF [38], CaF [39], chloride by AlCl [40], InCl [41], bromide by AlBr [42], CaBr [43], sulfur by CS [44], phosphorus by PO [45].

Table 2: Comparision of AAS techniques

	FAAS	**GFAAS**	**Hydride Generation AAS**	**Cold Vapour AAS**
Elements	68	50	As, Se, Sb, Bi, Pb, Sn	Hg
Limit of Detection	+	++++	++++	++++
Precision	++++	+	+	+
Interferences	+++	+	++++	++++
Analysis Time	++++	+	++	++
Sample Preparation	+++	+++	++	++
Operation Skills	+++	++	++	++
Operation Costs	++++	++	++	++

+Bad, ++:Moderate, +++:Good, ++++:Very Good

Inductively Coupled Plasma Optical Emission Spectrometry

Inductively coupled plasma-optical (or atomic) emission spectrometry (ICP-OES or ICP-AES) is an analytical technique used for determination of trace metals. This is a multi-element technique which uses a plasma source to excite the atoms in samples. These excited atoms emit light of a characteristic wavelength, and a dedector measures the intensity of the emitted light, which is related with the concentration. Samples are heated through 10000 °C to atomize effectively which is an important advantage for ICP technique.

Another advantage is multi element analysis. With ICP technique, 60 elements can be analysed in single sample run less than a minute simultaneously, or in a few minutes sequentially. Besides instrument is only optimized for one time for e set of metal analysis. High operating temperature lowers the interferences. Determinations can be accomplised in wide lineer range and refractory elements can be determined at low concentrations (B, P, W, Zr, U). On the other hand consumption of inert gas is very much higher than AAS techniques which cause high operating costs.

ICP instruments can be 'axial' and 'radial' according to their plasma configuration. In radial configuration, the plasma source is viewed from the side. Emissions from axial plasma are viewed from horizontally along its length, which reduces background signals resulting in lower detection limits. Some instruments have both viewing modes [46]. The block diagram of ICP-OES is depicted in Figure 2. Generally, radio frequency (RF) powered torch as a source, polychromators as a wavelength selector, photomultiplier (PMT) or charge capacitive discharged arrays (CCD) as a dedectors are used.

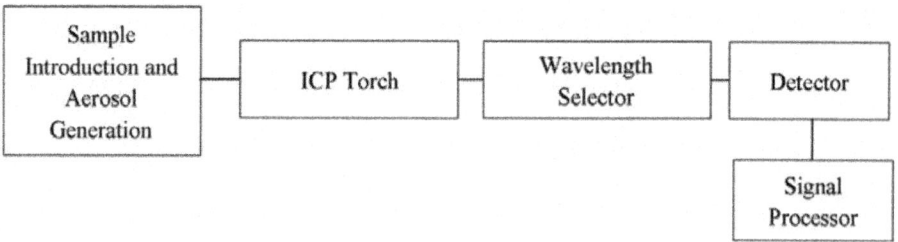

Figure 2: Block Diagram of ICP-OES.

Enrichment/separation procedures have been applied prior to ICP analyses as well. Atanassove et al. used sodium diethyldithiocarbamate to co-precipitate for the pre-concentration of Se, Cu, Pb, Zn, Fe, Co, Ni, Mn, Cr and Cd to detect by ICP-AES [47]. Zougagh et al. determined Cd in water ICP-AES with on-line adsorption preconcentration using DPTH-gel and TS-gel microcolumns [48]. Zogagh et al. developed a simple, sensitive, low-cost and rapid, flow injection system for the on-line preconcentration of lead by sorption on a microcolumn packed with silica gel funtionalized with methylthiosalicylate (TS-gel) then the metal is directly retained on the sorbent column and subsequently then eluted from it by EDTA and elution determined by ICP-AES [49].

Inductively Coupled Plasma Mass Spectrometry

Inductively coupled plasma mass spectrometry (ICP-MS) is a multi-element technique which uses plasma source to atomize the sample, and then ions are detected by mass spectrometer. Mass spectrometer separate ions according to their mass to charge ratio. This technique has excellent detection limits, in ppt (part per thousand) range. Samples generally introduced as an aerosol, liquid or solid. Solid samples are dissolved prior to analysis or by a laser solid samples are converted directly to aerosol. All elements can analyze in a minute, simultaneously. But it needs high skilled operator, because method development is moderately difficult from other techniques. There are various types of ICP-MS instruments; HR-ICP-MS (high resolution inductively coupled plasma mass spectrometry and MC-ICP-MS (multi collector inductively coupled plasma mass spectrometry). HR-ICP-MS, has both magnetic sector and electric sector to separate and focus ions. By these instruments eliminatenation or reduction of the effect of interferences due to mass overlap is accomplished but operation cost, time and complexity will increase. MC-ICP-MS, are designed to perform high-precision isotope ratio analysis. They have multiple detectors to collect every isotope of a single element but the major disadvantage of system is that all the isotopes should be in a narrow mass range which eliminates these instruments from routine analysis [46]. The block diagrame of ICP-MS is depicted in Figure 3. The main difference from ICP-OES is that quadropole mass spectrometers are used instead of wavelength selectors to detect the analytes.

Krishna et al. used moss (Funaria hygrometrica), immobilized in a polysilicate matrix as substrate for speciation of Cr(III) and Cr(VI) in various water samples and determined by ICP-MS and FAAS [41]. Hu et al. simultaneously separated and speciated inorganic As(III)/As(V) and Cr(III)/Cr(VI) in natural waters by capillary microextraction with mesoporous Al_2O_3 before determination with ICP-MS [51]. Chen et al. speciated of chromium in waste water using ion chromatography coupled inductively coupled plasma mass spectrometry [52].

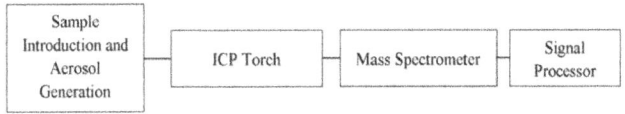

Figure 3: Block diagram of ICP-MS instrument.

Both ICP-OES and ICP-MS are not free of interferences. ICP-OES suffers from spectral interferences due to wavelength overlap of different elements. Similarly, in ICP-MS, the combination of different elements forms diatomic molecules which give the same (or indistinguishable) signal as that the analyte. In Table 3, a comparision of AAS and ICP techniques is given.

Table 3: Comparision of AAS and ICP Techniques

	FAAS	GFAAS	ICP-OES	ICP-MS
Analysis Time	++	+	+++	+++
Cost of Instrument	+++	++	++	+
Solid Sample Analysis	-	+++	-	-
Operating Cost	+	++	++	++
LOD	+	++	++	+++
Lineer Range	+	+	+++	+++
Precision	+++	+	++	++

-: Cannot Accomplished, +: Bad, ++: Medium, +++: Good

Laser Induced Breakdown Spectroscopy (LIBS)

Laser Induced Breakdown Spectroscopy (LIBS) is a type of atomic emission spectroscopy which uses a highly energetic laser pulse as the excitation source. It is based on analysing of atomic emission lines close to the surface of sample generated by laser pulse where the very high field intensity initiates an avalenche ionisation of the sample elements, giving rise to the breakdown effect. Spectral and time-resolved analysis of this emission are suitable to identify atomic species originally present at the sample surface [53]. It can determine various metals but only limitation is the power of laser, sensitivity and wavelength range of the spectrometer. Generally this technique is used for solid samples because there are many accurate methods for liquid samples which does not require preparatory steps. Addition to this, using LIBS for liquid samples may cause many problems due to the complex laser-plasma generation mechanisms in liquids [54]. Also splashing, waves, bubbles and aerosols caused by the shockwave accompanying the plasma formation effects precision and analytical performance. In order to overcome these problems, there are various procedures for liquid samples like analysing the surface of a static liquid body, the surface of a vertical flow of a liquid, the surface of a vertical flow of a liquid or of infalling droplets, the bulk of a liquid or dried sample of the liquid deposited on a solid substrate [55]. Though the results obtained were satisfactory, but it is obvious that such experimental tricks contradict with one of the most attractive advantages of LIBS, namely working on an unprepared sample, which facilitate in-situ and real-time measurements [56]. The block diagram of LIBS is depicted in Figure 4. Laser generates spark and plasma light is collected by a fiber optic and directed into a spectrograph. A sample output spectrum can be seen from figure 4.

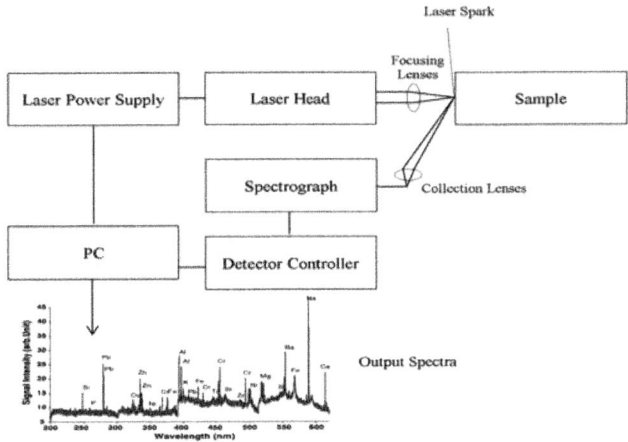

Figure 4: Scheme of an LIBS instrument (Spectra from Reference 48).

Gondal and Hussain, accomplished to determine many toxic trace elements in paint manufacturing plant waste water by LIBS. The results of LIBS method showed accuracy with the results found by ICP in the range of 0.03-0.6 %, which shows that this method can easily be used for trace element analysis [57]. Rai and Rai have also determined Cr in waste water collected from Cr-electroplating industry [58].

Anodic Stripping Voltammetry (ASV)

Anodic Stripping Voltammetry (ASV) is an analytical technique that specifically detects heavy metals in various matrices. Its sensitivity is 10 to 100 times more than ETAAS for some metals. Since its limit of detection is low, it may not require any preconcentration step. It also allows determining 4 to 6 metals simultaneously with inexpensive instrumentation. ASV technique consists of three steps. First step is electroplating of certain metals in solution onto an electrode which concentrates the metal. Second, stirring is stopped and then finally metals on the electrode are stripped off which generates a current that can be measured. This current is characteristic for each metal and by its magnitude quantification can be done. The stripping step can be either linear, staircase, square wave, or pulse [59].

Sonthalia et al. used anodic stripping for determination of various metals (Ag, Cu, Pb, Cd and Zn) in several waste water samples. Boron-doped diamond thin film is used [60]. McGaw and Swain compared the performance of boron-doped diamond (BDD) with Hg-coated glassy carbon (Hg-GC) electrode for the anodic stripping voltammetry (ASV) for determination of heavy metal ions

(Zn^{2+}, Cd^{2+}, Pb^{2+}, Cu^{2+}, Ag^{+}). Generally Hg has been used as the electrode for ASV but there is an ongoing search for alternate electrodes and diamond is one of these. Produced BDD showed comparable results with Hg electrodes [61]. Bernalte et al. determined mercury by screen-printed gold electrodes with anodic stripping voltammetry [62]. Mousavi et al. developed a sensitive and selective method for the determination of lead (II) with a 1,4-bis(prop-2-enyloxy)-9,10-anthraquinone (AQ) modified carbon paste electrode [63]. Kong et al. produced a method for the simultaneous determination of cadmium (II) and copper (II) during the adsorption process onto Pseudomonas aeruginosa was developed. The concentration of the free metal ions was successfully detected by square wave anodic stripping voltammetry (SWASV) on the mercaptoethane sulfonate (MES) modified gold electrode, while the P. aeruginosa was efficiently avoided approaching to the electrode surface by the MES monolayer [64]. Giacomino et al. investigated parameters affecting the determination of mercury by anodic stripping voltammetry using a gold electrode. Potential wave forms (linear sweep, differential pulse, square wave), potential scan parameters, deposition time, deposition potential and surface cleaning procedures were examined for their effect on the mercury peak shape and intensity and five supporting electrolytes were tested. The best responses were obtained with square wave potential wave form and diluted HCl as supporting electrolyte [65].

The literature is full of papers on the application of various methods for the determination of metals in waste water. Innumerous procedures, preconcentration/separation techniques, digestion techniques for several samples have been proposed.

REMOVAL OF HEAVY METALS FROM WASTE WATER

Cadmium, zinc, copper, nickel, lead, mercury and chromium are often detected in industrial wastewaters, which originate from metal plating, mining activities, smelting, battery manufacture, tanneries, petroleum refining, paint manufacture, pesticides, pigment manufacture, printing and photographic industries, etc. The toxic metals, probably existing in high concentrations (even up to 500 mg/L), must be effectively treated/removed from the wastewaters. If the wastewaters were discharged directly into natural waters, it will constitute a great risk for the aquatic ecosystem, whilst the direct discharge into the sewerage system may affect negatively the subsequent biological wastewater treatment [66].

In recent years, the removal of toxic heavy metal ions from sewage, industrial and mining waste effluents has been widely studied. Their presence in streams and lakes has been responsible for several types of health problems in animals, plants and human beings. Among the many methods available to reduce heavy metal concentration from wastewater, the most common ones are chemical precipitation, ion-exchange, adsorption, coagulation, cementation, electro-dialysis, electro-winning, electro-coagulation and reverse osmosis (See in Figure 5) [4, 67-70].

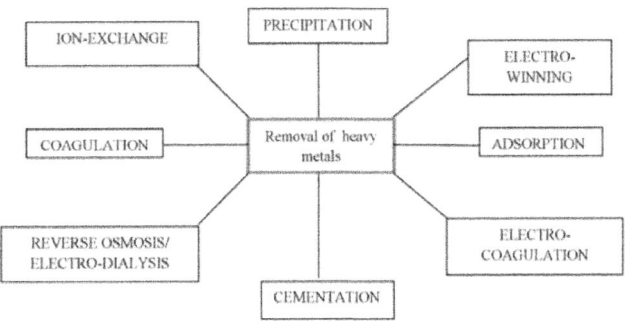

Figure 5: Some conventional methods for the removal of heavy metals.

Some conventional methods are explained below [4, 71-76];

*   Precipitation is the most common method for removing toxic heavy metals up to parts per million (ppm) levels from water. Since some metal salts are insoluble in water and which get precipitated when correct anion is added. Although the process is cost effective its efficiency is affected by low pH and the presence of other salts (ions). The process requires addition of other chemicals, which finally leads to the generation of a high water content sludge, the disposal of which is cost intensive. Precipitation with lime, bisulphide or ion exchange lacks the specificity and is ineffective in removal of the metal ions at low concentration.

*   Ion exchange is another method used successfully in the industry for the removal of heavy metals from effluents. Though it is relatively expensive as compared to the other methods, it has the ability to achieve ppb levels of clean up while handling a relatively large volume. An ion exchanger is a solid capable of exchanging either cations or anions from the surrounding materials. Commonly used matrices for ion exchange are synthetic organic ion exchange resins. The disadvantage of this method is that it cannot handle concentrated metal solution as the matrix gets easily fouled by organics and other solids in the wastewater. Moreover ion exchange is nonselective and is highly sensitive to pH of the solution.

*   Electro-winning is widely used in the mining and metallurgical

industrial operations for heap leaching and acid mine drainage. It is also used in the metal transformation and electronics and electrical industries for removal and recovery of metals. Metals like Ag, Au, Cd, Co, Cr, Ni, Pb, Sn and Zn present in the effluents can be recovered by electro-deposition using insoluble anodes.

- Electro-coagulation is an electrochemical approach, which uses an electrical current to remove metals from solution. Electro-coagulation system is also effective in removing suspended solids, dissolved metals, tannins and dyes. The contaminants presents in wastewater are maintained in solution by electrical charges. When these ions and other charged particles are neutralized with ions of opposite electrical charges provided by electrocoagulation system, they become destabilized and precipitate in a stable form.

- Cementation is a type of another precipitation method implying an electrochemical mechanism in which a metal having a higher oxidation potential passes into solution e.g. oxidation of metallic iron, Fe(0) to ferrous Fe(II) to replace a metal having a lower oxidation potential. Copper is most frequently separated by cementation along with noble metals such as Ag, Au and Pb as well as As, Cd, Ga, Pb, Sb and Sn can be recovered in this manner.

- Reverse osmosis and electro-dialysis involves the use of semi-permeable membranes for the recovery of metal ions from dilute wastewater. In electro-dialysis, selective membranes (alternation of cation and anion membranes) are fitted between the electrodes in electrolytic cells, and under continuous electrical current the associated.

Most of these methods suffer from some drawbacks such as high capital and operational costs and problem of disposal of residual metal sludge. Ionexchange is feasible when an exchanger has a high selectively for the metal to be removed and the concentrations of competing ions are low. The metal may then be recovered by incinerating the metal-saturated resin and the cost of such a process naturally limits its application to only the more valuable metals. In many cases, however, the heavy metals are not valuable enough to warrant the use of special selective exchangers/resins from an economic point of view. Cost efective alternative technologies or sorbents for treatment of metals contaminated waste streams are needed. Natural materials that are available in large quantities, or certain waste products from industrial or agricultural operations, may have potential as inexpensive sorbents. Due to their low cost, after these materials have been expended, they can be disposed of without expensive regeneration. Cost is an important parameter for comparing the sorbent materials. However, cost information is seldom reported, and the expense of individual sorbents varies depending on the degree of processing required and local availability. In general, a sorbent can be assumed as 'low cost" if it requires little processing, is abundant in nature, or is a by-product or waste material from another industry. Of course improved sorption capacity may compensate the cost of additional processing. This has encouraged research into using low-cost adsorbent materials to purify water contaminated with metals. Another major disadvantage with conventional treatment technologies is the production of toxic chemical sludge and its disposal/treatment is not eco-friendly. Therefore, removal of toxic heavy metals to an environmentally safe level in a cost effective and environment friendly manner assumes great importance.

In light of the above, biological materials and some adsorption materials have emerged as an economic and eco-friendly option. Adsorption is one the physico-chemical treatment processes found to be effective in removing heavy metals from aqueous solutions. According to literature, an adsorbent (sorbent) can be considered as cheap or low-cost if it is abundant in nature, requires little processing and is a byproduct of waste material from waste industry [66-68, 76-78].

Of course improved sorption capacity may compensate the cost of additional processing. Some of the reported low-cost sorbents such as bark/tannin-rich materials, lignin, chitin/chitosan, dead biomass, seaweed/algae/alginate, xanthate, zeolite, clay, fly ash, peat moss, bone gelatin beads, leaf mould, moss, iron-oxide-coated sand, modified wool and modified cotton. Important parameters for the sorbent effectiveness are effected by pH, metal concentration, ligand concentration, competing ions, and particle size [4, 66-68].

Another type of sorbent is plant waste [68]. Plant wastes are inexpensive as they have no or very low economic value. Most of the adsorption studies have been focused on untreated plant wastes such as papaya wood, maize leaf, teak leaf powder, lalang (Imperata cylindrica) leaf powder, rubber (Hevea brasiliensis) leaf powder, Coriandrum sativum, peanut hull pellets, sago waste, saltbush (Atriplex canescens) leaves, tree fern, rice husk ash and neem bark, grape stalk wastes, etc. Some of the advantages of using plant wastes for wastewater treatment include simple technique, requires little processing, good adsorption capacity, selective adsorption of heavy metal ions, low cost, free availability and easy regeneration. However, the application of untreated plant wastes as adsorbents can also bring several problems such as low adsorption capacity, high chemical oxygen demand (COD) and biological chemical demand (BOD) as well as total organic carbon (TOC) due to release of soluble organic compounds contained in the plant materials. The increase of the COD,

BOD and TOC can cause depletion of oxygen content in water and can threaten the aquatic life. Therefore, plant wastes need to be modified or treated before being applied for the decontamination of heavy metals. A comparison of adsorption efficiency between chemically modified and unmodified adsorbents was also reported in literature [67].

In a conclusion, a wide range of low-cost adsorbents obtained from naturel and chemical sorbent or chemically modified plant wastes has been studied and most studies were focused on the removal of heavy metal ions such as Cd, Cu, Pb, Zn, Ni and Cr(VI) ions. The most common chemicals used for treatment of plant wastes are acids and bases. Chemically modified plant wastes vary greatly in their ability to adsorb heavy metal ions from solution. Chemical modification in general improved the adsorption capacity of adsorbents probably due to higher number of active binding sites after modification, better ion-exchange properties and formation of new functional groups that favours metal uptake. Although chemically modified plant wastes can enhance the adsorption of heavy metal ions, the cost of chemicals used and methods of modification also have to be taken into consideration in order to produce 'low-cost' adsorbents.

Another option is using biological materials [4, 66-68, 76, 77]. Biomaterials of microbial and plant origin interact effectively with heavy metals. Metabolically inactive dead biomass due to their unique chemical composition sequesters metal ions and metal complexes from solution, which obviates the necessity to maintain special growth-supporting conditions. Metal-sorption by various types of biomaterials can find enormous applications for removing metals from solution and their recovery. Rather than searching thousands of microbial species for particular metal sequestering features, it is beneficial to look for biomasses that are readily available in large quantities to support potential demand. While choosing biomaterial for metal sorption, its origin is a major factor to be taken into account, which can come from (a) microorganisms as a by-product of fermentation industry, (b) organisms naturally available in large quantities in nature and (c) organisms cultivated or propagated for biosorption purposes using inexpensive media. Different non-living biomass types have been used to adsorb heavy metal ions from the environment. Seaweed, mold, bacteria, crab shells and yeast are among the different kinds of biomass, which have been tested for metal biosorption or removal. Advantages and disadvantages of biosorption by non-living biomass are as follows [67, 75-77]:

Advantages of biosorption;

- Growth-independent, non-living biomass is not subject to toxicity limitation of cells. No requirement of costly nutrients required for the growth of cells in feed solutions. Therefore, the problems of disposal of surplus nutrients or metabolic products are not present.

- Biomass can be procured from the existing fermentation industries, which is essentially a waste after fermentation.

- The process is not governed by the physiological constraint of living microbial cells.

- Because of non-living biomass behave as an ion exchanger; the process is very rapid and takes place between few minutes to few hours. Metal loading on biomass is often very high, leading to very efficient metal uptake.

- Because cells are non-living, processing conditions are not restricted to those conducive for the growth of cells. In other words, a wider range of operating conditions such as pH, temperature and metal concentration is possible. No aseptic conditions are required for this process.

- Metal can be desorbed readily and then recovered if the value and amount of metal recovered are significant and if the biomass is plentiful, metal-loaded biomass can be incinerated, thereby eliminating further treatment.

Disadvantages of biosorption;

- Early saturation can be problem i.e. when metal interactive sites are occupied, metal desorption is necessary prior to further use, irrespective of the metal value.

- The potential for biological process improvement (e.g. through genetic engineering of cells) is limited because cells are not metabolizing. Because production of the adsorptive agent occurs during pre-growth, there is no biological control over characteristic of biosorbent. This will be particularly true if waste biomass from a fermentation unit is being utilized.

- There is no potential for biologically altering the metal valency state. For example less soluble forms or even for degradation of organometallic complexes.

Metabolic independent processes can mediate the biological uptake of heavy metal cations. Biosorption offers an economically feasible technology for efficient removal and recovery of metal(s) from aqueous solution. The process of biosorption has many attractive features including the selective removal of metals over a broad range of pH and temperature, its rapid kinetics of adsorption and desorption and low capital and operation cost. Biosorbent can easily be produced using inexpensive growth media or obtained as a by-product from industry. It is desirable to develop biosorbents with a wide range of metal affinities that can remove a variety of metal cations. These will be particularly useful for industrial effluents, which carry more than one type of metals. Alternatively a mixture of non-living biomass consisting of more than one type of microorganisms can be employed as biosorbents. Bacterial biomass, algal biomass, fungal biomass were applied to removal of metals in the waste waters. The use of immobilized biomass rather than native biomass has been recommended for large-scale application but various immobilization techniques have yet to be thoroughly investigated for ease, efficency and cost effectivity [67, 77].

Biosorption processes are applicable to effluents containing low concentrations of heavy metals for an extended period. This aspect makes it even more attractive for treatment of dilute effluent that originates either from an industrial plant or from the primary wastewater treatment facility. Thus biomass-based technologies need not necessarily replace the conventional treatment routes but may complement them. At present, information on different biosorbent materials is inadequate to accurately define the parameters for process scale up and design perfection including reliability and economic feasibility. To provide an economically viable treatment, the appropriate choice of biomass and proper operational conditions has to be identified. To predict the difference between the uptake capacities of the biomass, the experimental results should be tested against an adsorption model. The development of a packed bed or fluidized-bed biosorption model would be helpful for evaluating industrial-scale biosorption column performance, based on laboratory scale experiments and to understand the basic mechanism involved in order to develop better and effective biosorbent.

CONCLUSION

An unfortunate consequence of industrialization and industrial production is the generation and release of toxic waste products which are polluting our environment. Many trace and heavy metals (Cd, Pb, Mn, Cu, Zn, Cr, Fe and Ni) and their compounds have been found that are toxic. Many of them are used in several industrial activities including metallurgy, tanneries, petroleum refining, electroplating, textiles and in pigments. Their presence in environment has been responsible for several types of health problems in animals, plants and human beings. If the wastewaters were discharged directly into natural waters, it will constitute a great risk for the aquatic ecosystem, whilst the direct discharge into the sewerage system may affect negatively.

The analysis of wastewater for trace and heavy metal contamination is an important step in ensuring human and environmental health. Wastewater is regulated differently in different countries, but the goal is to minimize the pollution introduced into natural waterways. In recent years, metal production emissions have decreased in many countries due to legislation, improved cleaning technology and altered industrial activities. Today and in the future, dissipate losses from consumption of various metal containing goods are of most concern. Therefore, wastewater may need to be measured for a variety of metals at different concentrations, in different wastewater matrices. A variety of inorganic techniques can be used to measure trace elements in waste water including atomic absorption spectrometry (AAS), inductively coupled plasma optical emission spectrometry (ICP-OES) and ICP mass spectrometry (ICP-MS). Depending upon the number of elements that need to be determined and the number of samples that need to be run, the most suitable technique for business requirements can be chosen.

Several industrial wastewater streams may contain heavy metals such as Cd, Sb, Cr, Cu, Pb, Zn, Co, Ni, etc. The toxic metals, probably existing in high or even in low concentrations, must be effectively treated/removed from the wastewaters. The various treatment methods employed to remove heavy or trace metals, adsorption and chemical precipitation process is the most common treatment technology. The conventional heavy metal removal process has some inherent shortcomings such as requiring a large area of land, a sludge dewatering facility, skillful operators and multiple basin configuration. In recent years, some new processes have been developed for the heavy metal removal from wastewater, like biosorption, neutralization, precipitation, ion exchange etc. The use of al these techniques for removal of the heavy metals offers several advantages and limitations compared to each other. The important parameters for the selection of removal technique of heavy metal from waste water are waste type, the growth of the wastewater field, cheap or low-cost removal material, operational costs and problem of disposal of residual metal sludge. The significance of developing new treatment/removal methods for heavy metal from waste or waste water samples has been widely recognized especially in the fields of environmental sciences.

REFERENCES

1. European Commission 2002 DG ENV. E3, Project ENV.E3/ETU/2000/0058, Heavy Metals in Waste Final Report

2. K. S. Hui, C. Y. H. Chao, S. C. Kot, 2005 Removal of mixed heavy metal ions in wastewater by zeolite 4A and residual products from recycled coal fly ash Journal of Hazardous Materials B 127 89 101

3. M. Karvelas, A. Katsoyiannis, C. Samara, 2003 Occurrence and fate of heavy metals in the wastewater treatment process Chemosphere 53 1201 1210

4. S. S. Ahluwalia, D. Goyal, 2007 Microbial and plant derived biomass for removal of heavy metals from wastewater Bioresource Technology 98 98 2243

5. J. H. Duffus, 2002 Heavy Metals"-A Meaningless Term? (IUPAC Technical Report) Pure Appl. Chem 74 5 793 807

6. A. Malik, 2004 Metal bioremediation through growing cells Environment International 30 261 278

7. B. Volesky, S. Schiewer, 2000 Biosorption, Metals Encyclopedia of Bioprocess Technology (Fermentation, Biocatalysis and Bioseparation) 1 433 453

8. http://www.lenntech.com/periodic/elements/cu.htm

9. http://www.ehow.com/list_7446916_sources-nickel-wastewater.html#ixzz1xxFBgbjy

10. A. D. Eaton, L. S. Clesceri, E. W. Rice, A. E. Greenberg, 2005 Standard Methods for the Examination of Water and Waste Water 21 ed. New York

11. B. A. Zarcinas, B. Cartwright, L. R. Spouncer, 1987 Nitric acid digestion and multi-element analysis of plant material by inductively coupled plasma spectrometry Communications in Soil Science and Plant Analysis 18 131 146

12. M. Hoenig, A. M. Kersabiec, 1996 Sample preparation steps for analysis by atomic spectroscopy methods: present status Spectrochimica Acta Part B 51 1297

13. N. R. Bader, 2011 Sample Preparation for Flame Atomic Absorpstion Spectroscopy: An Overview Rasayan Journal of Chemistry 4 1 49 55

14. ASTM Standard D1971-11. 2003 Standard Practices for Digestion of Water Samples for Determination of Metals by Flame Atomic Absorption, Graphite Furnace Atomic Absorption, Plasma Emission Spectroscopy, or Plasma Mass Spectrometry ASTM International West Conshohocken, PA DOI: 10.1520/D1971-11

15. D. A. Skoog, F. J. Holler, T. A. Nieman, 1998 Principles of instrumental analysis Philadelphia Saunders College Pub

16. Environmental Sampling and Analysis for Metals Csaba Csuros and Maria Csuros CRC Press 2002

17. M. E. Mahmoud, I. M. M. Kenawy, M. A. H. Hafez, R. R. Lashein, 2010 Removal, preconcentration and determination of trace heavy metal ions in water samples by AAS via chemically modified silica gel N-(1-carboxy-6-hydroxy) benzylidenepropylamine ion exchanger Desalination 250 62 70

18. A. Afkhami, T. Madrakian, H. Siampour, 2006 Flame atomic absorption spectrometric determination of trace quantities of cadmium in water samples after cloud point extraction in Triton X-114 without added chelating agents Journal of Hazardous Materials B 138 269 272

19. A. A. Mohamed, A. T. Mubarak, Z. M. H. Marstani, K. F. Fawy, 2006 A novel kinetic determination of dissolved chromium species in natural and industrial waste water Talanta 70 460 467

20. M. E. Mahmoud, M. M. Osman, O. F. Hafez, A. H. Hegazi, E. Elmelegy, 2010 Removal and preconcentration of lead (II) and other heavy metals from water by alumina adsorbents developed by surface-adsorbed-dithizone Desalination 251 123 130

21. R. J. Cassella, O. I. B. Magalhaes, M. T. Couto, E. L. S. Lima, M. Angelica, F. S. Neves, F. M. B. Coutinho, 2005 Synthesis and application of a functionalized resin for flow injection/F AAS copper determination in waters Talanta 67 121 128

22. J. S. Carletto, K. Cravo Di Pietro Roux, H. Franca Maltez, Martendal, E. Carasek, 2008 Use of 8-hydroxyquinoline-chitosan chelating resin in an automated on-line preconcentration system for determination of zinc(II) by F AAS Journal of Hazardous Materials 157 88 93

23. S. Gunduz, S. Akman, M. Kahraman, 2011 Slurry analysis of cadmium and copper collected on 11-mercaptoundecanoic acid modified TiO2 core-Au shell nanoparticles by flame atomic absorption spectrometry Journal of Hazardous Materials 186 212 217

24. J. L. Burguera, M. Burguera, C. Rondon, P. Carrero, M. R. Brunetto, Y. Petit, de Pena, 2000 Determination of beryllium in natural and waste waters using on-line flow-injection preconcentration by precipitation:dissolution for electrothermal atomic absorption spectrometry Talanta 52 27 37

25. A. Baysal, S. Akman, F. Calisir, 2008 A novel slurry sampling analysis of lead in different water samples by electrothermal atomic absorption spectrometry after coprecipitated with cobalt/pyrrolidine dithiocarbamate complex Journal of Hazardous Materials 158 454 459

26. N. M. M. Coelho, A. Cósmen, Silva. C. da, da. Moraes, Silva, 2002 Determination of As(III) and total inorganic arsenic by flow injection hydride generation atomic absorption spectrometry Analytica Chimica Acta 460 227 233

27. J. Dedina, D. L. Tsalev, 1995 Hydride Generation Atomic Absorption Spectrometry Wiley New York

28. J. Y. Cabon, C. L. Madec, 2004 Determination of major antimony species in seawater by continuous flow injection hydride generation atomic

absorption spectrometry Analytica Chimica Acta 504 209 215

29. M. Yersel, A. Erdem, A. E. Eroğlu, T. Shahwan, 2005 Separation of trace antimony and arsenic prior to hydride generation atomic absorption spectrometric determination Analytica Chimica Acta 534 293 300

30. A. N. Anthemidis, G. A. Zachariadis, J. A. Stratis, 2005 Determination of arsenic(III) and total inorganic arsenic in water samples using an on-line sequential insertion system and hydride generation atomic absorption spectrometry Analytica Chimica Acta 547 237 242

31. H. Erdoğan, Ö. Yalçınkaya, A. R. Türker, 2011 Determination of inorganic arsenic species by hydride generation atomic absorption spectrometry in water samples after preconcentration/separation on nano ZrO_2/B_2O_3 by solid phase extraction Desalination 280 391

32. D. Karadeniz, N. Korkmaz, O. Y. Ertas, Ataman, 2002 A novel silica trap for lead determination by hydride generation atomic absorption spectrometry Spectrochimica Acta Part B 57 571 580

33. S. Kagaya, Y. Kuroda, Y. Serikawa, K. Hasegawa, 2004 Rapid determination of total mercury in treated waste water by cold vapor atomic absorption spectrometry in alkaline medium with sodium hypochlorite solution Talanta 64 554 557

34. N. Pourreza, K. Ghanemi, 2009 Determination of mercury in water and fish samples by cold vapor atomic absorption spectrometry after solid phase extraction on agar modified with 2-mercaptobenzimidazole Journal of Hazardous Materials 161 982 987

35. S. Sahan, U. Sahin, 2012 An automated on-line minicolumn preconcentration cold vapour atomic absorption spectrometer: Application to determination of cadmium in water samples Talanta 88 701 706

36. H. Gleisner, B. Welz, J. W. Einax, 2010 Optimization of fluorine determination via molecular absorption of gallium mono-fluoride in a graphite furnace using a high-resolution continuum source spectrometer Spectrochimica Acta Part B 65 864 869

37. N. Ozbek, S. Akman, 2012 Method development for the determination of fluorine in water samples via the molecular absorption of strontium monofluoride formed in an electrothermal atomizer Spectrochimica Acta Part B 69 32 37

38. N. Ozbek, S. Akman, 2012 Method development for the determination of fluorine in toothpaste via molecular absorption of aluminum mono fluoride using a high-resolution continuum source nitrous oxide/acetylene flame atomic absorption spectrophotometer Talanta 94 246 250

39. S. Morés, G. C. Monteiro, F. S. Santos, E. Carasek, B. Welz, 2011 Determination of fluorine in tea using high-resolution molecular absorption spectrometry with electrothermal vaporization of the calcium mono-fluoride CaF Talanta 85 2681 2685

40. U. Heitmann, H. Becker-Ross, S. Florek, M. D. Huang, M. Okruss, 2006 Determination of non-metals via molecular absorption using high-resolution continuum source absorption spectrometry and graphite furnace atomization J. Anal. Atom. Spectrom 21 1314 1320

41. M. D. Huang, H. Becker-Ross, S. Florek, U. Heitmann, M. Okruss, 2006 Determination of halogens via molecules in the air-acetylene flame using high-resolution continuum source absorption spectrometry, Part II: Chlorine Spectrochimica Acta Part B 61 959 964

42. M. D. Huang, H. Becker-Ross, S. Florek, U. Heitmann, M. Okruss, 2008 High-resolution continuum source electrothermal absorption spectrometry of AlBr and CaBr for the determination of bromine Spectrochimica Acta Part B 63 566

43. T. Limburg, J. W. Einax, 2012 Determination of bromine using high-resolution continuum source molecular absorption spectrometry in a graphite furnace Microchemical Journal DOI: 10.1016/j. microc.2012.05.016

44. A. Baysal, S. Akman, 2011 A practical method for the determination of sulphur in coal samples by high-resolution continuum source flame atomic absorption spectrometry Talanta 85 2662 2665

45. R. C. de Campos, L. T. C. Correia, F. Vieira, T. D. Saint'Pierre, A. C. Oliveira, R. Goncalves, 2011 Direct determination of P in biodiesel by high-resolution continuum source graphite furnace atomic absorption spectrometry Spectrochimica Acta Part B 66 352 355

46. www.thermo.com

47. D. Atanassova, V. Stefanova, E. Russeva, 1998 Co-precipitative pre-concentration with sodium diethyldithiocarbamate and ICP-AES determination of Se, Cu, Pb, Zn, Fe, Co, Ni, Mn, Cr and Cd in water Talanta 47 1237 1243

48. M. Zougagh, A. Garcıa, J. M. de Torres, Pavon. Cano, 2002 Determination of cadmium in water by ICP-AES with on-line adsorption preconcentration using DPTH-gel and TS-gel microcolumns Talanta 56 753 761

49. M. Zougagh, A. Garcıa, E. de Torres, Alonso. J. M. Vereda, Pavón. Cano, 2004 Automatic on line preconcentration and determination of lead in water by ICP-AES using a TS-microcolumn Talanta 62 503 510

50. M. V. B. Krishna, K. Chandrasekaran, S. V. Rao, D. Karunasagar, J. Arunachalam, 2005 Speciation of Cr(III) and Cr(VI) in waters using immobilized moss and determination by ICP-MS and FAAS Talanta 65 135 143

51. W. Hu, F. Zheng, B. Hu, 2008 Simultaneous separation and speciation of inorganic As(III)/As(V) and Cr(III)/Cr(VI) in natural waters utilizing capillary microextraction on ordered mesoporous Al2O3 prior to their on-line determination by ICP-MS Journal of Hazardous Materials 151 58 64

52. Z. Chen, M. Megharaj, R. Naidu, 2007 Speciation of chromium in waste water using ion chromatography inductively coupled plasma mass spectrometry Talanta 72 394 400

53. R. Barbini, F. Colao, R. Fantoni, V. Lazic, A. Palucci, F. Capitelli, H. J. L. van der Steen, 2000 Laser Induced Breakdown Spectroscopy For Semi-Quantitative Elemental Analysis In Soils And Marine Sediments Proceedings Of Earsel-Sig-Workshop Lidar Dresden/Frg June 16 - 17

54. J. B. Stiger, H. P. M. De Haan, R. Guichert, C. P. A. Deckers, M. L. Daane, 2000 Environ. Pollut. 107 451

55. L. St-Onge, E. Kwong, M. Sabsabi, E. B. Vadas, 2004 Rapid analysis of liquid formulations containing sodium chloride using laser-induced breakdown spectroscopy Journal of Pharmaceutical and Biomedical Analysis 36 277 284

56. B. Charfi, M. A. Harith, 2002 Panoramic laser-induced breakdown spectrometry of water Spectrochimica Acta Part B 57 1141 1153

57. M. A. Gondal, T. Hussain, 2007 Determination of poisonous metals in wastewater collected from paint manufacturing plant using laser-induced breakdown spectroscopy Talanta 71 73 80

58. N. K. Rai, A. K. Rai, 2008 LIBS-An efficient approach for the determination of Cr in industrial wastewater Journal of Hazardous Materials 150 835 838

59. A. W. Bott, 1992 Stripping Voltammetry Current Separations 12(3 141 147

60. P. Sonthalia, E. Mc Gaw, Y. Show, G. M. Swain, 2004 Metal ion analysis in contaminated water samples using anodic stripping voltammetry and a nanocrystalline diamond thin-film electrode Analytica Chimica Acta 522 35 44

61. E. A. Mc Gaw, G. M. Swain, 2006 A comparison of boron-doped diamond thin-film and Hg-coated glassy carbon electrodes for anodic

stripping voltammetric determination of heavy metal ions in aqueous media Analytica Chimica Acta 575 180 189

62. E. Bernalte, C. Marin, E. Sanches, Gil. Pinilla, 2011 Determination of mercury in ambient water samples by anodic stripping voltammetry on screen-printed gold electrodes Analytica Chimica Acta 689 60 64

63. M. F. Mousavi, A. Rahmani, S. M. Golabi, M. Shamsipur, H. Sharghi, 2001 Differential pulse anodic stripping voltammetric determination of lead(II) with a 1, 4bis(prop-2-enyloxy)-9,10-anthraquinone modified carbon paste electrode Talanta 55 305 312

64. B. Kong, B. Tang, X. Liu, X. Zeng, H. Duan, S. Luo, W. Wei, 2009 Kinetic and equilibrium studies for the adsorption process of cadmium(II) and copper(II) onto Pseudomonas aeruginosa using square wave anodic stripping voltammetry method Journal of Hazardous Materials 167 455 460

65. A. Giacomino, O. Abollino, M. Malandrino, E. Mentasti, 2008 Parameters affecting the determination of mercury by anodic stripping voltammetry using a gold electrode Talanta 75 266 273

66. W. S. Wan Ngah, M. A. K. M. Hanafiah, 2008 Removal of heavy metal ions from wastewater by chemically modified plant wastes as adsorbents: A review Bioresource Technology 99 3935 3948

67. A. Demirbas, 2008 Heavy metal adsorption onto agro-based waste materials: A review Journal of Hazardous Materials 157 220 229

68. S. E. Bailey, T. J. Olin, M. R. Bricka, D. D. Adrian, 1999 A Review Of Potentially Low-Cost Sorbents For Heavy Metals Water Research 33 11 2469 2479

69. K. C. Sekhar, C. T. Kamala, N. S. Chary, Y. Anjaneyulu, 2003 Removal of heavy metals using a plant biomass with reference to environmental control The International Journal of Mineral Processing 68 37 45

70. B. G. Lee, R. M. Rowell, 2004 Removal of heavy metals ions from aqueous solutions using lignocellulosic fibers Journal of Natural Fibers 1 97 108

71. D. Mohan, C. U. Pittman, 2007 Arsenic removal from water/wastewater using adsorbents-A critical review Journal of Hazardous Materials 142 1-2 1 53

72. S. R. Wickramasinghe, B. Han, J. Zimbron, Z. Shen, M. N. Karim, 2004 Arsenic removal by coagulation and filtration: comparison of groundwaters from the United States and Bangladesh Desalination 169 231 244

73. D. S. Wang, H. X. Tang, 2001 Modified inorganic polymer flocculants-PFSi: its precipitation, characterization and coagulation behavior Water Research 35 3473 3581

74. M. Kang, M. Kawasaki, S. Tamada, T. Kamei, Y. Magara, 2000 Effect of pH on the removal of arsenic and antimony using reverse osmosis membranes Desalination 131 293 298

75. N. Ahalya, T. V. Ramachandra, R. D. Kanamadi, 2003 Biosorption of heavy metals Research Journal of Chemistry and Environment 7 71 79

76. Y. Sag, T. Kutsal, 2001 Recent Trends in the Biosorption of Heavy Metals: A Review Biotechnology and Bioprocess Engineering 6 376 385

77. B. Volesky, 2007 Biosorption and me Water Research 41 4017 4029

78. A. R. Turker, 2007 New sorbents for solid-phase extraction for metal enrichment Clean 35 548 557

Chapter 9

SELECTIVE REMOVAL OF HEAVY METAL IONS FROM WATERS AND WASTE WATERS USING ION EXCHANGE METHODS

Zbigniew Hubicki and Kołodyńska
Maria Curie-Skłodowska University, Poland

INTRODUCTION

Environmental pollution by toxic metals occurs globally through military, industrial, and agricultural processes and waste disposal (Duffus, 2002). Fuel and power industries generate 2.4 million tons of As, Cd, Cr, Cu, Hg, Ni, Pb, Se, V, and Zn annually The metal industry adds 0.39 million tons/yr of the same metals to the environment, while agriculture contributes 1.4 million tons/yr, manufacturing contributes 0.24 million tons/yr and waste disposal adds 0.72 million tons/yr. Metals, discharged or transported into the environment, may undergo transformations and can have a large environmental, public health, and economic impact (Brower et al. 1997; Nriagu & Pacyna, 1988; Gadd & White, 1993).

Among different techniques used for removal of high concentrations of heavy metals, precipitation-filtration, ion exchange, reverse osmosis, oxidation-reduction, solvent extraction, as well as membrane separation should be mentioned (Hubicki,et al. 1999; Dąbrowski et al. 2004). However, some of the wastes contain substances such as organics, complexing agents and alkaline earth metals that may decrease the metal removal and result in unacceptable concentrations of heavy metals in the effluents. The pollutants of concern include cadmium, lead, mercury, chromium, arsenic, zinc, cobalt and nickel as well as copper. They have a number of applications in basic engineering works, paper and pulp industries, leather tanning, petrochemicals, fertilizers, etc. Moreover, they have also negative impact on human health.

Cadmium: is a metal of great toxicological concern. An important source of human exposure to cadmium is food and water, especially for the population living in the vicinity of industrial plants, from which cadmium is emitted to the air. In the case of exposure to occupational cadmium compounds, they are absorbed mainly by inhalation. Through the gastrointestinal tract less than 10% cadmium is absorbed. An important source of human exposure to cadmium is food and water. In natural water its typical concentration lies below 0.001 mg/dm^3, whereas, the upper limit recommended by EPA (Environmental Protection Agency) is less than 0.003 mg/dm^3. The maximum limit in drinking water is 0.003 mg/dm^3.

Cadmium accumulates in kidneys, pancreas, intestines and glands altering the metabolism of the elements necessary for the body, such as zinc, copper, iron, magnesium, calcium and selenium. Damage to the respiratory tract and kidneys are the main adverse effects in humans exposed to cadmium compounds. In humans exposed to fumes and dusts chronic toxicity of cadmium compounds is usually found after a few years. The main symptom of emphysema is that it often develops without preceding bronchitis. The second basic symptom of chronic metal poisoning is kidney damage. It includes the loss and impairment of smell, pathological changes in the skeletal system (osteoporosis with spontaneous fractures and bone fractures), pain in the extremities and the spine, difficulty in walking, the formation of hypochromic anemia. The most known 'Itai-Itai' disease caused by cadmium exposure is mixed osteomalacia and osteoporosis. However, an important source of cadmium in soils are phosphate fertilizers. Large amounts of cadmium are also introduced to soil together with municipal waste. The high mobility of cadmium in all types of soils is the reason for its rapid integration into the food chain. Daily intake of cadmium from food in most countries of the world is 10-20 mg.

Lead: is a toxic metal, which accumulates in the vital organs of men and animals and enters into the body through air, water and food. According to the WHO (World Health Organization) standards, its maximum limit in drinking water is 0.05 mg/dm^3 but the maximum discharge limit for lead in wastewater is 0.5 mg/dm^3. Its cumulative poisoning effects are serious haematological damage, anaemia, kidney malfunctioning, brain damage etc. Chronic exposure to lead causes severe lesions in kidney, liver, lungs and spleen.

Lead is used as industrial raw material in the manufacture of storage batteries, pigments, leaded glass, fuels, photographic materials, matches and explosives. Lead being one of very important pollutants comes from wastewaters from refinery, wastewaters from production of basic compounds containing lead, wastewaters with the remains of after production solvents and paints. Large toxicity of lead requires that its contents are reduced to the minimum (ppb level). To this end there are applied chelating ions with the functional phosphonic and aminophosphonic groups. Also weakly basic anion exchangers in the free base form can be used for selective removal of lead(II) chloride complexes from the solutions of pH in the range 4-6. Also a combined process of cation exchange and precipitation is often applied for lead(II) removal form wastewaters (Pramanik et al. 2009). The average collection of lead by an adult was estimated at 320-440 mg/day. Acute poisoning with inorganic lead compounds occurs rarely. In the case of acute poisoning in man, the symptoms are burning in the mouth, vomiting, abdominal cramps, diarrhea, constipation progressing to systolic, blood pressure and body temperature. At the same time there is hematuria, proteinuria, oliguria, central nervous system damage. Alkyl lead compounds are more toxic than inorganic lead connections. Tetraethyl lead toxicity manifested primarily in lead damage of the nervous system. Toxic effects of lead on the central nervous system are observed more in children. In adults, the effects of lead toxicity occur in the peripheral nervous system. Symptoms of chronic poisoning may vary. The acute form of poisoning known as lead colic is the general state of various spastic internal organs and neurological damage in the peripheral organs. Long-term lead poisoning can lead to organic changes in the central and peripheral nervous systems. Characteristic symptoms include pale gray skin colour and the lead line on the gums (blue-black border).

In nature, natural circulation of mercury vapour has a significant influence on the content of the soil and water. Elemental mercury in the rain water creates compounds by oxidation to divalent mercury. Both the chemical reaction, and under the influence of biological factors, and especially the activity of bacteria in the sediments of water bodies methyl and dimethyl mercury compounds are formed. Mercury, a fixed component of the waste water treatment that may be used for soil fertilization is a major threat to the inclusion of the metal in nutritional products. Drinking water may contain up to 300 ng Hg/dm^3, in highly industrialized areas it can reach up to 700 ng/dm^3. Daily consumption of mercury from food in the general population is less than 20 $\mu g/day$. 80% of mercury absorbed by the respiratory system is retained in the body. In the case of ingestion of inorganic mercury salts, salivation, burning in the throat, vomiting, bloody diarrhea, necrosis of the intestinal mucosa and kidney damage, leading to anuria and uremii can occur. The concentration of mercury vapour over 1 mg/m^3 damages lung tissue and causes severe pneumonia. The classic symptoms of metallic mercury vapour poisoning are manifested by tremor, mental disorders, inflammation of the gums. Its maximum limit in drinking water is 0.0005 mg/dm^3.

Chromium: occurring as Cr(III) or Cr(VI) in natural environments, is an important material resource, an essential micronutrient or toxic contaminant. Cr(III) is required for normal development of human and animal organisms but Cr(VI) activates teratogenic processes, disturbs DNA synthesis and can give rise to mutagenous changes leading to malignant tumours (WHO, Report 1998). Natural sources of chromium include weathered rocks, volcanic exhalations and biogeochemical processes and, in the man-polluted environment, mainly wastes after processing and utilization of chromium compounds. Chromium is an important and widely applied element in industry. The hexavalent and trivalent chromium is often present in electroplating wastewater. Other sources of chromium pollution are leather tanning, textile, metal processing, paint and pigments, dyeing and steel fabrication. To remove toxic chromium compounds from sewages there are used such methods as: precipitation, coagulation, solvent extraction and various kinds of membrane processes, ion flotation, adsorption and ion exchange (Bajda, 2005). The maximum limit in drinking water is 0.05 mg/dm^3. The Polish drinking groundwater chromium content ranges on the average from 0.07 to 2 mg/dm^3. 0.02 mg/dm^3 is accepted as the permissible content of chromium in groundwater. The daily dose taken by the adult can be 50-200 mg/day (or 60-290 mg/day). Cr(III) cation predominates in most tissues except the liver. Chromium is associated with nucleic acids and is the subject to the concentration in liver cells. It plays an important role in the metabolism of glucose, certain proteins and fats, is part of enzymes and stimulates the activity of others. All compounds of chromium, with the exception of chromate, are rapidly cleared from the blood. Chromium also accumulates in the liver and kidneys. High concentrations of chromium, observed in the lungs of people exposed to this metal, indicate that at least part of chromium is stored in this organ in the form of insoluble compounds. The binding of chromium with the elements of the blood and transport of chromium by the blood depends mainly on its valence. Hexavalent chromium readily crosses the membranes of red blood cells and after reduction to trivalent chromium is bound to hemoglobin. The reduction of hexavalent to trivalent chromium, occurring within cells, considered as the activation of the carcinogenic chromium, increases because the probability of interaction of trivalent chromium on the DNA. Clinical signs of acute toxicity of chromium compounds are characterized by severe abdominal pain, vomiting and bloody diarrhea, severe kidney damage with hematuria leading to anuria, observed gastrointestinal ulceration. Chromium compounds and chromic acid are especially dangerous and cause serious damage to internal organs. Chronic exposure leads to chronic disorders in the body.

Arsenic: is present in over 160 minerals. It is readily bioaccumulative and therefore its concentration in polluted waters may reach 430 mg/dm^3 in plants and 2.5 mg/dm^3 in fish. The upper limit of arsenic recommended by US EPA, EU and WHO is 0.01 mg/dm^3. However, many countries have retained the earlier WHO guideline of 0.05 mg/dm^3 as their standard.

Arsenic accumulates in tissues rich in keratin, like hair, nails and skin. Arsenic and its inorganic compounds can cause not only cancer of the respiratory system and skin, but also neoplastic lesions in other organs. Arsenic compounds enter the body from the gastrointestinal tract and through skin and respiratory system. Arsenic compounds have affinity for many enzymes and can block their action, and above all disturb the Krebs cycle. Inorganic arsenic compounds are more harmful than organic and among them AsH_3 and As_2O_3 should be mentioned. 70-300 mg of As_2O_3 is considered to be the average lethal dose for humans. The dose of 10-50 ppb for 1 kg of body weight can cause circulatory problems, resulting in necrosis and gangrene of limbs. The dominant effects of arsenic in humans are changes in the skin and mucous membranes as well as peripheral nerve damage. There are xerosis soles and palms, skin inflammation with ulceration. In addition, there is perforation of the nasal septum. The values of the maximum allowable concentration (NDS) in Poland set for inorganic arsenic compounds are 0.3 mg/m^3 and 0.2 mg/m^3 for AsH_3.

In nature, **zinc** occurs in the form of minerals. An important source of zinc pollution is the burning of coal, petroleum and its products. Incineration of municipal solid waste can introduce about 75% zinc to urban air. Also, municipal wastewater generally contains significant amounts of zinc. The use of municipal and industrial waste in agriculture results in the accumulation of zinc in the surface layers of soil. Another source of this metal in soils are some preparations of plant protection products, as well as phosphatic fertilizers. The degree of toxicity of zinc is not big, but it depends on the ionic form, and changes under the influence of water hardness and pH. The daily average download of zinc by an adult is estimated at about 10-50 mg /day. The toxic dose is 150-600 mg. It is necessary for the proper functioning of living organisms and it is involved in the metabolism of proteins and carbohydrates. High doses of zinc cause damage to many biochemical processes followed by its deposition in the kidneys, liver, gonads. Kidney play an important role in maintaining zinc homeostasis in the body. Zinc is relatively non-toxic to humans and animals. Hazard zinc mainly connected with secondary copper deficiency does not give specific symptoms.

Nickel: is a moderately toxic element as compared to other transition metals. It is a natural element of the earth's crust; therefore its small amounts are found in food, water, soil, and air. Nickel occurs naturally in the environment at low levels. Nickel concentrations in the groundwater depend on the soil used, pH, and depth of sampling. The average concentration in the groundwater in the Netherlands ranges from 7.9 µg/dm^3 (urban areas) to 16.6 µg/dm^3 (rural areas). Acid rain increases the mobility of nickel in the soil and thus might increase nickel concentrations in the groundwater. In the groundwater with a pH below 6.2, nickel concentrations up to 0.98 mg/dm³ have been measured, whereas the upper limit recommended by FAO (Food & Agricultural Organization of the United Nation) for nickel in water is 0.02 mg/dm³. According to the Polish standards the maximum discharge limit for nickel in waste water is 2-3 mg/dm³. The maximum limit in drinking water in Europe is 0.01 mg/dm³. Although it has been suggested that nickel may be essential to plants and some animal species as well as in human nutrition, this metal causes damage to humans. Nickel occurs in seams of coal in the amount of 4-60 mg/kg. Crude oil contains about 50-350 mg/kg of the metal. The most dangerous is tetracarbonyl nickel occurring mostly in nickel refineries. The content of this metal in industrial and municipal wastewater ranges 20-3924 mg/kg. An important source of nickel pollution is its emissions to the air, the combustion of coal and liquid fuels, primarily by diesel engines. It is assumed that the concentration of nickel in the waters of the rivers should be about 1 μg/dm³, while in most rivers of Europe it is as high as 75 μg/dm³. Large amounts of nickel are given to surface waters from municipal wastewater in which the concentration exceeds 3000 ppm s.m. The permissible concentration should be 20 μg/dm³. Nickel readily accumulatives particularly in phytoplankton or other aquatic plants. The daily absorption of nickel by humans ranges 0.3-0.5 mg. In humans, the absorption of nickel from the gastrointestinal tract is less than 10%. Nickel taken with food and water is poorly absorbed and rapidly excreted from the body. It accumulates mainly in bones, heart, skin and various glands. Nickel inhalation of atmospheric air is largely accumulated in the lungs. Practically fatal or acute poisoning with nickel or its salts is not found. The most toxic compound is carbonyl nickel. An excess of inhaled nickel causes damage to the mucous membranes. Moreover, its symptoms are allergic disorders (protein metabolism disorder in plasma, changes in the chromosomes and changes in bone marrow and cancer. It is known that inhalation of nickel and its compounds can lead to serious problems, including, among others, respiratory system cancer. Moreover, nickel can cause skin disorder which is a common occupational disease in workers who handle its large amounts. Also dermatitis is the most common effect of chronic dermal exposure to nickel. Chronic inhalation exposure to nickel in humans also results in detrimental respiratory effects.

Copper: is generally found in the earth's crust, usually in the form of sulphides. Municipal and industrial waste waters are an important source of pollution of rivers and water reservoirs. Copper accumulating plants may be the cause of poisoning. Copper is present in all types of water, and its content is subject to large variations (Barceloux, 1999). The natural content of copper in the river water ranges 0.9-20 $\mu g/dm^3$ and for saline waters 0.02-0.3 $\mu g/dm^3$. Copper is an essential nutritional element being a vital part of several enzymes. It is one of the components of human blood. The estimated adult dietary intakes are between 2 and 4 mg/day. The demand for copper is increased in pregnant women, children and the elderly. Good dietary sources of copper include animal liver, shellfish, dried fruit, nuts and chocolate. In some cases drinking water may also provide significant levels of copper. Copper in the body is involved in oxidation-reduction processes, acts as a stimulant on the amount and activity of hemoglobin, in the process of hardening of collagen, hair keratinization, melanin synthesis as well as affects on lipid metabolism and properties of the myelin sheath of nerve fibers. In animal cells it is mainly concentrated in the mitochondria, DNA, RNA, and the nucleus. Copper readily forms a connection with various proteins, especially those of sulphur. Although copper is an essential metal, it can, in some circumstances, lead to toxic effects including liver damage and gastrointestinal disturbances. Such as Wilson's disease (also known as hepatolenticular degeneration), Indian Childhood Cirrhosis (ICC) which are characterised by an accumulation of copper-containing granules within liver cells. Ingestion of high levels of copper salts is known to cause gastrointestinal upsets. Additionally, absorption of copper compounds by inhalation causes congestion of the nasal mucosa, gastritis, diarrhea and toxic symptoms such as chronic lung damage. Copper compounds act on the intact skin, causing it to itch and inflammation. They can cause conjunctivitis, ulceration and corneal opacity, nasal congestion and as well as sore throat and nasal septum. The upper limit recommended by WHO for copper is less than 1.3 mg/dm^3. The maximum limit in drinking water is 0.05 mg/dm^3 (Fewtrell et al. 1996).

ION EXCHANGE (IX)

Ion exchange may be defined as the exchange of ions between the substrate and surrounding medium. The most useful ion exchange reaction is reversible. When the reaction is reversible, the ion exchanger can be reused many times. Generally resins are manufactured in the spherical, stress and strain free form to resist physical degradation. They are stable at high temperatures and applicable over a wide pH range. Ion exchange resins, which are completely insoluble in most aqueous and organic solutions, consist of a cross linked polymer matrix to which charged functional groups are attached by covalent bonding (Sherrington, 1998). The ion exchangers which contain cations or anions as counter-ions are called cation exchangers or anion exchangers, respectively. The usual matrix is polystyrene cross linked for structural stability with 3 to 8 percent of divinylbenzene (3-8 % DVB) (Kunin, 1958;Helfferich, 1962).

The resins of higher cross linking (12-16% DVB) are more costly, both to make and to operate and they are specially developed for heavy duty industrial applications. These products are more resistant to degradation by oxidizing agents such as chlorine, and withstand physical stresses that fracture lighter duty materials. Typical ion exchangers are produced with a particle size distribution in the range 20-50 mesh (for separation of anions from cations or of ionic species from nonionic ones). For more difficult separations, materials of smaller particle size or lower degrees of cross linking are necessary. Moreover, when the separation depends solely upon small differences in the affinity of the ions, a particle size of 200-400 mesh is required and when the selectivity is increased by the use of complexing agents, the particle size in the 50-100 mesh is adequate. The ion exchangers finer than 100 mesh are employed for analytical purposes and for practical applications on the commercial scale the materials finer than 50 mesh are used.

Depending on the type of functional groups of exchanging certain ions, the ion exchangers with strongly acidic e.g., sulphonate $-SO_3H$, weakly acidic e.g., carboxylate $-COOH$, strongly basic e.g., quaternary ammonium $-N^+R_3$ and weakly basic e.g., tertiary and secondary amine $-N^+R_2H$ and $-N^+RH_2$ should be mentioned. The strong acidic cation exchangers are well dissociated over a wide pH range and thus reaching its maximum sorption capacity. On the other hand, weak acidic cation exchangers containing, for example, carboxylic functional groups reach the maximum sorption capacity at pH> 7.0 as presented in Fig.1.

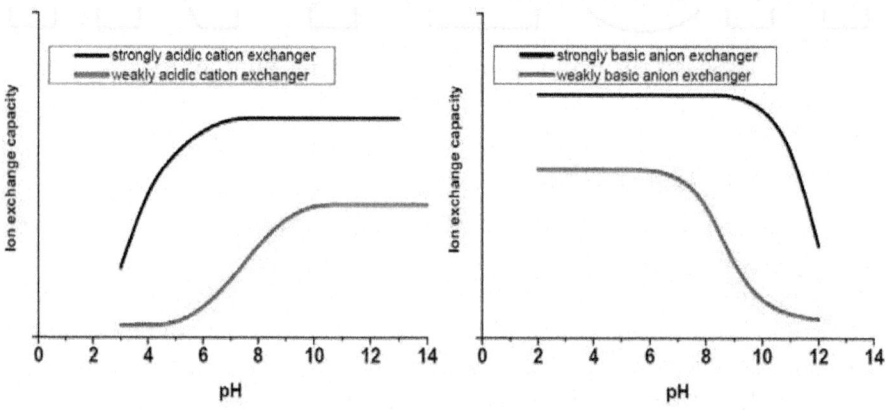

Figure 1: The sorption capacity of ion exchangers depending on pH.

Additionally, ion exchangers possess: the iminodiacetate functional groups $(-N\{CH_2COOH\}_2)$, phenol $(-C_6H_4OH)$, phosphonic $(-PO_3H_2)$ and phosphine $(-PO_2H)$ functional groups. These groups are acidic in nature and are dissociated with the exchange of H^+ or Na^+ ions for other cations from the solution.

Negative charge of the functional groups is offset by an equivalent number of mobile cations so-called counter ions. Counter ions can be exchanged for other ions from the solution being in the contact with the resin phase.

There are also amphoteric exchangers, which depending on the pH of the solution may exchange either cations or anions. More recently these ion exchangers are called bipolar electrolyte exchange resins (BEE) or zwitterionic ion exchangers (Nesterenko & Haddad, 2000). The aminocarboxylic amphoteric ion exchangers AMF-1T, AMF-2T, AMF-2M, ANKB-35 as well as the carboxylic cation exchanger KB-2T were, for example used for recovery of Ni(II) from the $Mn(NO_3)_2–H_2O$ system (Kononowa et al. 2000).

The individual ions present in the sample are retained in varying degrees depending on their different affinity for the resin phase. The consequence of this phenomenon is the separation of analyte ions, such as metal ions, however, the nature and characteristics of the resin phase determine the effectiveness of this process (Fritz, 2005). The affinity series which for various types of ion exchangers are as follows:

Cation Exchangers with the Sulphonic Functional Groups

It is well known that the affinity of sulphonic acid resins for cations varies with the ionic size and charge of the cation. The affinity towards cation increases with the increasing cation charge:

$Na^+< Ca^{2+}<Al^{3+}<Th^{4+}$,

and in the case of different cations with the same charge the affinity increases with the increasing atomic number:

$Li^+< H^+< Na^+< NH_4^+ < K^+< Rb^+< Cs^+< Ag^+< Tl^+$

$Mg^{2+} < Ca^{2+} < Sr^{2+} < Ba^{2+}$

$Al^{3+} < Fe^{3+}$.

Generally, the affinity is greater for large ions with high valency.

For the strong acidic cation exchanger the affinity series can be as follows:

$Pu^{4+} >> La^{3+} > Ce^{3+} > Pr^{3+} > Nd^{3+} > Sm^{3+} > Eu^{3+} > Gd^{3+} > Tb^{3+} > Dy^{3+} > Ho^{3+} > Er^{3+} > Tm^{3+}> Yb^{3+} > Lu^{3+} > Y^{3+} > Sc^{3+} > Al^{3+} >> Ba^{2+} > Pb^{2+} > Sr^{2+} > Ca^{2+} > Ni^{2+} > Cd^{2+} > Cu^{2+} > Co^{2+} > Zn^{2+} > Mg^{2+} > UO_2^{2+} >> Tl^+ > > Ag^+ > Cs^+ > Rb^+ > K^+ > NH_4^+ > Na^+ > H^+ > Li^+$

and for Lewatit SP-112 it is as: $Ba^{2+} > Pb^{2+} > Sr^{2+} > Ca^{2+} > Ni^{2+} > Cd^{2+} > Cu^{2+} > Co^{2+} > Zn^{2+} > Fe^{2+} >Mg^{2+} > K^+ >NH_4^+ > Na^+ >H^+$.

Cation Exchangers with the Carboxylic Functional Groups

Cation exchangers with the carboxylic functional groups show the opposite the affinity series for alkali and alkaline earth metal ions. Noteworthy is the fact that the cations exhibit a particularly high affinity for H^+. The affinity of this type of cation is therefore as follows:

$H^+ > Mg^{2+} > Ca^{2+} > Sr^{2+} > Ba^{2+} > Li^+ > Na^+ > K^+ > Rb^+ > Cs^+$.

Anion Exchangers with the Quaternary Ammonium Functional Groups

The charge of the anion affects its affinity for the anion exchanger in a similar way as for the cation exchangers:

citrate > tartrate > PO_4^{3-} > AsO_4^{3-} > ClO_4^- > SCN^- > I^- > $S_2O_3^{2-}$ > WO_4^{2-} > MoO_4^{2-} > CrO_4^{2-} > $C_2O_4^{2-}$

> SO_4^{2-} > SO_3^{2-} > HSO_4^- > HPO_4^{2-} > NO_3^- > Br^- > NO_2^- > CN^- > Cl^- > HCO_3^- > $H_2PO_4^-$ > CH_3COO^-

> IO_3^- > $HCOO^-$ > BrO_3^- > ClO_3^- > F^- > OH^-

for *Dowex 1* (type 1):

ClO_4^- > I^- > HSO_4^- > NO_3^- > Br^- > NO_2^- > Cl^- > HCO_3^- > CH_3COO^- > OH^- > F^-,

for *Dowex 2* (type 2):

ClO_4^- > I^- > HSO_4^- > NO_3^- > Br^- > NO_2^- > Cl^- > HCO_3^- > OH^- > CH_3COO^- > F^-

Anion Exchangers with the Tertiary and Secondary Amine Functional Groups

Only with the exception of the OH- ion, the affinity of the anion exchangers with the tertiary and secondary functional groups is approximately the same as in the case of anion exchangers with the quaternary ammonium functional groups. These medium and weakly basic anion exchangers show very high affinity for OH^- ions.

Anion exchange materials are classified as either weak base or strong base depending on the type of exchange group. These are two general classes of strong base anion exchangers e.g. types 1 and 2 depending on chemical nature. The synthesis of the weak base anion exchangers with the tertiary amine groups is usually provided by the chloromethylation of PST-DVB followed by the amination by secondary amine (Drăgan & Grigoriu, 1992). Weak base resins act as acid adsorbers, efficiently removing strong acids such as sulphuric and hydrochloric ones. They are used in the systems where strong acids predominate, where silica reduction is not required, and where carbon dioxide is removed in degasifiers. Preceding strong base units in demineralizing processes, weak base resins give more economical removal of sulphates and chlorides. The selectivity for the bivalent ions such as SO_4^{2-} depends strongly on the basicity of the resin, the affinities of various functional groups following the order: primary > secondary > tertiary > quaternary. Therefore among the factors affecting the sorption equilibrium the most important are: first of all nature of functional groups and the concentration of the solution (Boari et al. 1974). At low concentration the resin prefers ions at higher valency and this tendency increases with solution diluting. It should be also mentioned that obtaining resins with the primary amine functional groups is difficult by chemical reactions on polystyrene-divinylbenzene copolymers. Weakly basic anion exchangers can be used, for example for zinc cyanide removal from the alkaline leach solutions in the Merrill Crowe process (Kurama & Çatlsarik, 2000).

Gel and Macroporous Resins

The development in polymerization technique has provided novel matrices for a series of new ion exchangers. They differ from the earlier corresponding copolymers that are characterized by being essentially cross linked gels of polyelectrolytes with pore structure defined as the distance between polymeric chains.

It is well known that the fouling of the resin by organic compounds and mechanical stress imposed by plant operating at high flow rates are the most important problems encountered in the use of the ion exchange resins (De Dardel & Arden, 2001). To overcome these problems the ion exchangers with a high degree of cross linking containing artificial open pores in the form of channels with diameters up to 150 nm were introduced (Fig. 2).

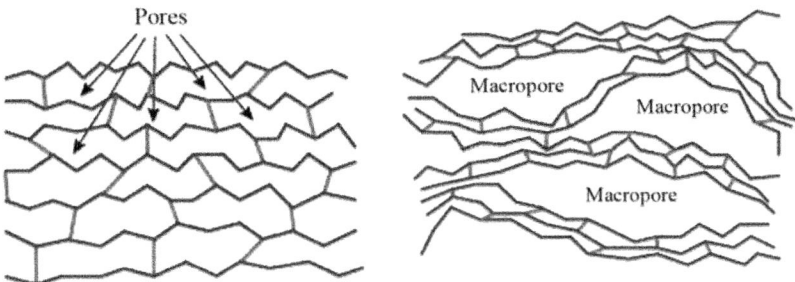

Figure 2: The structure of gel and macroporous ion exchangers (http://dardel.info/IX/index.html).

The first macroporous ion exchanger was a carboxylic resin made by Rohm and Haas, which covered a wide variety of acrylic compounds copolymerized with polyvinyl cross linking agents to make insoluble, infusible weakly acidic resins. By 1948 AmberliteTM IRC-50, made by the copolymerization of methacrylic acid and divinylbenzene was in production and possessed the 'sponge structure' (Abrams & Milk, 1997). According to the definition by Stamberg and Valter (1970) the macroporous resin should be characterised by measurable inner surface by any suitable method resulting from pores 5 nm, even in the completely dried state. In contrast, the gel materials did not show any porosity in the dry state. Then the term 'macroreticular' (sometimes abbreviated to MR) was selected to distinguish resins with a particular type of porosity obtained by application of precipitating diluents such as t-amyl alcohol. In 1979 Amber-Hi-Lites stated that 'macroreticular' resins are those made by a copolymerization technique which brings about precipitation during the polymerization, thus resulting in a product which has two phases, a gel phase in the form of microspheres formed during the phase separation and the pore phase surrounding the microspheres (Kunin, 1979). Later when quantitative porosity measurements were used it was shown that other methods of preparation gave products similar to those declared as 'macroreticular'. Therefore classification of resins should be based on their properties and function (Ion exchange resins and adsorbents, 2006).

During last decades the great progress was made by the development of the macroporous ion exchange resins. It should be mentioned that macroporous resins can also perform as adsorbents because of their pore structure. For organic ion exchange resins based on cross linked polystyrene the porosity was originally selected by the degree of cross linkage. These gel type resins are able to sorb organic substances from water according to their degree of porosity and the molecular weight of the adsorbate. They not only allow for large molecules or ions to enter the sponge like structure but also to be eluted during the regeneration. Therefore they perform two functions: ion exchange by means of the functional groups and the reversible adsorption and elution due to the macroporous structure. They are also resistant against organic fouling which results in a longer resin life compared with the conventional gel type ion exchangers as well as the quality of the treated water is much better because of the adsorption of organic species by the macroporous structure. The SEM scan of the macroporous anion exchanger Lewatit MonoPlus MP 500 is presented in Fig. 3.

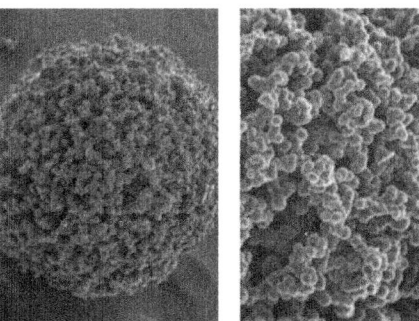

Figure 3: SEM scans of macroporous resins.

Ion exchange applications can be performed by either column (flow continuous) and batch technique. In column operations the ion exchange resin is placed in the vertical column to form a bed. Once the application is completed, the resin can be regenerated to use in another cycle. In batch operations the resin is shacked in a vessel with the solution to be treated. After the application is completed, the resin can be regenerated in place or transferred to a column for regeneration.

While the main aims in the production of conventional ion exchangers were focused on obtaining a high ion exchange capacity and improved chemical resistance and thermal and mechanical strength, in the case of monodisperse ion exchange resins, these efforts directed towards improvement of kinetic parameters. Heterodisperse ion exchangers are usually characterized by a standard grain size of 0.3-1.2 mm and uniformity coefficient (UC) within the limits of 1.5-1.9. In the case of monodisperse ion exchange resins during the manufacturing process the grain size from 0.6 mm and uniformity coefficient within the limits 1.1-1.2 is usually achieved. In addition, monodisperse ion exchangers, due to the uniform packing of the column, show more than 12% higher ion exchange capacity, faster kinetics of exchange and a much higher mechanical strength, which is extremely important from the economical point of view. As the particle size of the ion exchanger material and its uniformity are the most important parameters influencing the hydraulics and kinetics of the ion exchange therefore the monodisperse ion exchangers provided better flow characteristics in column applications in comparison to the conventional heterodisperse ion exchangers (the flow rate decreases with the decreasing particle size, however, smaller particles have larger outer surface, but cause larger head loss in the column processes) (Scheffler, 1996; Krongauz & Kocher, 1997). The visualization of the monodisperse and hetrodisperse ion exchangers is presented in Fig. 4a-b.

For example the research carried out by Zainol & Nicol (2009a) shows that in the the sorption process of Ni(II) and other metal ions the monodisperse resin (Lewatit MonoPlus TP 207) proved to be superior to the conventional heterodisperse ones in terms of loading capacity for Ni(II) and also the kinetics of adsorption. This makes it a preferred choice for different applications.

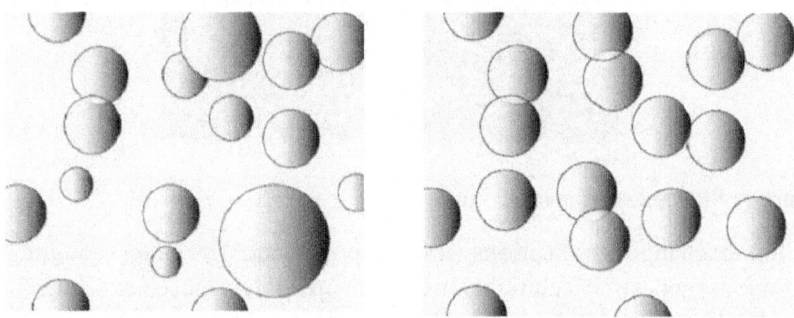

Figure 4: The monodisperse and hetrodisperse ion exchangers.

The influence of temperature on the equilibrium properties of ion exchange resins was studied extensively. The decrease of the capacity of the cation exchange resins based on the polystyrene matrix due to the operation temperature is not a significant problem. However, the relatively slight decomposition gives enough decomposition products to cause significant problems elsewhere. This may be decomposition of the bone polystyrene matrix, resulting in styrene sulphonic acid derivatives or as a substitution of the sulphonic group giving sulphate. Further decomposition of styrene sulphonic acid derivatives will also result in sulphate as one of the end products (desulphonation). The amount of sulphate produced is sometimes high. The information on the stability of the ion exchange resins mainly deals with the anion exchange resins. The mechanism of the degradation of quaternary ammonium salts and tertiary anions is well-known (Reynolds, 1982; Fernandez-Prini, 1982; Fisher, 2002). The effect of temperature on the properties of chelating ion exchangers was also described in the paper by Ivanov (1996).

APPLICATION OF ION EXCHANGERS FOR HEAVY METAL IONS REMOVAL

Ion exchange technique can remove traces of ion impurities from water and process streams and give a product of desired quality. Ion exchangers are widely used in analytical chemistry, hydrometallurgy, antibiotics, purification and separation of radioisotopes and find large application in water treatment and pollution control (Clifford, 1999; Luca et al. 2009). The list of metals which are recovered and purified on an industrial scale by means of ion exchange include: uranium, thorium, rare earth elements (REEs), gold, silver, platinum metals (PGM), chromium, copper, zinc, nickel, cobalt and tungsten.

In some of these cases, the scale of operations is relatively small, for instance in the rare earth elements or noble metals, but the values of recovered metals are very high. Ion exchange process is particularly suitable for purification of metal ions with a high value and low processing. The alternative is also a process of large-scale recovery of trace amounts of metals from waste streams, such as cadmium and mercury, chromium, or copper and zinc. The use of ion exchange processes in hydrometallurgy is high and every year continues to grow. It is associated mainly with the progress of what is observed in the synthesis of new selective chelating ion exchangers containing complexing ligands (Minczewski, et al. 1982;

CHELATING AND SPECIAL ION EXCHANGERS

Typical disadvantage of lack a of the selectivity towards heavy metal ions and alkali and alkaline earth metal ions of most widely used functionalized ion exchangers such as Chelex 100 is overcome by introducing chelating ligands capable of removing selective metal ions. It exhibits high affinity for heavy metal ions: $Cu^{2+} > Hg^{2+} > Pb^{2+} > Ni^{2+} > Zn^{2+} > Cd^{2+} > Co^{2+} > Fe^{2+} > Mn^{2+} > Ca^{2+} > Mg^{2+} > Sr^{2+} > Ba^{2+} >>>$ alkali ions $> H^+$, whereas for sulphonic ones the analogous affinity series can be as presented earlier (for Lewatit SP 112).

Generally, the functional group atoms capable of forming chelate rings usually include oxygen, nitrogen and sulphur. Nitrogen can be present in a primary, secondary or tertiary amine, nitro, nitroso, azo, diazo, nitrile, amide and other groups. Oxygen is usually in the form of phenolic, carbonyl, carboxylic, hydroxyl, ether, phosphoryl and some other groups. Sulphur is in the form of thiol, thioether, thiocarbamate, disulphide groups etc. These groups can be introduced into the polymer surface by copolymerization of suitable monomers, immobilization of preformed ligands, chemical modification of groups originally present on the polymer surface. However, the last two are most often used (Warshawsky, 1987). Chelating resins with such type of ligands are commonly used in analysis and they can be classified according to Fig.1. (Kantipuly et al. 1990). The choice of an effective chelating resin is dictated by the physicochemical properties of the resin materials. These are the acid-base properties of the metal species and the resin materials, the polarizability, selectivity, sorptive capacity, kinetic and stability characteristics of the resin. The sorption capacity of chelating ion exchangers depends mainly on the nature of functional groups and their content as well as solution pH as for their selectivity it depends on the relative position of functional groups, their spatial configuration, steric effects, and sometimes their distance from the matrix and to a lesser extent on the properties of the matrix. Their use allows the recovery of valuable metals from ores and sludge, sea water and industrial effluents. They are used as flotation agents, depressants, flocculants and collectors.

It is worth emphasizing that these resins are invaluable wherever it is necessary to concentrate or remove elements present in very low concentrations.

With a range of well known chelating ion exchangers only a few types are produced on an industrial scale. Among the most important ones these with the functional groups: amidoxime, dithiocarbamate, 8-hydroxychinoline, iminodiacetate, aminophosphonic, bispicolylamine, diphosphonic, sulphonic and carboxylic acid groups, thiol, thiourea as well as isothiourea should be selected (Sahni & Reedijk, 1984; Busche et al. 2009). Among them the chelating ion exchangers possessing methylglucoamine, bis(2pirydylmethyl)amine also known as bispikolilamine, thiol etc. are used for special applications such as removal of precious metal ions, heavy metal ions from the acidic medium, boron and special oxoanions removal. A separate group are ion exchangers of solvent doped type used for In, Zn, Sn, Bi, etc. separation. The advantages of ion exchangers from these groups include good selectivity, preconcentration factor, binding energy and mechanical stability, easy regeneration for multiple sorption-desorption cycles and good reproducibility in the sorption characteristics.

CHELATING ION EXCHANGERS WITH THE HYDROXAMIC AND AMIDOXIME FUNCTIONAL GROUPS

The choice of hydroxamic acids is based on their application in mineral processing as collectors in flotation of haematite, pyrolusite or bastnaesite ores. The copolymer of malonic acid dihydroximate with styrene-divinylbenzene was used for uranium(VI) removal from sea water (Park & Suh, 1996). In the paper by Ahuja (1996) it was found that glycin hydroximate resin shows maximum adsorption for Fe(III), Cu(II) and Zn(II) at pH 5.5; for W(VI), U(VI), Co(II) and Ni(II) at pH 6.0 as well as for Cd(II) at pH 6.5. It can be recommended for separation of Cu(II) from Co(II) and Ni(II) at pH 5.5. However, the iminodiacetic– dihydroximate resin can be applied for U(VI), Fe(II), Cu(II) separation according to the affinity series: $UO_2^{2+} > Fe^{3+} > Cu^{2+} > Zn^{2+} > Co^{2+} > Cd^{2+} > Ni^{2+} > Zn^{2+}$. Hydroxamic acids exist in the two tautomeric forms: and metal ions are coordianated by the hydroxamide functional group (a).

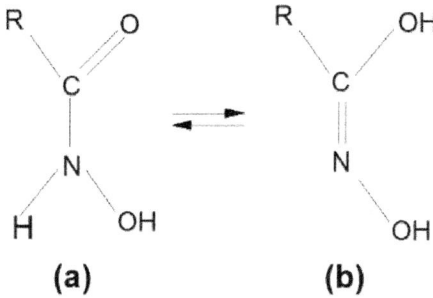

(a)   **(b)**

Chelating resins with the amidoxime functional groups such as Duolite ES-346, and Chelite N can be applied for the concentration of solutions containing Ag(I), Al(III), Cd(II), Co(II), Cr(III), Cu(II), Fe(III), Hg(II), Mn(II), Ni(II), Mo(VI), Pb(II), Ti(IV), U(VI), V(V) and Zn(II) in the presence of alkali and alkaline earth metal ions (Samczyński & Dybczyński, 1997; Dybczyński et al. 1988). Alkali and alkaline earth metal ions are poorly retained by these resins. Duolite ES-346 is commonly used to extract uranium(VI) from seawater and As(III) from aqueous solutions. It can be also applied for Pd(II) removal (Chajduk-Maleszewska & Dybczyński, 2004). It is characterized by high selectivity towards Cu(II) ions due to the presence of amidoxime groups and small quantities of hydroxamic acid (RCONHOH):

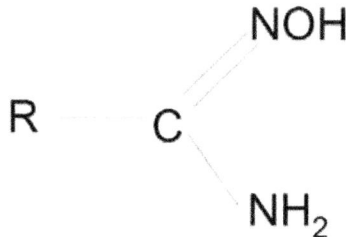

where: R is the resin matrix.

It was found that for the amidoxime resins the selectivity series can be as follows: Cu(II) > Fe(III) > As(III) > Zn(II) > Ni(II) > Cd(II) > Co(II) > Cr(III) > Pb(II). Interesting results were obtained by observing the impact of acidity on the behaviour of this ion exchanger. At low pH values (<3) there was a decrease in the chelating ability of Duolite ES 346 for heavy metal ions as well as degradation of its functional groups according to the reactions (Ferriera et al. 1998):

However, the second reaction is the representative of the degradation of amidoxime groups under less acidic conditions (pH < 3.0). The increase in pH causes the weakening of the hydrogen ions competition for active sides resulting in an increase in the complexation of metal ions such as Cu(II). The fact that the degradation of the functional groups of Duolite ES-346 occurs under the influence of strong mineral acids is a serious problem which can significantly reduce the chelating capacity of the resin. However, this effect was made use of the recovery of adsorbed ions on the resin ion exchange.Corella et al. (1984) demonstrated that poly(acrylamidoxime) can be successfully used for the preconcentration of trace metals from aqueous solutions.

Also salicylic acid is a ligand which can selectively complex with Zn(II), Pb(II), Fe(III). Using the salicylic acid loaded resin for preconcentration of Zn(II) and Pb(II), it was proved that the preconcentration factors are much higher than those for bis(2-hydroxyethyl)dithiocarbamate (Saxena et al. 1995). However, the phenol-formaldehyde resin with the salicylaldoxime and salicylaldehyde functional chelating groups shows high selectivity for Cu(II) ions (Ebraheem & Hamdi, 1997).

The affinity order of metal complexes of salicylaldoxime is as follows: $Fe^{3+} > Cu^{2+} > Ni^{2+} > Zn^{2+} > Co^{2+}$.

CHELATING ION EXCHANGERS WITH THE DITHIO-CARBAMATE FUNCTIONAL GROUPS

The high affinity for transition metal ions is also exhibited by the classes of ion exchangers with the dithiocarbamate functional groups (including commercially available Nisso ALM 125), in which sulphur is the donor atom. Ion exchangers of this type have high affinity for the Hg(II), Pb(II), Cd(II) ions as well as precious metal ions, however, they do not adsorb alkali and alkaline earth metals. It was shown that dithiocarbamates obtained with the share of primary amines are less stable than those obtained with the share of secondary amines, and the binding of metal ions to the functional group of the donor atom increases in the number of $Fe^{2+} < Ni^{2+} < Cu^{2+}$. The sorption efficiency is dependent on the presence of ions in the solution such as SCN^{-}. In the paper by McClain and Hsieh (2004) the selective removal of Hg(II), Cd(II) and Pb(II) was presented. This resin is also effective for separation and concentration of Mn(II), Pb(II), Cd(II), Cu(II), Fe(III) and Zn(II) from complex matrices (Yebra-Biurrun et al. 1992). The copolymer of poly(iminoethylo)dithiocarbamate was used for sorption of VO^{2+}, Fe(II), Fe(III), Co(II), Ni(II) and Cu(II) (Kantipuly et al. 1990)

CHELATING ION EXCHANGERS WITH THE 8-HYDROXYQUINOLINE FUNCTIONAL GROUPS

A simple method for immobilization of 8-hydroxyquinoline in a silica matrix is described byLührmann (1985). The sorbent was used in the sorption of Cu(II), Ni(II), Co(II), Fe(III), Cr(III), Mn(II), Zn(II), Cd(II), Pb(II) and Hg(II) at pH from 4 to 6. It was shown that the sorption capacity varies in the range from 0.2 to 0.7 mM /g, and the partition coefficients from 1×10^3 to 9×10^4. Ryan and Weber showed (1985) that this type of sorbent has better sorption properties with respect to Cu(II) than Chelex 100 with the iminodiacetate functional groups. Th sorbents based on 8-hydroxyquinoline can be used, e.g. for concentration of the trace metal ions Mn(II), Co(II), Ni(II), Cu(II), Zn(II), Cd(II), Pb(II) and Cr(III) from sea water (Pyell & Stork, 1992).

CHELATING ION EXCHANGERS WITH THE IMINODIACETIC FUNCTIONAL GROUPS

Recently the attention has been paid to the ion exchangers with the amino- or iminoacids groups. The presence of two carboxyl groups and the tertiary nitrogen atom provides strong preference for chromium(III) and copper(II) (Marhol & Cheng, 1974). Therefore, for the commercial chelating ion exchangers such as Chelex 100, Dowex A 1, CR-20, Lewatit TP 207, Lewatit TP 208, Purolite S 930, Amberlite IRC 748 (formely Amberlite IRC 718) or Wofatit MC-50 the sorption process of metal ions proceeds according to the order: $Cr^{3+} > Cu^{2+} > Ni^{2+} > Zn^{2+} > Co^{2+} > Cd^{2+} > Fe^{2+} > Mn^{2+} > Ca^{2+} >> Na^{+}$. This type of ion exchangers also exhibis high affinity for Hg(II) and Sb(V) ions. It should be noted that depending on the pH value they may occur in the following forms (Zainol & Nicol. 2009a):

$$RCH_2HN^+ \diagup^{CH_2COOH}_{\diagdown CH_2COOH}$$

pH=2,21

$$RCH_2HN^+ \diagup^{CH_2COOH}_{\diagdown CH_2COO^-}$$

pH=3,99

$$RCH_2HN^+ \diagup^{CH_2COO^-}_{\diagdown CH_2COO^-}$$

pH=7,41

$$RCH_2N \diagup^{CH_2COO^-}_{\diagdown CH_2COO^-}$$

pH=12,3

At pH <2.0 the nitrogen atom and the two carboxylic groups are protonated. In this case the chelating ion exchanger with the iminodiacetic functional groups behaves as a weakly basic anion exchanger. At pH ~ 12, the nitrogen atom and the two carboxylic groups undergo deprotonation – the ion exchanger behaves as a typical weakly acidic cation exchanger. For pH medium values, the iminodiacetic resin behaves as an amphoteric ion exchanger. The iminodiacetate groups provide electron pairs so that the binding force for the alkaline earth metals is 5000 times as large as that for alkali metals like Ca(II), which react with divalent metals to form a stable coordination covalent bond. Therefore, the affinity series determined for the iminodiacetic ion exchanger can be presented in the order: $Hg^{2+} > UO_2^{2+} > Cu^{2+} > Pb^{2+} > Ni^{2+} > Cd^{2+} > Zn^{2+} > Co^{2+} > Fe^{2+} > Mn^{2+} > Ca^{2+} > Mg^{2+} > Ba^{2+} > Sr^{2+} >> Li^+ > Na^+ > K^+$.

Amberlite IRC 748 in the K(I) form was also used for removal of Ca(II), Mg(II) from the potassium chromate solution (Yua et al. 2009). The optimum pH obtained for Ca(II) and Mg(II) adsorption onto Amberlite IRC 748 from the potassium chromate solution is 9.8 and 9.5, respectively. It was also noted that an increase of temperature and resin dosage resulted in their higher adsorption and the equilibrium conditions were attained within 480 min. The experimental data were relatively well interpreted by the Langmuir isotherm and the monolayer adsorption capacities of Ca(II) and Mg(II) were equal to 47.21 mg/g and 27.70 mg/g, respectively. This is of great importance because manufacturing of chromium trioxide by electrolyzing chromate salts, as a green process with the zero emission of waste, is studied widely now (Li et. al 2006). It should be also pointed out that separation factors between Mg(II) and Ca(II) and other divalent metal ions on an iminodiacetate resin are much smaller than those expected from the stability constants of their IDA complexes in solutions. Such phenomena were qualitatively described as the 'polymer effect' or operation of ion exchange as well as complexation reactions. Pesavento et al. (1993) gave a quantitative explanation for these anomalies on the basis of the Gibbs-Donnan model. Ca(II) and Mg(II) ions are adsorbed forming the $R(Hida)_2M$ complexes in acidic media and R(ida)M in neutral and alkaline systems whereas Ni(II) or Cu(II) etc. forms the R(ida)M complexes:

Commercially available chelating resins with the iminodiacetate functional group (Amberlite IRC 748, Lewatit TP 207, Lewatit TP 208, Purolite S 930, Lewatit MonoPlus TP 207) have been evaluated for their suitability for the adsorption of Ni(II) and other metal ions (Al(III), Ca(II), Co(II), Cr(III), Cu(II), Fe(II/III), Mg(II), Mn(II) and Zn(II)) from the tailings of a pressure acid leach process for nickel laterites. The Amberlite IRC 748 and TP MonoPlus 207 resins were found to be the most suitable in terms of loading capacity for nickel and kinetics of adsorption. Although all the five resins studied have the same functional groups their performance is not identical. The observed differences are possibly caused by variations in the synthesis procedure which results in variations in the structure of the matrix, degree of cross linking, density of functional groups, proportion of iminodiacetate groups and also the particle size (Zainol & Nicol, 2009a)

Additionally, the research carried out by Biesuz et al. (1998) shows that in the case of Ni(II) and Cd(II) sorption the structure of formed complexes is different. Ni(II) forms complexes of R(ida)M type, whereas Cd(II) $R(idaH)_2M$. However, in the paper by Zagorodni & Muhammed (1999) it was stated that the complexes $R(Hida)_2M$ should be extremely weak or even impossible. The adsorption equilibrium of Ni(II), Co(II), Mn(II) and Mg(II) on Amberlite IRC 748 has been discussed in (Zainol & Nicol, 2009b). The resin proves to have high selectivity for Ni(II) and Co(II) which suggests that these metals can be easily separated from Mg(II) and Mn(II) at pH 4 and 5. The following order of selectivity of the resin was also found: Ni(II) > Co(II) > Mn(II) > Mg(II).

The kinetics of Cd(II) sorption from separate solutions and from the mixtures with the nonionic surfactant Lutensol AO-10 (oxyethylated alkohols) in the hydrogen form of chelating iminodiacetic ion exchanger has also been investigated (Kaušpėdienė et al. 2003). It was stated that the sorption of Cd(II) from separate solutions and from the mixture with AO-10 is controlled by the intraparticle diffusion in acidic (pH 5) and alkaline media (pH 7.6). The presence of AO-10 leads to a decrease in the rate of intraparticle diffusion. The iminodiacetate resin has a large collective adsorption with Cr(III) ion. The Cr(III) form bearing waste water can be removed at any pH in the range 3-6 at 2h of the phase contact time. Therefore for treatment of leather tanning, electroplating, textile and dyeing waste water the application of this resin is economical (Gode & Pehlivan, 2003)

Adsorption of trivalent metal ions on iminodiacetate resins was not studied as extensively as that of divalent metal ions. The known selectivity order of trivalent metal ions on an iminodiacetate resin can be presented as: $Sc^{3+} > Ga^{3+} > In^{3+} > Fe^{3+} > Y^{3+} > La^{3+} > Al^{3+}$ (Yuchi et al. 1997).

Also since the end of the 1960s fibrous adsorbents with the iminodiacetic acid groups have been studied. For example, the capacity of a commercially available iminodiacetic acid fiber named Ionex IDA-Na was established to be 0.9-1.1 mmol/g for Cu(II). The fibrous materials containing iminodiacetate groups were developed by the group of Jyo et al. (2004). Although the metal ion selectivity of the present fiber was close to that of iminodiacetic acid resins, the metal adsorption rate of chloromethylstyrene-grafted polyethylene coated polypropylene filamentary fiber is much higher than that of commercially available granular exchangers of this type having cross-linked polystyrene matrices. In the column mode adsorption of Cu(II), breakthrough capacities of Cu(II) were independent of the flow rates of feeds up to 200-300/h. The main reasons for the extremely fast adsorption rate of sorbent can be ascribed to the diameter of the fiber being much less than those of the resins as well as to the fact that the functional groups were introduced onto non cross linked grafted polymer chains. Their chemical and physical stabilities are comparable to those of commercially available iminodiacetic acid resins.

CHELATING ION EXCHANGERS WITH THE PHOS-PHONIC AND AMINOPHOSPHONIC FUNCTIONAL GROUPS

Among various types of ion exchangers with the acidic ligands, those having phosphonate functionality are of particular interest since they are selective towards heavy metal cations. Development of this type of ion exchangers started in the late 1940s with phosphorylation of poly(vinyl alcohol) using various phosphorylating agents (Trochimczuk & Streat, 1999; Trochimczuk 2000). Besides phosphate, phosphinic and phosphonic resins, containing $-OPO(OH)_2$, $-PO(OH)$ and $-PO(OH)_2$ functional groups, respectively, they also contain methylenediphosphonate, ethylenediphosphonate and carboxyethyl phosphonate ones (Marhol et al. 1974;

Kabay, 1998a; Ogata et al. 2006). In all cases they display good selectivity towards metal ions even at very low pH (except for ethylenediphosphonate and carboxyl containing resins, being less acidic, more selective at the pH value from 1 to 2).

Chelating ion exchangers with the phosphonic functional groups are characterized by extremely high selectivity towards Th(IV) and U(IV,VI) as well as Cu(II), Cd(II), Zn(II), Ni(II), Ag(I), Au(III) and Fe(III) ions. Commercially available resins containing the phosphonic groups are Diaion CRP200 and Diphonix® Resin. In the case of Diphonix® Resin besides the diphosphonic functional groups in the structure of the ion exchanger, there are also carboxylic and sulphonic functional groups whose presence determines better hydrophilic properties. Diphonix® Resin as well as Diphonix A with the functional phopshonic and ammonium (type 1) or pyridyne (type 2) groups have been of significant interest lately (Chiarizia et al. 1993; Horwitz et al. 1993; Chiarizia et al. 1994; Chiarizia et al. 1996,Alexandratos, 2009). Diphonix® Resin was developed by the Argonne National Laboratory and University of Tennessee. It is synthesized by a patented process involving copolymerization of tetraalkylvinylidene diphosphonate with styrene, divinylbenzene, and acrylonitrile followed by sulphonation with concentrated sulphuric acid. Finding a method for effective copolymerization of vinylidene-1,1-diphosphonate (VDPA) ester was a major achievement because of the steric hindrance imposed on the vinylidene group by the diphosphonate group. This difficulty was overcome by using another relatively small monomer, acrylonitrile, as a carrier to induce polymerization of vinylidene-1,1-diphosphonate (Horwitz et al. 1994; Horwitz, et al. 1995). The protonation constants of Diphonix Resin® which are pK_1 and $pK_2 < 2.5$ $pK_3=7.24$ and $pK_4=10.46$ appear almost equal to the protonation constants of the starting material VDPA which are $pK_1=1.27$, $pK_2 = 2.41$, $pK_3=6.67$ and $pK_4=10.04$ (Nash et al. 1994).

In the past few years there were many publications on the separation of lanthanides and actinides on the chelating resins with the phosphonic groups. Lanthanides in minerals occur in small amounts, usually in the form of mixtures, often isomorphic, so that their extraction and separation create many problems. To this end also Diphonix® Resin can be used especially at low pH. It is characterized by high affinity for U(VI), Pu(IV), Np(IV), Th(IV), Am(III) and Eu(III). It was found that from 1 M HNO_3 solutions the distribution coefficient of Diphonix® Resin for U(VI) ions is 70,000 compared to 900 for sulphinic acid resin (Alexandratos, 2007) and the recovery coefficient for Eu(III) under the same conditions is 98.3, whereas for the sulphonic acid resin 44.9 (Ripperger & Alexandratos, 1999). In the paper by Phillips et al. it was demonstrated that Diphonix Resin® can be successfully used for removal of uranium from the solutions of pH > 5 including high concentration of NO_3^- ions as it is less sensitive to interference by such ions as carbonates, nitrates(V), sulphates(VI), Fe(III), Ca(II) and Na(I) (Philips et al. 2008). It can be also used for removal of V(V), Cr(III), Mn(II), Co(II), Ni(II), Zn(II), Cd(II), Hg(II) and Pb(II) form waters and waste waters; V(V), Cr(III), Mn(II), Co(II), Ni(II), Cu(II), Zn(II), Cd(II), Hg(II) and Pb(II) from drinking water; Mn(II), Co(II) and Ni(II) from waste waters of the oil industry; Cr(III) from acidic solutions, Fe(III) from the solutions containing complexing agents in the process of removing scale and radionuclides from radioactive waste waters. Smolik et al. (2009) investigated separation of zirconium(IV) from hafnium(IV) sulphuric acid solutions on Diphonix Resin®. It was found that the best medium for separation of hafnium(IV) and zirconium(IV) is 0.5 M sulphuric acid. A decrease in temperature lowers the degree of metals separation, while lower flow rates through the column increases zirconium(IV) from hafnium(IV) separation. Recent studies have shown that Diphonix Resin® can also be used for removal of Cd(II) and Cr(III) from the phosphoric acid solutions through column tests. Kabay et al. (1998b) found that the acid concentration strongly determines the resin behaviour with respect to the sorption/elution of Cd(II) and Cr(III). In the paper by Cavaco et al. it was pointed but that Diphonix Resin® has strong affinity for Cr(III) ions and high selectivity towards Fe(III) and Ni(II) (Cavaco et al. 2009). The mechanism of sorption on Diphonix Resin® can be written as (Hajiev et al. 1989):

$$R-(PO_3H_2)_2^{2-} + M^{2+} \rightleftarrows R-(PO_3H_2)^{2-}\rightarrow M^{2+}$$

where: R is the resin matrix.

However, according to the literature, Diphonix Resin® has the best selectivity for transition metals such as Fe(III), Cu(II) and Ni(II) over Cr(III). High affinity of Diphonix Resin® for Fe(III) compared to the mono- and divalent ions e.g. Ca(II) was reported in several papers. Owing to its very good separation capability, Diphonix Resin® was also applied in the project FENIX Iron Control System to remove iron from the spent copper electrolyte in Western Metals Copper Ltd. (Queensland, Australia). In this plant, copper(I) sulphate(VI) was used as a reducing agent at the reaction temperature of 85 °C to increase the elution of Fe(III):

$$Fe_2(SO_4)_3 + 6HR \rightleftarrows 2Fe(R)_3 + 3H_2SO_4$$

$$CuSO_4 + Cu \rightleftarrows Cu_2SO_4$$

$$2Fe(R)_3 + 3H_2SO_4 + Cu_2SO_4 \rightleftarrows 2FeSO_4 + 2CuSO4 + 6HR$$

where: R is the resin matrix.

In the paper by Lee & Nicol (2007) it was proved that sorption capacities of Diphonix Resin® for Fe(III) and Co(II) ions in the sulphate(VI) system at pH 2 are equal to 130 mg/g and 90 mg/g, respectively.

The obvious disadvantage of this ion exchanger is therefore the fact that it is difficult to remove Fe(III) ions. To this end 1-hydroxyethane-1,1-diphosphonic acid (HEDP) is used. In the case of Cr(III), Mn(II), Co(II), Ni(II), Cu(II), Zn(II), Cd(II), Hg(II), Sn(II) and Pb(II) ions 2M H_2SO_4 can be also applied.

In the group of chelating ion exchangers containing phosphonic and aminophosphonic functionalities the resins with aminoalkylphosphonic functional groups, such as Duolite C-467, Duolite ES-467, Lewatit OC 1060, Purolite S 940, Purolite S 950 and Chelite P occupy a significant position. In the sorption of heavy metal ions on this kind of chelating ion exchangers the following affinity series is obtained: $Pb^{2+} > Cu^{2+} > UO_2^{2+}$, Zn^{2+}, $Al^{3+} > Mg^{2+} > Sr^{2+} > Ca^{2+} > Cd^{2+} > Ni^{2+} > Co^{2+} > Na^+ > Ba^{2+}$. These ion exchangers as well as the previously mentioned phosphonic ones exhibit poor affinity for Ca(II) and Mg(II). The effectiveness of sorption of the above mentioned metal ions, however, decreases with the decreasing pH. It is worth mentioning that depending on pH value, the aminoalkylphosphonic groups may occur in the following forms:

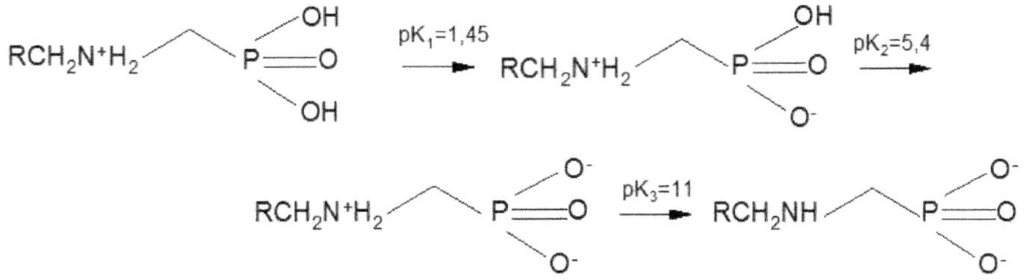

and therefore the selectivity of metal ions sorption depends on the degree of ionization of phosphonic groups. In the case of acidic solutions due to protonation of the nitrogen atom of aminophosphonic group there are formed combinations with the following structure:

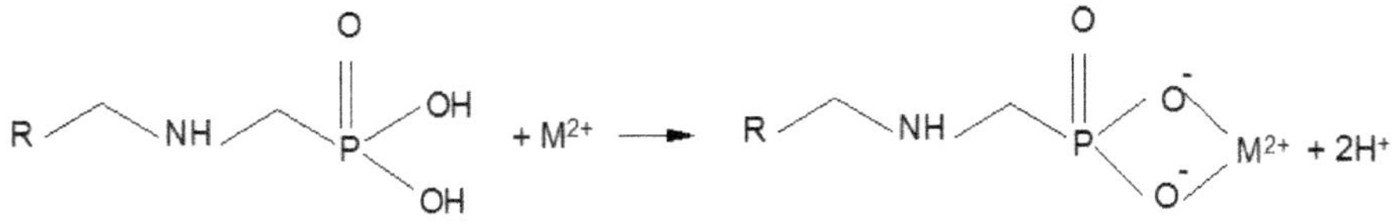

One of the most favourable modes of chelation of the phosphonic acid group is the formation of a four-membered ring through determination of two P-OH groups.

Additionally, in the case of the aminoalkylphosphonic groups, due to the fact that between the aromatic ring of the matrix and the nitrogen atom there is also presented the alkyl group, the increase of the electron density on the nitrogen atom of the amino group is expected. It affects the growth of its protonation. Therefore, this preferred zwitterion form can be as follows:

$$RCH_2N^+H_2 \qquad\qquad P \begin{array}{c} OH \\ O \\ O^- \end{array}$$

However, the possibility of coordination of the secondary nitrogen atom at lower pH seems to be impossible with respect to its protonated nature and also for steric reasons. Therefore the only potential donor and binding sites of Duolite ES-467 are the oxygen atoms of the phosphonic groups at lower pH values. The chelating, aminomethylphosphonic functional group is also potentially a tridentate ligand having two bonding sites at a phosphonic acid groups and one coordination site at the secondary nitrogen atom (Kertman, 1997; Nesterenko et al. 1999). Formation of a four membered ring through bonding of one of the OH groups and coordination of the oxygen atom has also been reported. These structures are presented below:

$$R-NH-P\begin{array}{c}O\\OH\\OH\end{array} + Cu^{2+} \longrightarrow R-NH \cdots Cu \cdots P=O + 2H^+$$

Chelating ion exchangers with the aminoalkylphosphonic functional groups, like picolylamine resins - Dowex M 4195 exhibits moderate selectivity for Cu(II) over Fe(III) in the acidic sulphate(VI) solutions compared to the iminodiacetic acid resins which show no or limited selectivity depending on pH. The stability constants for divalent metal ions with aminomethylphosphonic acid have been found in the order: $Ca^{2+} < Mg^{2+} < Co^{2+} < Ni^{2+} < Cu^{2+} > Zn^{2+}$ (Sahni et al. 1985). In the paper by Milling and West (1984), it was found that Duolite ES-467 possesses a higher capacity for copper(II) ions compared to nickel(II) and iron(III) and that the capacity decreases with the decreasing pH and metal ion concentration in the solution.

Besides Duolite ES-467, Purolite S 950 has been proved to have a high affinity for various heavy metal ions and it is successfully applied in metallurgical and wastewater treatment processes. In the paper by Koivula et al. (2000) Purolite S-950 was used for purification of effluents from metal plating industry containing Zn(II), Ni(II), Cu(II) and Cd(II) ions. Among others, it was stated that Purolite S-950 showed lower sorption capacity equal to 1.2 eq/dm³ for zinc chloride compared to zinc solutions containing KCl and NH₄Cl (1.3 eq/dm³). Under analogous conditions the sorption capacity for Cd(II) was 1.1 eq/dm³. Recovery of Ni(II) and Co(II) from organic acid complexes using Purolite S 950 was also studied by Deepatana & Valix (2006). They found that sorption capacities for nickel sulphate(VI) for Dowex M4195 (94.51 mg/g), Amberlite IRC 748 (125.03 mg/g) and Ionac SR-5 (79.26 mg/g) are much higher than those for Purolite S-950 in the case of sorption of Ni(II) complexes with citric acid (18.42 mg/g), malic acid (14.45 mg/g) and lactic acid (19.42 mg/g) mainly due to the steric hindrance. For Co(II) ions analogous results were obtained (citric (5.39 mg/g for citric acid; 7.54 mg/g for malic acid and 10.48 mg/g for lactic acid). The elution efficiencies of these complexes from Purolite S-950 resins were high (82–98%) therefore it would appear that the adsorption process involves weak interactions. However, in the case of the sorption of Cu(II) and Zn(II) ions from the sulphate solutions at pH 1.9 on the aminomethylphosphonic resin Lewatit R 252K and the iminodiacetic resin Lewatit TP 207 it was found that separation factors were much lower for Lewatit R 252K (83.0 at 10 °C and 30.0 at 80 °C) than for Lewatit TP 207 (1.67 at 10 °C and 1.4 at 80 °C) (Muraviev et al. 1995).

CHELATING ION EXCHANGERS WITH THE METHYL-GLUCAMINE FUNCTIONAL GROUPS

Selective ion exchange resins also include chelating ion exchangers containing N-methyl (polyhydroxohexyl)amine functional groups also called methylglucamine. Commercially available ion exchangers of this type are: Amberlite IRA 743, Duolite ES-371, Diaion CRB 02, Dowex BSR 1, Purolite S 108 and Purolite S110.

These ion exchangers show high selectivity for boron (in the form of trioxyboric acid H_3BO_3) (Alexandratos, 2007; Alexandratos, 2009). The boron sorption process proceeds according to the scheme:

Besides boron the following components of waste water should be also taken into account: Na(I), K(I), Ca(II), Mg(II), Cl⁻, SO_4^{2-}, HCO_3^-, CO_3^{2-} and effects should be also considered. Ion exchangers of this type can be used in the removal of Cr(VI) and As(V) (Dambies et al. 2004; Gandhi et al. 2010) although the mechanism of sorption of chromate ions(VI) involves both electrostatic interactions with the protonate amino group and the reduction of Cr(VI) to Cr(III):

As for arsenate removal the process should be conducted from aqueous solutions at neutral pH. The percent removal of arsenate from the aqueous solution of 100 mg/ dm^3 arsenate and 560 mg/dm^{3}sulphate on NMDG resin is 99% and the reaction is unaffected by the presence of phosphate ions and the solution pH above 9.0, indicating that it can be regenerated with the alkaline solution. It was determined that the key variable in its selectivity is that the resin has to be protonated prior to contact with the aqueous solution (Alexandratos, 2007).

CHELATING ION EXCHANGERS WITH THE BIS (2-PYRIDYLMETHYL)AMINE FUNCTIONAL GROUPS

The ion exchange resins with the bis(2-pyridylmethyl)amine (bpa) functional groups also known as bispicolylamine are capable of selective sorption of transition metals, particularly Cu(II) ions due to the presence of donor atoms (nitrogen atoms) which are capable of coordination reaction with Cu(II). Due to this fact, such chelating ion exchange resins can combine ion exchange and complexing reactions and then exhibit high selectivity for metal ions. Dowex M 4195 possessing such functional groups is commercially available. It was synthesized in the early 1970s by Dow Chemical Co. and formerly known as Dowex XFS-4195 or DOW 3N. Also two others: Dowex M4196 (formerly Dowex XFS-4196) N-(2-hydroxyethyl)picolylamine or Dowex XFS-43084 (DOW 2N) with N-(2-hydroxypropyl)picolylamine were recognized (Jones & Pyper, 1979; Grinstead, 1984).

Bis(2-pyridylmethyl)amine (bpa) is an uncharged tridentate ligand having the ability to form charged complexes with most divalent metals. The 1:1 complexes with the metal ions of [M(bpa)]$^{2+}$ type are stable (Hirayama & Umehara, 1996). Based on the pK$_a$ values of bis(2-pyridylmethyl)amine (pK$_1$=0.5, pK$_2$=2.2, pK$_3$=3.4), it can be stated that at low pH values three nitrogen atoms would be protonated, while in the middle range of pH only one. For instance, Cu(II) ions (the coordination number is equal to 4) with the bis(2-pyridylmethyl)amine group and water molecule coordinate to it giving a square planar structure. In the next stage the H_2O molecule can be replaced by the anion, which is able to coordinate Cu(II) by a ligand exchange reaction:

$$R-N(bpa)_2 + H^+ \rightleftarrows R-NH^+(bpa)_2$$
$$2R-HN^+(bpa)_2 + M^{2+} \rightleftarrows [R-HN^+(bpa)_2]_2 \rightarrow M^{2+}$$

The complexes formed in the resin phase possess the following structure:

The obtained complex ion exchanger provides a new mode for the recognition of ions in the chromatographic analysis. Dowex M-4195 is a weak base ion exchanger and 1 M H_2SO_4 is in the protonated form (pKa = 3.2). It is also resistant to osmotic shock. Diniz et al. (2000, 2002, 2005) showed that the affinity series of metal ions determined in the one-component system for Dowex M 4195 is as follows: Cu(II) > Ni(II) > Co(II) > Pb(II) > Fe(III) > Mn(II) and it is slightly different from that in the multicomponent system: Cu(II) > Ni(II)

> Pb(II) > Fe(III) > Co(II) > Mn(II). The affinity of the transition metal cations for Dowex M 4195 in most cases was in agreement with the Irving-Williams order (Irving and Williams, 1953): Fe(II) < Co(II) < Fe(III) < Ni(II) < Cu(II) > Zn(II). It can be used for purification of chloride solutions after leaching of Mn(II) containing trace amounts of Co(II), Pb(II), Ni(II) and Cu(II). It can be also used for gold recovery (Tuzen, 2008).

On a commercial scale Dowex M 4195 has been used, among others, for separation of Ni(II) ions in the presence of Co(II) at INCO's Port Colborne refinery in Canada and Zambia Chambishi Cobalt Plant (Diniz et al. 2005) for purifying cobalt electrolytes. The efficiency of sorption of both ions is affected not only by pH, but by also by the concentration of sulphate(VI) ions and temperature. It is worth mentioning that separation of the twin pair Co(II)-Ni(II) is one of the most difficult tasks in inorganic chemistry. Contrary to Lewatit TP 207 and Amberlite IRC 718 with the iminodiacetate functional groups, Dowex M-4195 is characterized by the maximum sorption capacity towards Cu(II) ions in the pH range 1-4 (Melling and West, 1984). The sorption process can be presented in the following reaction:

$$RH_n + Cu^{2+} \rightleftarrows R\text{--}Cu^{2+} + nH^+$$

where: R is the resin matrix, n is the stoichiometric ratio, for n≠2 the SO_4^{2-} ions sorption occurs.

Partial washing out of copper(II) ions proceeds by means of 4M H_2SO_4, whereas the total one by means of NH_3 H_2O. In the case of sorption of Cu(II) ions sorption in the presence of Fe(III) ions, the ion exchange mechanism must be assumed. Sorption of both ions is affected not only by pH, but also by concentration of sulphate(VI) ions and temperature. Fe(III) ions sorption increases significantly with the temperature rise from 293 K to 303 K, whereas it does not change for Cu(II) ions.

The ion exchangers with the picolylamine functional groups can be the basis for obtaining the polymeric ligand exchanger (PLE) with the structure presented above (Zhao et al. 1998; Kołodyńska 2009c):

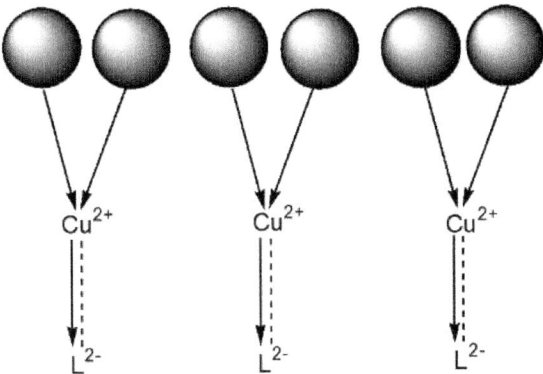

Such kind of ion exchanger consists of the cross linked polystyrene-divinylbenzene matrix, covalently attached bispicolylamine functional groups and Lewis acid cations (such as Cu^{2+}, Ni^{2+}, Fe^{3+}, Co^{2+} according to the series: $Cu^{2+} > Ni^{2+} > Fe^{3+} > Co^{2+}$) coordinated to the functional groups and without neutralization of their positive charge so that the anion exchanger is obtained. It is expected to show high affinity not only for phosphates(V) HPO_4^{2-}, arsenate(V) $HAsO_4^{2-}$ and chromate(VI) CrO_4^{2-} ions but also oxalates ox^{2-}, perchlorates ClO_4^-, tartaric acid as well as simultaneous and selective removal of heavy metal and chromate ions (contrary to other ion exchangers) (Saygi et al. 2008; Dimick, 2008; Du et al. 2008). It was found that for phosphates(V) HPO_4^{2-} removal, the sorption efficiency is much higher than that for the strongly basic macroporous anion exchanger Amberlite IRA 958 and proceeds according to the reaction:

$$R-M^{2+}(2Cl^-) + HPO_4^{2-} \rightleftarrows R-M^{2+}(HPO_4^{2-}) + 2Cl^-$$

The desorption process can be written as follows:

$$R-M^{2+}(HPO_4^{2-}) + 2Cl^- + H^+ \rightleftarrows R-M^{2+}(2Cl^-) + H_2PO_4^-$$

$$R-M^{2+}(HPO_4^{2-}) + 2Cl^- \rightleftarrows R-M^{2+}(2Cl^-) + HPO_4^{2-}$$

Four chelating ion exchange polymeric resins were tested to remove Ni(II) and Co(II) from synthetic solutions on the commercially available ion exchangers Dowex M4195, Amberlite IRC 748, Ionac SR-5 and Purolite S930. Among the selected resins, Dowex M4195 showed the best results for Ni(II) and Co(II) selective sorption from acid liquors in the whole pH range and with small influence of other elements. Even at pH 1 Dowex M4195 was the most effective (Mendes & Martins, 2004).

CHELATING ION EXCHANGERS WITH THE THIOL, THI-OUREA AND ISOTHIOUREA FUNCTIONAL GROUPS

In the group of ion exchangers with the thiol functional groups (Chelite S, Duolite ES-465, Imac GT 73) Imac TMR resin is very important. It is the macroporous ion exchanger with the PS-DVB matrix, which besides the thiol ones, also possesses the sulphone groups. Imac TMR is used for selective sorption of Hg(II) ions from the process solutions as well as Ag(I), Au(III), Pt(IV) and Pd(II). The sorption process of Hg(II) with the saline solution proceeds according to the reaction:

$$R{-}SH + Hg^{2+} \longrightarrow (R{-}S)_2Hg^{2+} + 2H^+$$

$$R{-}SH + HgCl^+ \longrightarrow R{-}SHgCl + H^+$$

Also ion exchangers with the isothiourea functional groups (Ionac SR 3, Lewatit TP 214, Purolite S 920, Srafion NMRR) exhibit high affinity for Hg(II) ions. They are also selective for noble metal ions. Depending on the pH, the isothiourea groups occur in the following forms:

For the first form, coordination bond formation is possible, whereas for the second one the sorption process proceeds in accordance with the anion exchange mechanism:

RETARDION 11A8

Dowex Retardion 11A8 is an example of a very interesting ion exchanger of the type 'snake in a cage' with the quaternary ammonium and carboxylic functional groups (amphoteric resin) (Dybczyński, 1987). It is produced by polymerizing acrylic acid monomer inside an anion exchange resin. Polyacrylic chains are (snake) alternate with the PS-DVB matrix (cage) and therefore they are trapped inside the cross linked ion exchange resin and cannot diffuse out. As a result, cationic and anionic sites are so closely associated that they partially neutralize their electric charges. Mobile ions, such as chlorides, nitrates(V) are attracted and retained on these unique sites until they are eluted with hot water.

Dowex Retardion 11A8 can be used for selective separation of Cd(II) ions in the presence of other heavy metal ions (Samczyński & Dybczyński, 2002). Cd(II) ions are sorbed from 2 M HCl and 2 M NH_4OH with 0.1 M NH_4Cl systems according to the reactions:

$$(n-2)RCl + [CdCl_n]^{2-n} \longrightarrow R([CdCl_n]^{2-n}) + (n-2)Cl^-$$

$$2RCOONH_4 + [Cd(NH_3)_n]^{2+} \longrightarrow (RCOO)_2[Cd(NH_3)_{n-x}] + 2NH_4^+ + nNH_3$$

In the case of separation of Ga(III), In(III), Tl(III), Pt(II), Pd(II) and Na(I), Ni(II), Cu(II) and Zn(II) mixtures in acidic media the resin acts mainly as an anion exchanger. For the elements that can exist as both cations and anions in solution (e.g., Ni, Co, Cu and Zn), the amphoteric properties of Retardion11A8 permit more specific isolation of certain elements from complex mixtures than would be possible with the use of monofunctional ion exchangers (Dybczyński & Sterlińska, 1974).

IMPREGNATED ION EXCHANGERS

The impregnated resins obtained by physically loaded organic reagents on a solid inert support material such as Amberlite XAD resins are an attractive material for separation and preconcentration of heavy metal ions (Prabhakaran & Subramanian, 2003). They are characterized by good porosity, uniform pore size distribution, high surface area as a chemical homogeneous, non-ionic structure. For instance, it was found that Amberlite XAD-2 functionalized with dithiocarbamate ligand, 1,8-dihydroxynaphthalene-3,6-disulphonic acid (chromotropic acid), 2(2-thiazolylazo)-p-cresol, 1-(2-pyridylazo)-2-naphthol, calmagite, xylenol orange (Abollino et al. 1998; Ferreira & Brito, 1999,Ferreira et al. 1999; Ferreira et al. 2000a; Ferreira et al. 2000b; Tewari & Singh, 1999; Tewari & Singh, 2000; Tewari & Singh, 2001; Tewari & Singh, 2002) can be used for selective sorption and preconcentration of heavy metal ions. Amberlite XAD-4 loaded with sodium diethyl dithiocarbamate; 2,3-dihydroxy benzoic acid (DHBA), ammonium pyrrolidine dithiocarbamate (APDC) and piperidine dithiocarbamate (pipDTC) were used for preconcentration and determination of metal ions in various matrices (Uzun et al. 2001; Hosseini et al. 2006; Ramesh et al. 2002). However, the most promising polymeric support with a larger surface area is Amberlite XAD 16. Amberlite XAD-16 loaded with quercitin (Sharma & Pant, 2009) is characterized by good adsorbent properties for large amounts of uncharged compounds (Tokalıoğlu et al. 2010). The solid phase extraction (SPE) process with the application of such materials is characterized by important advantages such simplicity, flexibility, economical, rapid, higher enrichment factors, absence of emulsion and low cost because of lower consumption of reagents.

In general, sorption selectivity of a resin can be affected by both sorbate-sorbent and sorbate-solvent interactions. It has been well recognized that resin matrix and functional groups can strongly affect ion exchange capacity and selectivity. Therefore in the presented paper the chelating ion exchangers Diphonix Resin® containing diphosphonic, sulphonic and carboxylic acid groups and Dowex M 4195 with the bis(2-pyridylmethyl) amine functional group were used for the sorption of Cu(II), Zn(II), Co(II), Pb(II) complexes with Baypure CX 100 (IDS) and Cu(II), Zn(II), Cd(II), Pb(II) complexes with Trilon M (MGDA). The presence of the sulphonic functional groups determines better hydrophilic properties of Diphonix Resin® compared to the traditional monofunctional ion exchangers.

EXPERIMENTAL

In the paper the results of the sorption of heavy metal ions such as Cu(II), Zn(II), Cd(II) and Pb(II) in the presence of the complexing agents of a new generation Baypure CX 100 (IDS) and Trilon M (MGDA) on commercially available chelating ion exchangers are presented.

The essential physicochemical properties of these chelating agents are given in Table 1.

CHARACTERISTICS OF THE CHELATING ION EX-CHANGE RESINS

The chelating ion exchange resins Dowex M 4195 and Diphonix Resin® were tested. Their short characteristics are presented in Table 2.

Table 1: Physicochemical properties of IDA and MGDA.

Properties	IDS	MGDA
Structure		
Form supplied	liquid	liquid
Molecular weight	337.1	271.0
Appearance	colourless to light yellow	clear yellowish
pH	10.3-11.4	11.0
Density [g mL⁻¹]	1.32-1.35 g/mL	1.31 g/mL

Solubility in H$_2$O	in any ratio	in any ratio
Solubility in NaOH	in any ratio	in any ratio
Biodegradability [%]	"/> 80%	"/> 68%
Termal stability	in any range	in any range

Table 2: Physicochemical properties of Dowex M 4195 and Diphonix Resin®.

Properties	Dowex M 4195	Diphonix Resin®
Matrix	PS-DVB	PS-DVB
Structure	macroporous	gel
Functional groups	bis(2-pyridylmethyl) amine bis-picolylamine	diphosphonic sulphonic carboxylic
Commercial form	weak base, partially H$_2$SO$_4$ salt	H+
Appearance	brown to green, opaque	beige, opaque
Total capacity	1.3 [eq/dm^3]	5,6 [mol/kg]
Moisture content	40-60 %	58.3 %
Bead size	0.300-1.200 [mm]	0.074-0.150 [mm]
Density	0.67 [g/cm^3]	1.05-1.11 [g/cm^3]
Max temp. range	353 K	313 K
Operating pH range	2 – 6	0 – 12

Before the experiments, the resins were washed with hydrochloric acid (0.1 M) or sulphuric acid (0.5 M) to remove impurities from their synthesis. After pre-treatment they were washed with deionised water.

The solutions of Cu(II), Zn(II), Cd(II) and Pb(II) complexes with Baypure CX 100 and Trilon M with the desirable concentrations were prepared by mixing appropriate metal chlorides or nitrates with the complexing agents solutions, respectively. For the studies the obtained solutions were used without pH adjustment. The pH values of the solutions of Cu(II), Zn(II), Cd(II) and Pb(II) complexes with IDS were as follows: 6.7, 6.5, 69 and 7.3, respectively. For the Cu(II), Zn(II), Cd(II) and Pb(II) complexes with MGDA these values were equal to 8.3, 9.8, 10.5 and 10.4. The other chemicals used were of analytical grade.

In batch experiments, 50 cm^3 of sample solution and ion exchanger (0.5 g) were put into a conical flask and shaken at different time intervals using the laboratory shaker Elpin type 357, (Elpin-Plus, Poland). After the pH of solutions was stabilized and equilibrated, the ion exchangers were filtered. The experiments were conducted in three parallel series. The reproducibility of the measurements was within 5%. Adsorption isotherms were obtained with different initial concentrations varying from 1×10^{-3} M to 2.5×10^{-2} M of metal ions and ligands while keeping the constant amount of resins at room temperature (295 K). The equilibrium between the solid and liquid phases was modelled by the Langmuir and Freundlich equations as presented earlier (Kołodyńska, 2010a; Kołodyńska 2010b; Kołodyńska 2010c). Kinetic studies were carried out at different time intervals varying from 1 to 120 min keeping the constant amount of resins at room temperature (295 K). The shaking speed was 180 rpm to maintain resin particles in suspension.

The amount of heavy metal complexes sorbed onto the resins was calculated by the difference between the amounts added and already present in the solution and that left in the solution after equilibrium.

The pH values were measured with a PHM 84 pH meter (Radiometer, Copenhagen) with the glass REF 451 and calomel pHG 201-8 electrodes. The concentrations of heavy metals were measured with the AAS spectrometer Spectra 240 FS (Varian, Australia).

RESULTS

As for the removal of toxic metal ions many different methods are available. Among them, the most commonly used are ion exchange, adsorption, reduction and precipitation. In many cases, the environmentally most compatible and cost-effective solutions include combination of two or more of these processes. From different waste waters those containing heavy metal ions and complexing agents require special attention.

COMPLEXING AGENTS

For over fifty years synthetic chelating agents from the group of aminopolycarboxylic acids (APCAs) have been the basis in many technological processes. Ethylenediaminetetraacetic acid (EDTA), nitrilotriacetic acid (NTA) and diethylenetriaminepentaacetic acid (DTPA) are the best known traditional complexing agents. They are commonly applied in many branches of industry forming stable, water soluble complexes with various metal cations or as a masking agent. Nowadays there are a number of alternative products on the market which claim to be as effective as EDTA and NTA. Among them, IDS and MGDA should be listed.

Iminodisuccinic acid (IDS) also known as Baypure CX 100 is a medium-strong chelator consisting of: iminodisuccinic acid sodium salt > 32%, aspartic acid sodium salt < 7%, fumaric acid sodium salt < 3.5 %, hydroxysuccinic acid < 0.9 %, maleic acid sodium salt < 0.9 % (IDS Na-salz, 1998; Vasilev et al. 1996; Vasilev et al. 1998; Reinecke et al. 2000, Kołodyńska et al. 2009;

Kołodyńska, 2009a). Iminodisuccinic acid sodium salt can form quintuple-bonded complexes with metal ions. In this case, complexing occurs via the nitrogen and all four carboxyl groups. As a result of the octahedric structure of the complete complex, a water molecule is required for the sixth coordination point (Kołodyńska, 2009b). In the paper by Hyvönen it was found that for low pH conditions (less than 3), the tendency for M(II)/M(III) ions to form complexes with IDS may be assumed as: Cu(II)>Fe(III)>Zn(II)>Mn(II), whereas for pH >7 it can be as follows: Cu(II)>Zn(II)> Mn(II)>Fe(III) (Hyvönen et al. 2003; Hyvönen & Aksela, 2010). IDS is able to replace EDTA when rather moderate chelating agents are sufficient for masking alkaline earth or heavy metal ions. As a substitute for EDTA it is used in a variety of applications, including detergent formulations, corrosion inhibitors, production of pulp and paper, textiles, ceramics, photochemical processes, and as trace nutrient fertilizers in agriculture.

Methylglycinediacetic acid (MGDA) was patented by BASF and marketed under the brand name Trilon M. The active ingredient contained in Trilon M is the trisodium salt of methylglycinediacetic acid. The acid dissociation constants pK_a of MGDA are as follows: pK_1=1.6, pK_2=2.5 pK_3=10.5 (Jachuła et al. 2011; Jachuła et al. 2012). The most important property of Trilon M is the ability to form complexes (MGDA is a tetradentate chelating ligand where chelation involves three carboxylate groups and nitrogen atom) with metal ions, soluble in water in the large pH range 2-13. These complexes remain stable, especially in alkaline media and even at temperatures of up to 373 K. It is worth mentioning that MGDA chelating capacity was investigated by Tandy et al. (2004) in soil washing. It was found that 89-100% of MGDA can be degraded in 14 days, 90% of EDDS in 20 days while no EDTA was degraded in 30 days.

Fig. 5a-b shows the comparison of the logarithmic stability constants (log K) for the complexes of IDS and MGDA and selected metal ions with the stability constants for EDTA.

A high or moderately high value for log K of Cu(II), Zn(II), Cd(II) and Pb(II) and first of all Fe(III) with IDS and MGDA indicates that these chelating agents have a high affinity for particular metal ions and they provide a preliminary indication of whether the chelating agent is suitable for the specific application.

As these complexing agents are widely applied, removal of their complexes with heavy metals is essential, especially when typical chemical precipitation methods are ineffective, even if solutions with high metal concentrations are treated. Therefore, more advanced techniques are required for cleaning up such contaminants and retardation of heavy metal ions mobility. Among these, the ion exchange with application of selective resins appears to be a more promising method for the treatment of such solutions.

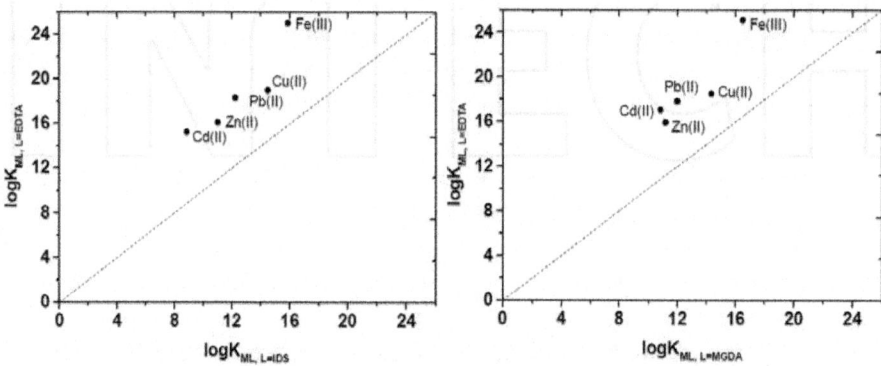

Figure 5: a-b. Comparison of conditional stability constants values of some complexes of metals with EDTA and IDS (a) as well as EDTA and MGDA (b).

Generally, chelating properties and selectivity of ion exchangers have been enhanced by: (i) immobilization of ligands with multiple coordinating sites such as bifunctional polymers or polyfunctional polymers possessing different functional groups, (ii) immobilization of low molecular weight complexing agents, (iii) by preparation of ion imprinted polymers (IIP), (iv) preparation of reactive ion exchangers (RIEX), (vi) immobilization of specific donor groups through application of Pearson's hard soft acid base theory, (vii) immobilization of macrocycles e.g. crown ethers, calixarenes, resorcinarenes etc. These approaches correspond to both chelating ion exchangers Dowex M 4195 and Diphonix Resin®. Additionally, their sorption selectivity can be affected by sorbate-sorbent and sorbate-solvent interactions. It has been well recognized that the resin matrix and the functional groups can strongly affect ion exchange capacity and selectivity (Clifford & Weber, 1983;Barron & Fritz, 1984; Li et al. 1998). Therefore, in the case of chelating ion exchangers, where the formation of coordination bonds is the basis of the sorption process, besides the parameters related to physicochemical properties of the resins, the effect of the presence of complexing agents should be also taken into account.

In the presence of the complexing agents, IDS and MGDA, there are formed:

$$M^{2+} + H_n ids^{n-4} \rightleftarrows [M(H_n ids)]^{n-2} \text{ where } n=1,2,3$$

and

$$M^{2+} + H_n mgda^{n-3} \rightleftarrows [M(H_n mgda)]^{n-2}, \text{ where } n=1,2.$$

Therefore using selective chelating ion exchangers the sorption effectiveness will be dependent on the decomposition of neutral or anionic species of [MH$_2$L], [MHL]$^-$ and [ML]$^{2-}$ type, where L=ids^{4-}, mgda^{3-}. Additionally, the 'sieve effect' is also important (Kołodyńska, 2010b; Kołodyńska 2010c; Kołodyńska 2011). In the case of the chelating resin Dowex M 4195 possessing the bis(2-pyridylmethyl)amine (bpa) functional groups, depending on the pH value the mechanism of sorption can be as presented earlier. Additionally, the ionic interaction mechanism between the protonated amines and the anionic complexes of the [ML]$^{2-}$ and [ML]$^-$ is also possible (Kołodyńska 2011). Therefore, appropriate reactions can be as follows:

$$2R-HN^+(bpa)_2\ Cl^- + [ML]^{2-} \rightleftarrows [R-HN^+(bpa)_2]_2[ML]^{2-} + 2Cl^-$$

$$R-HN^+(bpa)_2\ Cl^- + [MHL]^- \rightleftarrows [R-HN^+(bpa)_2][MHL]^- + Cl^-$$

or

$$R-HN^+(bpa)_2\ Cl^- + [ML]^- \rightleftarrows [R-HN^+(bpa)_2][ML]^- + Cl^-$$

where: R is the Dowex M 4195 skeleton (PS-DVB), L is the ids^{4-} or mgda^{3-} ligand.

The analogous mechanism of sorption in the case of Diphonix chelating ion exchanger should be considered.

Kinetic studies

For the kinetic data, a simple kinetic analysis was performed using the pseudo first order and the pseudo second order equations:

$$\log(q_e - q_t) = \log(q_e) - \frac{k_1 t}{2.303} \tag{1}$$

$$\frac{t}{q_t} = \frac{t}{q_e} + \frac{1}{k_2 q_e^2} \tag{2}$$

where: q_e is the amount of metal complexes sorbed at equilibrium (for the pseudo first order model also denoted as q_1 and q_2 for the pseudo second order model) (mg/g), qt is the amount of metal complexes sorbed at time t (mg/g), k_1, k_2 are the equilibrium rate constants (1/min), respectively.

The sorption of Cu(II), Zn(II), Cd(II) and Pb(II) complexes with IDS on Dowex M 4195 in the M(II)-L=1:1 system is presented in Fig.6a. The analogous data for the Cu(II), Zn(II), Cd(II) and Pb(II) complexes with MGDA sorption on Dowex M 4195 are presented in Fig.6b and for Diphonix Resin® in Figs.6c and 6d.

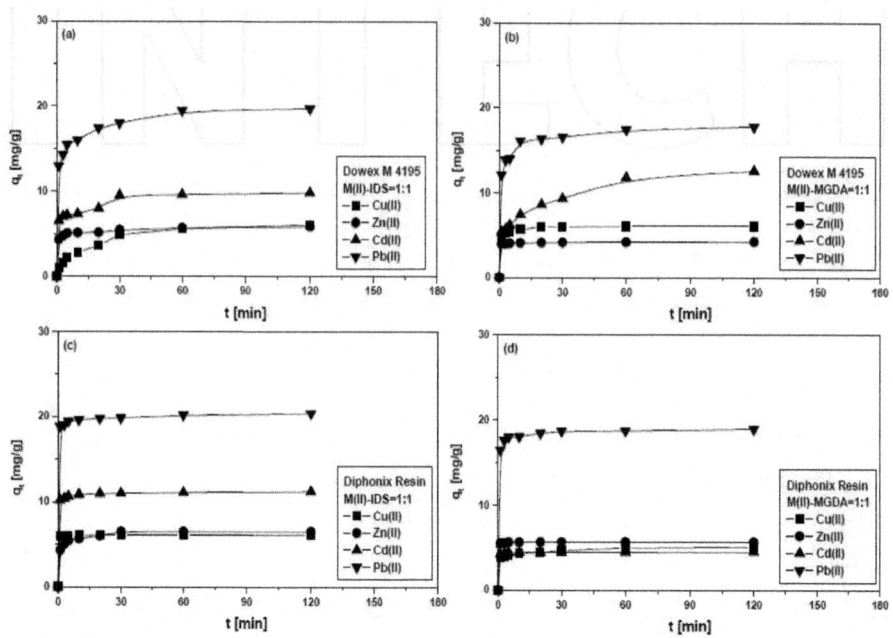

Figure 6: a-b. The effect of the phase contact time on the sorption capacities of Cu(II), Zn(II), Cd(II) and Pb(II) complexes with IDS on Dowex M 4195 (a) and Diphonix Resin® (c) as well as the Cu(II), Zn(II), Cd(II) and Pb(II) complexes with MGDA on Dowex M 4195 (b) and Diphonix Resin® (d) (c_0 1×10^{-3} mol/dm³, shaking speed 180 rpm, shaking time 1-120 min, room temperature).

The straight lines of t/q_t vs. t suggest the applicability of the pseudo second kinetic model to determine the q_e, k_2 and h parameters (from the intercept and the slope of the plots). These kinetic parameters are presented in Tables 3 and 4.

It was shown that the equilibrium was reached very quickly. More than 90% of metal ions were bound to Dowex M 4195 and Diphonix Resin® within 10-20 min of the phase contact time and therefore a slight increase until a plateau was reached after about 2 h was observed. The values of the theoretical q_e for the studied resins were in good agreement with those obtained experimentally ($q_{e,exp}$). On Dowex M 4195 about 95 %, 100 %, 99 % and 97.5 % of the Cu(II), Zn(II), Cd(II) and Pb(II) complexes with IDS and 94 %, 98 %, 96 % and 95 % complexes with MGDA are sorbed at this time, respectively. On Diphonix Resin® for the Cu(II), Zn(II), Cd(II) and Pb(II) complexes with IDS and MGDA the adequate values are as follows: 94 %, 89 %, 97 % and 98 % as well as 97 %, 86 %, 99 % and 96 %.

Table 3: The pseudo second order kinetic parameters for the sorption of Cu(II), Zn(II), Cd(II) and Pb(II) complexes with IDS and MGDA on Dowex M 4195

System	qe.exp [mg/g]	q2 [mg/g]	$k2$	h	$R2$
Cu(II)-IDS=1:1	5.63	5.61	1.012	5.789	0.9987
Zn(II)-IDS=1:1	5.91	5.88	1.007	4.897	0.9988
Cd(II)-IDS=1:1	9.81	9.89	0.987	12.456	0.9999
Pb(II)-IDS=1:1	19.10	19.00	0.845	16.789	0.9992
Cu(II)-MGDA=1:1	6.05	5.98	2.335	9.237	0.9999
Zn(II)-MGDA=1:1	4.17	4.03	1.017	16.783	0.9996
Cd(II)-MGDA=1:1	12.55	12.23	0.924	10.123	0.9999
Pb(II)-MGDA=1:1	17.77	17.46	0.688	7.525	0.9999

Table 4: The pseudo second order kinetic parameters for the sorption of Cu(II), Zn(II), Cd(II) and Pb(II) complexes with IDS and MGDA on Diphonix Resin®

System	qe.exp [mg/g]	q2 [mg/g]	$k2$	h	$R2$
Cu(II)-IDS=1:1	6.12	6.21	2.211	10.207	0.9999
Zn(II)-IDS=1:1	6.01	6.09	1.345	7.123	0.9991
Cd(II)-IDS=1:1	10.21	10.11	0.988	23.434	0.9999
Pb(II)-IDS=1:1	20.39	20.26	0.876	37.551	0.9998
Cu(II)-MGDA=1:1	5.66	5.61	3.469	11.111	0.9999
Zn(II)-MGDA=1:1	4.48	4.48	2.395	48.077	0.9999
Cd(II)-MGDA=1:1	10.23	10.24	0.024	2.475	0.9999
Pb(II)-MGDA=1:1	18.94	18.93	0.188	67.568	0.9999

These results indicate that the sorption process of metal ions in the presence of IDS and MGDA on Dowex M 4195 and Diphonix Resin® followed a pseudo second order kinetics, which meant that both the external mass transfer and intraparticle diffusion together were involved in the sorption process. The correlation coefficients (R2) obtained for the pseudo second order kinetic model are in the range 0.9991 -1.000 for all metal complexes. The pseudo first order parameters were not shown because the correlation coefficients for this model are low (0.7438-0.8745 for the IDS complexes and 0.919-0.986 for the MGDA complexes on Diphonix Resin®.

The breakthrough curves for Cu(II) ions in the presence of MGDA on Dowex M4195 from single metal ion solutions of a concentration 1×10^{-3} M are shown in Fig. 7. Typical 'S' shaped curves were obtained in the experiments. Analogous results were obtained on Diphonix Resin®. It should be mentioned the UV exposition does not have a significant effect on the decomposition of the complexes in the resin phase.

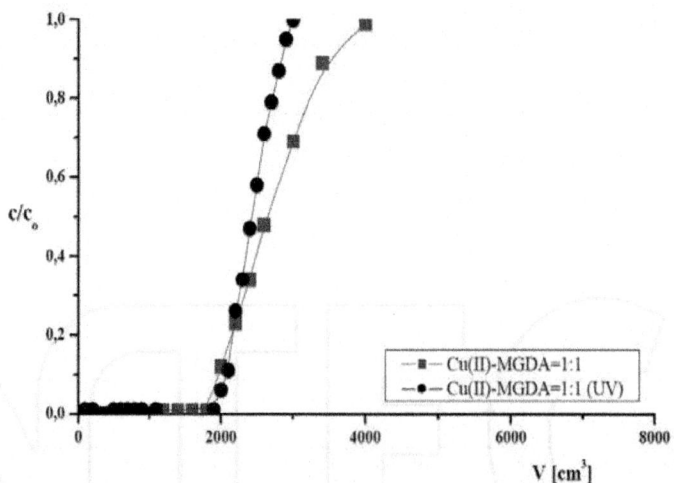

Figure 7: The breakthrough curves of Cu(II) complexes with MGDA on Dowex M 4195 without and with UV exposition (c_0 1×10^{-3} mol/dm³, bed volume 10 cm³, flow rate 0.6 cm³/min).

It is well known that the particle size of ion exchange resins influences the time required to establish equilibrium conditions and two types of diffusion must be considered in an ion exchange equilibrium e.g. the film diffusion (the movement of ions from a surrounding solution to the surface of an ion exchange particle) and the internal diffusion (the movement of ions from the surface to the interior of an ion exchange particle). Film diffusion is usually the controlling reaction in dilute solutions whereas the internal diffusion is controlling in more concentrated solutions. The particle size of an ion exchange resin affects both the film diffusion and the internal diffusion (Kołodyńska, 2011).

According to the manufacturer data the particle size of Dowex M 4195 is 0.300-1.200 mm. However, Diphonix Resin® available on the commercial scale is in the range 0.30-0.85 mm, 0.15-0.30 mm and 0.075-0.15 mm.

In the presented paper Diphonix Resin® with the particle size 0.075-0.150 mm was used to study the sorption process of Cu(II), Zn(II), CdII) and Pb(II) in the presence of IDS and MGDA. In the paper by Cavaco et al. (2009) it was found that for the range 0.15-0.30 mm, 50 % of the particles have diameters less than 0.223 mm. As follows from the obtained results, the bead size of the used chelating ion exchangers has also approximately the Gaussian distribution (Fig. 8 a-b). It was found that with the increase of bead dimensions, the volume fractions of disc-similar beads decrease and the beads are more spherical (Kołodyńska, 2011).

A decrease in the particle size thus shortens the time required for equilibration of particle size and pore characteristics have an effect on equilibrium concentration and influence sorption kinetics. Therefore this factor is essential, especially when the sorption of metal complexes, not metal ions is taken into account. In the case of large complexes the sieve effect is observed.

Kinetic sorption experiments were also carried out with the increased complexes concentrations from 1×10^{-3} mol/dm^3 to 2×10^{-2} mol/dm^3 and these results were presented in(Kołodyńska, 2011).

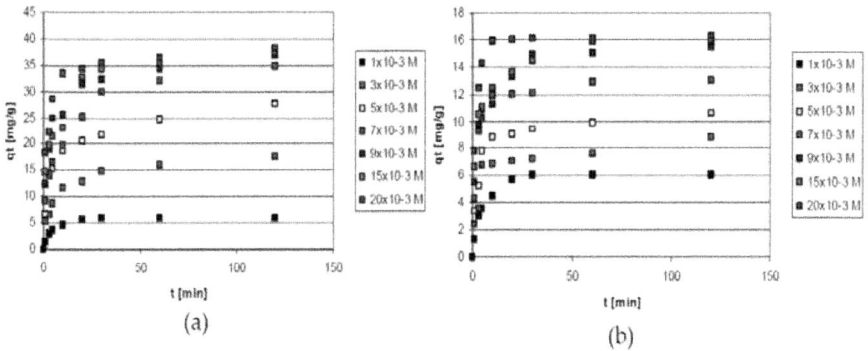

Figure 8: a-b. Comparison of the distribution of the bead size of Dowex M 4195 (a) and Diphonix Resin® (b) based on the Zingg classification.

It was found that with an increase of metal complexes concentrations a continuous increase in the amount adsorbed per unit mass of ion exchanger was observed till the equilibrium was achieved. For the pseudo second order kinetic model, the rate k_2 values decrease with the increasing initial concentrations, while h increases.

PH EFFECTS

The effect of pH was studied for the Cu(II), Zn(II), Cd(II) and Pb(II) in the M(II)-IDS=1:1 and M(II)-MGDA=1:1 systems at the pH varied from 2 to 12. The optimal sorption range of the Cu(II), Zn(II), Cd(II) and Pb(II) complexes with IDS practically does not change in the pH range from 4 to 10 both on Dowex M 4195 and Diphonix Resin® whereas, at high pH values, decrease in removal efficiency is observed. In the case of the Cu(II), Zn(II), Cd(II) and Pb(II) complexes with MGDA a slight decrease in sorption efficiency with the increasing pH was also shown.

ADSORPTION STUDIES

The Langmuir equation was applicable to the homogeneous adsorption system, while the Freundlich equation was the non-empirical one employed to describe the heterogeneous systems and was not restricted to the formation of the monolayer. The well-known Langmuir equation was represented as:

$$\frac{1}{q_e} = \frac{1}{bq_0c_e} + \frac{1}{q_0}$$

(3)

where: q_e is the equilibrium M(II) ions concentration on the ion exchanger, (mg/g), c_e is the equilibrium M(II) ions concentration in solution (mg/dm^3), q_0 is the monolayer capacity of ion exchanger (mg/g), b is the Langmuir adsorption constant (L/g) related to the free energy of adsorption.

The values of q_0 and b were calculated from the slope and the intercept of the linear plots c_e/q_e vs. c_e. On the other hand, the Freundlich equation was represented as:

$$q_e = K_F c_e^{bF}$$

(4)

where: KF and 1/n are the Freundlich constants corresponding to the adsorption capacity and the adsorption intensity.

The plot of ln q_e vs. ln c_e was employed to generate the intercept KF and the slope 1/n.

The exemplary results presented in Fig.9a-b indicate that for the studied range of concentration of Cu(II) complexes with MGDA (1×10^{-3} M - 2×10^{-2} M) the sorption capacity of Dowex 4195 and Diphonix Resin® increases.

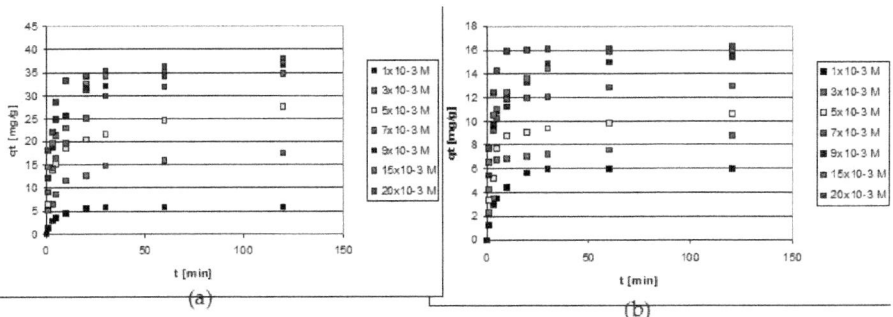

Figure 9: a-b. The effect of the concentration on the sorption capacities of Cu(II) complexes with MGDA on Dowex M 4195 (a) and Diphonix (b) (c_0 1×10^{-3} -20x10^{-3} mol/dm³, shaking speed 180 rpm, shaking time 1-120 min, room temperature).

The experimental data obtained for the sorption of Cu(II), Zn(II), Cd(II) and Pb(II) in the presence of IDS and MGDA on Dowex M 4195 and Diphonix Resin® were well represented by the Langmuir isotherm model (Table 5). The correlation coefficients of the linear plot of c_e/q_e vs. c_e obtained from them were high, ranging from 0.9512 to 0.9999 (Kołodyńska, 2011). The highest values of the Langmuir parameter q_0 were obtained in the case of Pb(II) complexes with IDS and MGDA on Dowex M 4195 and Diphonix Resin®. They are equal to 121.58 mg/g and 97.64 mg/g on Dowex M 4195 and 112.37 mg/g and 100.20 mg/g on Diphonix Resin®, respectively.

Table 5: The Langmuir and Freundlich isotherm parameter values for the sorption of Cu(II), Zn(II), Cd(II) and Pb(II) ions in the presence of IDS and MGDA on Dowex M4195

System	Langmuir				Freundlich		
	$q_{e, ex}p$	q^0	K_L	R^2	K_F	n	R^2
Cu(II)-IDS=1:1	38.23	37.56	0.023	0.9876	9.54	3.21	0.9865
Zn(II)-IDS=1:1	21.07	20.87	0.046	0.9923	6.23	2.45	0.9456
Cd(II)-IDS=1:1	88.00	87.65	0.008	0.9998	4.23	4.58	0.9687
Pb(II)-IDS=1:1	122.37	121.58	0.012	0.9989	1.23	5.69	0.9623
Cu(II)-MGDA=1:1	37.99	38.05	0.026	0.9932	12.48	4.23	0.9758
Zn(II)-MGDA=1:1	16.78	17.01	0.052	0.9983	7.56	3.69	0.9823
Cd(II)-MGDA=1:1	65.78	63.21	0.061	0.9996	6.23	6.11	0.9877
Pb(II)-MGDA=1:1	98.78	97.64	0.042	0.9999	2.48	9.25	0.9837

For the studied systems regeneration tests were conducted using HCl, HNO_3, H_2SO_4 and NaCl at 1M and 2M concentrations. Based on the series of five experiments using known amounts of Cu(II) complexes with IDS and MGDA sorbed, it was established that the overall recoveries of Cu(II) eluted from Dowex M 4195 and Diphonix Resin® by 2M HCl and H_2SO_4 were above 98 %, suggesting that the recovery is quantitative.

CONCLUSIONS

The presence of biodegradable complexing agents of a new generation that is IDS and MGDA affects the sorption process of Cu(II), Zn(II), Cd(II) and Pb(II) ions on Dowex M4195 and Diphonix Resin®. The effectiveness of sorption depends on the type of complexes and their stability that facilitates their decomposition in the resin phase. The batch equilibrium was relatively fast and reached equilibrium after about 10-20 min of the contact. The experimental data have been analyzed using the Langmuir and Freundlich models. The sorption of studied metal ions in the presence of IDS and MGDA Dowex M 4195 and Diphonix Resin® followed the pseudo second order kinetics. As follows from the experiment pH does not have a significant effect on the sorption of Cu(II), Zn(II), Cd(II) and Pb(II) ions in the presence of IDS and MGDA on the chelating ion exchanger under consideration. The affinity of the above analyzed heavy metal complexes with IDS and MGDA Dowex M 4195 and Diphonix Resin® were found to be as follows: Pb(II) > Cd(II) > Cu(II) > Zn(II) for IDS and MGDA. The studied complexing agents can be proposed as alternative chelating agents to EDTA or NTA for the removal of heavy metal ions from waters and wastewaters.

REFERENCES

1. O. Abollino, M. Aceto, M. C. Bruzzoniti, E. Mentasti, C. Sarzanini, 1998Determination of metals in highly saline matrices by solid-phase extraction and slurry-sampling inductively coupled plasma-atomic emission spectrometry, Analytica Chimica Acta, 375293298

2. I. M. Abrams, J. R. Milk, 1997A history of the origin and development of macroporous ion-exchange resins, Reactive and Functional Polymers, 35

3. M. Ahuja, A. K. Rai, P. N. Mathur, 1999Adsorption behaviour of metal ions on hydroximate resins, Talanta, 4319551963

4. S. D. Alexandratos, 2007New polymer-supported ion-complexing agents: Design, preparation and metal ion affinities of immobilized ligands, Journal of Hazardous Materials, 139467470

5. S. D. Alexandratos, 2009Ion exchange resins. A retrospective from Industrial and Engineering Chemistry Research, Industrial and Engineering Chemistry Research, 48388398

6. T. Bajda, 2005Chromatite Ca[CrO4] in soil polluted with electroplating effluents (Zabierzów, Poland), Science of the Total Environment, 336269274

7. D. G. Barceloux, 1999Copper, Journal of Toxicology- Clinical Toxicology, 37217230

8. R. E. Barron, J. S. Fritz, 1984Effect of functional group structure on the selectivity of low-capacity anion exchangers for monovalent anions, Journal of Chromatography, 2841325

9. R. Biesuz, M. Pesavento, A. Gonzalo, M. Valiente, 1998Sorption of proton and heavy metal ions on a macroporous chelating resin with an iminodiacetate active group as a function of temperature, Talanta, 47127136

10. G. Boari, L. Liberti, C. Merli, R. Passino, 1974Exchange equilibria on anion resins, Desalination, 15145166

11. J. B. Brower, R. L. Ryan, M. Pazirandeh, 1997Comparison of ion-exchange resins and biosorbents for the removal of heavy metals from plating factory, Environmental Science and Technology, 3129102914

12. B. Busche, R. Wiacek, J. Davidson, V. Koonsiripaiboon, W. Yantasee, R. S. Addleman, G. E. Fryxell, 2009Synthesis of nanoporous iminodiacetic acid sorbents for binding transition metals, Inorganic Chemistry Communications, 12312315

13. S. A. Cavaco, S. Fernandes, C. M. Augusto, M. J. Quina, L. M. Gando-Ferreira, 2009Evaluation of chelating ion-exchange resins for separating Cr(III) from industrial effluents, Journal of Hazardous Materials, 169516523

14. D. Clifford, W. J. Weber, Jr , 1983The determinants of divalent/monovalent selectivity in anion exchangers, Reactive Polymers, Ion Exchangers, Sorbents, 17789

15. D. A. Clifford, 1999Ion exchange and inorganic adsorption, In: Water Quality and Treatment Fifth Edition, Ray Lettermam, ed., McGraw Hill, Inc., New York, 9

16. E. Chajduk-Maleszewska, R. Dybczyński, 2004Effective separation and preconcentration of trace amounts of Pd on Duolite ES 346 resin and its use for the determination of Pd by NAA, Analytical Chemistry, 49281297

17. R. Chiarizia, E. P. Horwitz, S. D. Alexandratos, M. J. Gula, 1997Diphonix resin: a review of its properties and applications, Separation Science and Technology, 32135

18. R. Chiarizia, E. P. Horwitz, S. D. Alexandratos, 1994Uptake of metal

ions by a new chelating ion-exchange resin. Part 4. Kinetics, Solvent Extraction and Ion Exchange, 12211237

19. R. Chiarizia, K. A. D'Arcy, E. P. Horowitz, S. D. Alexandratos, A. W. Trochimczuk, 1996Uptake of metal ions by a new chelating ion exchange resin. Part 8. Simultaneous uptake of cation and anion species, Solvent Extraction and Ion Exchange, 14519542

20. M. B. Corella, S. Siggia, R. M. Barnes, 1984Synthesis and characterization of a poly(acryloamidoxime) metal chelating resin, Analytical Chemistry, 56967972

21. L. Dambies, R. Salinaro, S. D. Alexandratos, 2004Immobilized N-methyl-D-glucamine as an arsenate-selective resin, Environmental Science and Technology, 3861396146

22. A. Dąbrowski, Z. Hubicki, P. Podkościelny, E. Robens, 2004Selective removal of the heavy metal ions from waters and industrial wastewaters by ion-exchange method, Chemosphere, 5691106

23. F. De Dardel, T. V. Arden, 2001Ion exchangers. Principles and applications, Ullmann's Encyclopaedia of Industrial Chemistry, Sixth Edition, Wiley-VCH Verlag GmbH, 170

24. A. Deepatana, M. Valix, 2006Recovery of nickel and cobalt from organic acid complexes: Adsorption mechanisms of metal-organic complexes onto aminophosphonate chelating resin, Journal of Hazardous Materials, 137925933

25. P. D. Dimick, A. Kney, J. Tavakoli, S. E. Mylon, D. Zhao, 2008A comparison of metal-loaded DOW3N ion exchanger for removal of perchlorate from water, Separation Science and Technology, 4323432362

26. X. Ding, S. F. Mou, K. Liu, Y. Yan, 2000Improved scheme of chelation ion chromatography with a mixed eluent for the simultaneous analysis of transition metals at mg l-1 levels, Journal of Chromatography, 883127136

27. X. Ding, S. Mou, 2001Retention behavior of transition metals on a bifunctional ion-exchange column with oxalic acid as eluent, Journal of Chromatography, 920101107

28. C. V. Diniz, F. M. Doyle, A. H. Martins, 2000Uptake of heavy metals by chelating resins from acidic manganese chloride solution, Minerals Metallurgy Processing, 17217222

29. C. V. Diniz, F. M. Doyle, V. S. T. Ciminelli, 2002Effect of pH on the adsorption of selected heavy metal ions from concentrated chloride solutions by the chelating resin Dowex M-4195, Separation Science and Technology, 3731693185

30. C. V. Diniz, V. S. T. Ciminelli, F. M. Doyle, 2005The use of the chelating resin Dowex M-4195 in the adsorption of selected heavy metal ions from manganese solutions, Hydrometallurgy, 78147155

31. S. Drăgan, G. Grigoriu, 1992Ion exchangers, I. Anion exchanges with tertiary amine groups on poly(acrylonitrile-co-divinylbenzene) network, Die Angewandte Macromolekulare Chemie, 2002736

32. W. Du, B. Pan, P. Jiang, Q. Zhang, W. Zhang, B. Pan, Q. Zhang, Q. Zhang, 2008Selective sorption and preconcentration of tartaric acid using a copper(II)-bound polymeric ligand exchanger, Chemical Engineering Journal, 1396368

33. J. H. Duffus, 2002Heavy metals"- a meaningless term? Pure Applied Chemistry, 74793807

34. R. Dybczyński, 1987Selective separation of zinc from other elements on the amphoteric resin Retardion 11A8 and its use for the determination of zinc in biological materials by neutron activation analysis, Analyst, 112449453

35. R. Dybczyński, Z. Hubicki, K. Kulisa, 1988Ion exchange behaviour of 23 elements and amphoteric properties of chelating resin Duolite ES 346 containing amidoxime groups, Solvent Extraction and Ion Exchange, 6

36. R. Dybczyński, E. Sterlińska, 1974The use of the amphoteric ion-exchange resin retardion11A8 for inorganic separations, Journal of Chromatography, 102263271

37. K. A. K. Ebraheem, S. T. Hamdi, 1997Synthesis and properties of a copper selective chelating resin containing a salicylaldoxime group, Reactive and Functional Polymers, 34510

38. R. Fernandez-Prini, 1982Hydrothermal decomposition of ion-exchange resins, Power Industry Research, 2101108

39. L. M. Ferriera, J. M. Loureiro, A. E. Rodrigues, 1998Sorption of metals by an amidoxime chelating resin. Part I: Equilibrium, Separation Science and Technology, 3315851604

40. S. Fisher, 2002Spotlight on the cation resin, Power Plant Chemistry, 4407410

41. S. L. C. Ferreira, C. F. de Brito, 1999Separation and preconcentration of cobalt after sorption onto Amberlite XAD-2 loaded with 2-(2-thiazolylazo)-p-cresol, Analytical Science, 15189

42. S. L. C. Ferreira, C. F. De Brito, A. F. Dantas, Araújo. N. M. Lopo De, A. C. S. Costa, 1999Nickel determination in saline matrices by ICP-AES after sorption on Amberlite XAD-2 loaded with PAN, Talanta,

4811731177

43. S. L. C. Ferreira, V. A. Lemos, B. C. Moreira, A. C. S. Costa, R. E. Santelli, 2000aAn on-line continuous flow system for copper enrichment and determination by flame atomic absorption spectroscopy, Analytica Chimica Acta, 403259264

44. S. L. C. Ferreira, J. R. Ferreira, A. F. Dantas, V. A. Lemos, N. M. L. Araújo, A. C. S. Costa, 2000bCopper determination in natural water samples by using FAAS after preconcentration onto Amberlite XAD-2 loaded with calmagite, Talanta, 5012531259

45. L. Fewtrell, D. Kay, F. Jones, A. Baker, A. Mowat, 1996Copper in drinking water. An investigation into possible health effects, Public Healh, 110175178

46. Fritz J.S.2005Factors affecting selectivity in ion chromatography, Journal of Chromatography, 1085817

47. G. M. Gadd, C. White, 1993Microbial treatment of metal pollution- a working biotechnology? Trends in Biotechnology, 11353359

48. M. R. Gandhi, N. Viswanathan, S. Meenakshi, 2010Adsorption mechanism of hexavalent chromium removal using Amberlite IRA 743 resin, Ion Exchange Letters, 32535

49. F. Gode, E. Pehlivan, 2003A comparative study of two chelating ion-exchange resins for the removal of chromium(III) from aqueous solution, Journal of Hazardous Materials, 100231243

50. R. R. Grinstead, 1984New developments in the chemistry of XFS 4195 and XFS 43084 chelating ion exchange resins, In: Ion Exchange Technology, Society of Chemical Industry, New York, 509518

51. S. N. Hajiev, S. V. Kertman, U. A. Leykin, A. N. Amelin, 1989Thermochemical study of ion echange processess. V. Sorption of copper ions in complex-forming resins, Thermochimica Acta, 139327332

52. F. Helfferich, 1962Ion Exchange Resins, Mc-Grow Hill:New York.

53. N. Hirayama, W. Umehara, 1996Novel separation of inorganic anions using a charged complex ion-exchanger, Analytica Chimica Acta, 33414

54. E. P. Horwitz, R. Chiarizia, H. Diamond, R. C. Gatrone, S. D. Alexandratos, A. Q. Trochimczuk, D. W. Crick, 1993Uptake of metal ions by a new chelating ion-exchange resin. Part 1. Acid dependencies of actinide ions, Solvent Extraction and Ion Exchange, 11943966

55. E. P. Horwitz, S. D. Alexandratos, R. C. Gatrone, R. Chiarizia, 1994Phosphonic acid based ion exchangers, US Patent 5281631.

56. E. P. Horwitz, S. D. Alexandratos, R. C. Gatrone, R. Chiarizia,

1995Phosphonic acid based ion exchange resins. US Patent 5449462.

57. M. S. Hosseini, H. Raissi, S. Madarshahian, 2006Synthesis and application of a new chelating resin functionalized with 2,3-dihydroxy benzoic acid for Fe(III) determination in water samples by flame atomic absorption spectrometry, Reactive and Functional Polymers, 6615391545

58. Z. Hubicki, A. Jakowicz, A. Łodyga, 1999Application of the ions from waters and sewages, In: Adsorption and its applications in industry and environmental protection. Studies in surface science and catalysis, ed. A. Dąbrowski, Elsevier, Amsterdam, New York.

59. H. Hyvönen, M. Orama, H. Saarinen, R. Aksela, 2003Studies on biodegradable chelating ligands: complexation of iminodisuccinic acid (ISA) with Cu(II), Zn(II), Mn(II) and Fe(III) ions in aqueous solution, Green Chemistry, 5410414

60. H. Hyvönen, R. Aksela, 2010Complexation of 3-hydroxy-2,2'-iminodisuccinic acid (HIDS) with Mg2+, Ca2+, Mn2+, Fe3+, Fe2+, Co2+, Ni2+, Cu2+, and Zn2+ ions in aqueous solution, Journal of Coordination Chemistry, 6320132025

61. IDS Na-salz1998Eine neue umweltfreundliche alternative zu klassischen komplexierungsmitteln. Bayer AG brochure, Leverkusen.

62. Ion exchange resins and adsorbents2006Dow Chemical, Co. brochure.

63. H. Irving, R. J. P. Williams, 1953The stability of transition-metal complexes. Journal of Chemical Society, 16231923210

64. V. A. Ivanov, V. D. Timofeevskaya, V. I. Gorshkov, N. V. Drozdova, 1996The role of temperature in ion exchange processesof separation and purification, Journal of Radioanalytical and Nuclear Chemistry, 2082345

65. S. Jeyakumar, V. G. Mishra, M. K. Das, V. V. Raut, R. M. Sawant, K. L. Ramakumar, 2011Separation behavior of U(VI) and Th(IV) on a cation exchange column using 2,6-pyridine dicarboxylic acid as a complexing agent and its application for the rapid separation and determination of U and Th by ion chromatography, Journal of Separation Science, 3418

66. K. C. Jones, R. A. Pyper, 1979Copper recovery from acidic leach liquors by continuous ion-exchange and electrowinning, Journal of Metals, 41925

67. N. Kabay, M. Demircioğlu, S. Yaylı, E. Günay, M. Yüksel, M. Sağlam, M. Streat, 1998aRecovery of uranium from phosphoric acid solutions using chelating ion-exchange resins, Industrial and Engineering Chemistry Research, 3719831990

68. N. Kabay, M. Demircioğlu, H. Ekinci, M. Yüksel, M. Sağlam, M. Akçay,

M. Streat, 1998bRemoval of metal pollutants (Cd(II) and Cr(III)) from phosphoric acid solutions by chelating resins containing phosphonic or diphosphonic groups, Industrial and Engineering Chemistry Research, 3725412547

69.   C. Kantipuly, S. Katragadda, A. Chow, H. D. Gesser, 1990Chelating polymers and related supports for separation and preconcentration of trace metals, Talanta, 37491517

70.   D. Kaušpėdienė, J. Snukiškis, A. Gefenienė, 2003Kinetics of cadmium(II) sorption by an iminodiacetic ion exchanger in the presence of a nonionic surfactant, Desalination, 1546777

71.   S. V. Kertman, G. M. Kertman, A. N. Amelin, Yu. A. Leykin, 1997Heats of the immersion of Co2+ and Cu2+ contained chelating resins, Thermochimica Acta, 2974956

72.   R. Koivula, J. Lehto, L. Pajo, T. Gale, H. Leinonen, 2000Purification of metal plating rinse waters with chelating ion exchangers, Hydrometallurgy, 5693108

73.   D. Kołodyńska, H. Hubicka, Z. Hubicki, 2009Studies of application of monodisperse anion exchangers in sorption of heavy metal complexes with IDS, Desalination, 239216228

74.   D. Kołodyńska, 2009aPolyacrylate anion exchangers in sorption of heavy metal ions with the biodegradable complexing agent, Chemical Engineering Journal, 150280288

75.   D. Kołodyńska, 2009bIminodisuccinic acid as a new complexing agent for removal of heavy metal ions from industrial effluents, Chemical Engineering Journal, 152277288

76.   D. Kołodyńska, 2009cChelating ion exchange resins in removal of heavy metal ions from waters and wastewaters in presence of a complexing agent, Przemysł Chemiczny, 88182189in Polish).

77.   D. Kołodyńska, 2010aBiodegradable complexing agents as an alternative to chelators in sorption of heavy metal ions, Desalination and Water Treatment. Science and Engineering, 16146155

78.   D. Kołodyńska, 2010bDiphonix Resin® in sorption of heavy metal ions in the presence of biodegradable complexing agents of a new generation, Chemical Engineering Journal, 1592736

79.   D. Kołodyńska, 2010cThe effect of the treatment conditions on metal ions removal in the presence of complexing agents of a new generation, Desalination, 26315951169

80.   D. Kołodyńska, 2011The chelating agents of a new generation as an

alternative to conventional chelators for heavy metal ions removal from different waste waters, In: Expanding issues in desalination (ed., Robert Y. Ning) InTech, Publishers 2011, 339371

81. D. Kołodyńska, J. Jachuła, Z. Hubicki, 2012Removal of heavy metal complexes with MGDA by synthetic resin Diphonix, Environmental Engineering and Management Journal, in press.

82. J. Jachuła, D. Kołodyńska, Z. Hubicki, 2011Sorption of Cu(II) and Ni(II) ions in presence of novel chelating agent methylglycinediacetic acid by microporous ion exchangers and sorbents from aqueous solutions, Central European Journal of Chemistry, 95265

83. J. Jachuła, D. Kołodyńska, Z. Hubicki, 2012Methylglycinediacetic acid as a new complexing agent for removal of heavy metal ions from industrial wastewater, Solvent Extraction and Ion Exchange, in press.

84. O. N. Kononova, A. G. Kholmogorov, S. V. Kachin, O. V. Mytykh, Y. S. Kononov, O. P. Kalyakina, G. L. Pashkov, 2000Ion exchange recovery of nickel from manganese nitrate solutions, Hydrometallurgy, 54107115

85. V. V. Krongauz, C. W. Kocher, 1997Kinetics of ion exchange in monodisperse resin, Journal of Applied Polymer Science, 5912711283

86. R. Kunin, 1958Ion Exchange Resins, 3rd Ed; Wiley: New York.

87. R. Kunin, 1979Amber-Hi-Lites 161, Rohm and Haas Co.

88. H. Kurama, T. Çatlsarik, 2000Removal of zinc cyanide from a leach solution by an anionic ion-exchange resin, Desalination, 12916

89. M. S. Lee, M. J. Nicol, 2007Removal of iron from cobalt sulphate solutions by ion exchange with Diphonix resin and enhancement of iron elution with titanium(III), Hydrometallurgy, 86612

90. P. Li, Gupta. A. K. Sen, 1998Genesis of selectivity and reversibility for sorption of synthetic aromatic anions onto polymeric sorbents, Environmental Science and Technology, 3237563766

91. C. W. Li, T. Qi, F. A. Wang, Y. Zhang, Z. H. Yu, 2006Variation of cell voltage with reaction time in electrochemical synthesis process of sodium dichromate, Chemical Engineering and Technology, 29481486

92. C. Luca, C. D. Vlad, I. Bunia, 2009Trends in weak base anion exchangers resins, Revue Roumaine de Chimie, 54107117

93. L. M. Lührmann, N. Stelter, A. Kettrup, 1985Synthesis and properties of metal collecting phases with silica immobilized 8-hydroxyquinoline, Fresenius Journal of Analytical Chemistry, 3224752

94. M. Marhol, H. Beranová, K. L. Cheng, 1974Selective ion-exchangers containing phosphorus in their functional groups. I. Sorption and

separation of some bivalent and trivalent ions, Journal of Radioanalytical and Nuclear Chemistry, 21177186

95. M. Marhol, K. L. Cheng, 1974Some chelating ion-exchange resins containing ketoiminocarboxylic acids as functional groups, Talanta, 21751762

96. A. Mc Clain, Y. L. Hsieh, 2004Synthesis of polystyrene-supported dithiocarbamates and their complexation with metal ions, Journal of Applied Polymer Sciences, 922004218225

97. F. D. Mendes, A. H. Martins, 2004Selective sorption of nickel and cobalt from sulphate solutions using chelating resins, International Journal of Mineral Processing, 74359371

98. J. Melling, D. W. West, 1984Proceedings of the International Conference on Ion-Exchange, Society of Chemical Industry, Cambridge, England, 724

99. J. Minczewski, J. Chwastowska, R. Dybczyński, 1982Separation and preconcentration methods in inorganic trace analysis, Wiley-VCH, New York.

100. D. Muraviev, A. Gonzalo, M. Valiente, 1995Ion exchange on the resin with temperature-responsive selectivity. 1. ion exchange equilibrium of Cu2+ and Zn2+ on iminodiacetic and aminomethylphosphonic resin, Analytical Chemistry, 6730283035

101. K. L. Nash, P. G. Rickert, J. V. Muntean, 1994Uptake of metal ions by a new chelating ion exchange resin. Part 3: Protonation constants via potentiometric titration and solid state 31P NMR spectroscopy, Solvent Extraction and Ion Exchange, 12193209

102. P. N. Nesterenko, M. J. Shaw, S. J. Hill, P. Jones, 1999Aminophosphonate-functionalized silica: A versatile chromatographic stationary phase for high performance chelation ion chromatography, Microchemical Journal, 625569

103. P. N. Nesterenko, P. R. Haddad, 2000Zwitterionic ion-exchangers in liquid chromatography, Analytical Sciences, 16565574

104. J. O. Nriagu, J. M. Pacyna, 1988Quantitative assessment of worldwide contamination of air, water and soils by trace metals, Nature, 333134139

105. T. Ogata, K. Nagayoshi, T. Nagasako, S. Kurihara, T. Nonaka, 2006Synthesis of hydrogel beads having phosphinic acid groups and its adsorption ability for lanthanide ions, Reactive and Functional Polymers, 66625633

106. I. H. Park, J. M. Suh, 1996Preparation of uranyl ion adsorptivity

of macroreticular chelating resins containing a pair of neighboring amidoxime groups in a monomeric styrene units, Angewandte Makromolekulare Chemie, 239121132

107. M. Pesavento, R. Biesuz, M. Gallorini, A. Profumo, 1993Sorption mechanism of trace amounts of divalent metal ions on a chelating resin containing iminodiacetate groups, Analytical Chemistry, 6525222527

108. D. H. Phillips, B. Gu, D. B. Watson, C. S. Parmele, 2008Uranium removal from contaminated groundwater by synthetic resins, Water Research, 42

109. D. Prabhakaran, M. S. Subramanian, 2003A new chelating sorbent for metal ion extraction under high saline conditions, Talanta, 5912271236

110. S. Pramanik, S. Sarkar, H. Paul, P. Chattopadhyay, 2009A new polymer with 2-methoxy-1-imidazolylazobenzene functionality for determination of lead(II) and iron(III), Indian Journal of Chemistry, 48A3037

111. U. Pyell, G. Stork, 1992Preparation and properties of an 8-hydroxyquinoline silica gel synthesized via Mannich reaction, Fresenius Journal of Analytical Chemistry, 342281286

112. D. K. Ryan, J. H. Weber, 1985Comparison of chelating agents immobilized on glass with Chelex 100 for removal and preconcentration of trace copper(II), Talanta, 321985859863

113. A. Ramesh, K. R. Mohan, K. Seshaiah, 2002Preconcentration of trace metals on Amberlite XAD-4 resin coated with dithiocarbamates and determination by inductively coupled plasma-atomic emission spectrometry in saline matrices, Talanta, 57243252

114. F. Reinecke, T. Groth, K. P. Heise, W. Joentgen, N. Müller, A. Steinbüchel, 2000Isolation and characterization of an Achromobacter xylosoxidans strain B3 and other bacteria capable to degrade the synthetic chelating agent iminodisuccinate, FEMS Microbiological Letters, 1884146

115. Reynolds T.D.1982Unit operations and processes in environmental engineering, BC Engineering Division, Edition II, Boston, MA.

116. K. P. Ripperger, S. D. Alexandratos, 1999Polymer-supported phosphorus-containing ligands for selective metal ion complexation. In: Adsorption and its applications in industry and environmental protection. Studies in surface science and catalysis, ed. A. Dąbrowski, Elsevier, Amsterdam, New York.

117. K. O. Saygi, M. Tuzen, M. Soylak, L. Elci, 2008Chromium speciation by solid phase extraction on Dowex M 4195 chelating resin and determination by atomic absorption spectrometry, Journal of Hazardous Materials, 15310091014

118. S. K. Sahni, J. Reedijk, 1984Coordination chemistry of chelating resins and ion exchangers, Coordination Chemistry Reviews, 591139

119. S. K. Sahni, R. Van Bennekom, J. Reedijk, 1985A spectral study of transition-metal complexes on chelating ion exchange resin containing aminophosphonic acid groups, Polyhedron, 416431658

120. Z. Samczyński, R. Dybczyński, 1997Some examples of the use of amphoteric ion exchange resins for inorganic separations, Journal of Chromatography, 789157167

121. Z. Samczyński, R. Dybczyński, 2002The use of Retardion 11A8 amphoteric ion exchange resin for the separation and determination of cadmium and zinc in geological and environmental materials by neutron activation analysis, Journal of Radioanalytical and Nuclear Chemistry, 2542002335341

122. R. Saxena, A. K. Singh, D. P. S. Rathore, 1995Salicylic acid functionalized polystyrene sorbent Amberlite XAD-2. Synthesis and applications as a preconcentrator in the determination of zinc(II) and lead(II) by using atomic absorption spectrometry, Analyst, 120403405

123. A. Scheffler, 1996Lewatit-MonoPlus. The latest generation of monodisperse ion exchange resin with outstanding properties for optimising water treatment system, Technical Bulletin, Bayer AG.

124. R. K. Sharma, P. Pant, 2009Solid phase extraction and determination of metal ions in aqueous samples using Quercetin modified Amberlite XAD-16 chelating polymer as metal extractant, International Journal of Environmental Analytical Chemistry, 89503514

125. D. C. Sherrington, 1998Preparation, structure and morphology of polymer supports, Chemical Communications, 22752286

126. M. Smolik, A. Jakóbik-Kolon, M. Porański, 2009Separation of zirconium and hafnium using Diphonix® chelating ion-exchange resin, Hydrometallurg, 95350353

127. J. Stamberg, V. Valter, 1970Entfärbungsharze- Anwendungsprinzipien, charakterisierung und verwendung (Decolourizing resins- principles of application, characterization and use). Akademie-Verlag, Berlin, Akadbmie-Verlag, Berlin, 63

128. S. Tandy, K. Bossart, R. Mueller, J. Ritschel, L. Hauser, R. Schulin, B. Nowack, 2004Extraction of heavy metals from soils using biodegradable chelating agents, Environmental Science and Technology, 38937944

129. P. K. Tewari, A. K. Singh, 1999Amberlite XAD-2 functionalized with chromotropic acid: Synthesis of a new polymer matrix and its

applications in metal ion enrichment for their determination by flame atomic absorption spectrometry, Analyst, 12418471851

130. P. K. Tewari, A. K. Singh, 2000Amberlite XAD-7 impregnated with Xylenol Orange: a chelating collector for preconcentration of Cd(II), Co(II), Cu(II), Ni(II), Zn(II) and Fe(III) ions prior to their determination by flame AAS, Fresenius Journal of Analytical Chemistry, 367562567

131. P. K. Tewari, A. K. Singh, 2001Synthesis, characterization and applications of pyrocatechol modified Amberlite XAD-2 resin for preconcentration and determination of metal ions in water samples by flame atomic absorption spectrometry (FAAS), Talanta, 53823833

132. P. K. Tewari, A. K. Singh, 2002Preconcentration of lead with Amberlite XAD-2 and Amberlite XAD-7 based chelating resins for its determination by flame atomic absorption spectrometry, Talanta, 56735744

133. S. Tokalıoğlu, H. Ergün, A. Çukurovalı, 2010Preconcentration and determination of Fe(III) from water and food samples by newly synthesized chelating reagent impregnated Amberlite XAD-16 resin, Bulletin of Korean Chemical Society, 3119761980

134. A. W. Trochimczuk, M. Streat, 1999Novel chelating resins with aminothiophosphonate ligands, Reactive & Functional Polymers, 40205213

135. A. W. Trochimczuk, 2000Synthesis of functionalized phenylphosphinic acid resins through Michael reaction and their ion-exchange properties, Reactive and Functional Polymers, 44919

136. M. Tuzen, K. O. Saygi, M. Soylak, 2008Novel solid phase extraction procedure for gold(III) on Dowex M 4195 prior to its flame atomic absorption spectrometric determination, Journal of Hazardous Materials, 156591595

137. A. Uzun, M. Soylak, L. Elçi, 2001Preconcentration and separation with Amberlite XAD-4 resin; determination of Cu, Fe, Pb, Ni, Cd and Bi at trace levels in waste water samples by flame atomic absorption spectrometry, Talanta, 54197202

138. A. Warshawsky, 1987Chelating ion exchangers, in: Ion exchange and sorption processes in hydrometallurgy, M. Streat, D.A. Naden (Eds.), Wiley VCH, Chichester, 166225

139. V. P. Vasilev, A. V. Katrovtseva, I. P. Gorelov, N. V. Tukumova, 1996Stability of complexes of Ni(II) with iminodisuccinic acid, Zhurnal Neorganic Khimii, 4113201323

140. V. P. Vasilev, A. V. Katrovtseva, S. A. Bychkova, N. V. Tukumova,

1998Stability of complexes of Co(II) and Cu(II) with iminodisuccinic acid, Zhurnal Neorganic Khimii, 43808809

141. M. C. Yebra-Biurrun, Bermejo. García-Dopazo-Barrera, A. , M. P. Bermejo-Barrera, 1992Preconcentration of trace amounts of manganese from natural waters by means of a macroreticulap poly(dithiocarbamate) resin, Talanta, 39671674

142. Z. Yua, T. Qia, J. Qua, L. Wanga, J. Chu, 2009Removal of Ca(II) and Mg(II) from potassium chromate solution on Amberlite IRC 748 synthetic resin by ion exchange, Journal of Hazardous Materials, 167406417

143. A. Yuchi, T. Sato, Y. Morimoto, H. Mizuno, H. Wada, 1997Adsorption mechanism of trivalent metal ions on chelating resins containing iminodiacetic acid groups with reference to selectivity, Analytical Chemistry, 6929412944

144. A. A. Zagorodni, M. Muhammed, 1999Explanation of the Zn/Cu dual temperature separation on Amberlite IRC-718 ion exchange resin, Separation Science and Technology, 3420132021

145. Z. Zainol, M. Nicol, 2009bIon-exchange equilibria of Ni2+, Co2+, Mn2+ and Mg2+ with iminodiacetic acid chelating resin Amberlite IRC 748, Hydrometallurgy, 99175180

146. Z. Zainol, M. Nicol, 2009aComparative study of chelating ion exchange resins for the recovery of nickel and cobalt from laterite leach tailings, Hydrometallurgy, 96283287

147. D. Zhao, Gupta. A. K. Sen, L. Stewart, 1998Selective removal of Cr(VI) oxyanions with a new anion exchanger, Industrial and Engineering Chemistry Research, 3743834387

CITATION

CHAPTER 1

Javier Miguel Ochando-Pulido and Antonio Martinez-Ferez, On the Recent Use of Membrane Technology for Olive Mill Wastewater Purification, doi:10.3390/membranes5040513.

CHAPTER 2

Tsuyoshi Ochiai, Shoko Tago, Mio Hayashi, Hiromasa Tawarayama, Toshifumi Hosoya and Akira Fujishima, TiO2-Impregnated Porous Silica Tube and Its Application for Compact Air- and Water-Purification Units, doi:10.3390/catal5031498.

CHAPTER 3

Marjan S. Ranđelović, Aleksandra R. Zarubica and Milovan M. Purenović (2012). New Composite Materials in the Technology for Drinking Water Purification from Ionic and Colloidal Pollutants, Composites and Their Applications, Prof. Ning Hu (Ed.), ISBN: 978-953-51-0706-4, InTech, DOI: 10.5772/48390.

CHAPTER 4

Ramiro Escudero, Francisco J. Tavera and Eunice Espinoza (2013). Treating of Waste Water Applying Bubble Flotation, Water Treatment, Dr. Walid Elshorbagy (Ed.), ISBN: 978-953-51-0928-0, InTech, DOI: 10.5772/54227.

CHAPTER 5

Adina Elena Segneanu, Cristina Orbeci, Carmen Lazau, Paula Sfirloaga, Paulina Vlazan, Cornelia Bandas and Ioan Grozescu (2013). Waste Water Treatment Methods, Water Treatment, Dr. Walid Elshorbagy (Ed.), ISBN: 978-953-51-0928-0, InTech, DOI: 10.5772/53755.

CHAPTER 6

Chunli Zheng, Ling Zhao, Xiaobai Zhou, Zhimin Fu and An Li (2013). Treatment Technologies for Organic Wastewater, Water Treatment, Dr. Walid Elshorbagy (Ed.), ISBN: 978-953-51-0928-0, InTech, DOI: 10.5772/52665.

CHAPTER 7

Bulent Sen, Mehmet Tahir Alp, Feray Sonmez, Mehmet Ali Turan Kocer and Ozgur Canpolat (2013). Relationship of Algae to Water Pollution and Waste Water Treatment, Water Treatment, Dr. Walid Elshorbagy (Ed.), ISBN: 978-953-51-0928-0, InTech, DOI: 10.5772/51927.

CHAPTER 8

Asli Baysal, Nil Ozbek and Suleyman Akman (2013). Determination of Trace Metals in Waste Water and Their Removal Processes, Waste Water - Treatment Technologies and Recent Analytical Developments, Prof. Fernando Sebastián García Einschlag (Ed.), ISBN: 978-953-51-0882-5, InTech, DOI: 10.5772/52025.

CHAPTER 9

Zbigniew Hubicki and Dorota Kołodyńska (2012). Selective Removal of Heavy Metal Ions from Waters and Waste Waters Using Ion Exchange Methods, Ion Exchange Technologies, Prof. Ayben Kilislioglu (Ed.), ISBN: 978-953-51-0836-8, InTech, DOI: 10.5772/51040.

INDEX